Suzuki Trail Bik

Owners
Workshop
Manual

T0252091

by Chris Rogers
with an additional Chapter on the TS100 and 125 ERZ models
by Pete Shoemark

Models covered

Model	cc	Introduction
TS100 ERN.	98cc.	Introduced UK only May 1979
TS100 ERT.	98cc.	Introduced UK December 1980, USA October 1979
TS100 ERX.	98cc.	Introduced UK August 1980, USA September 1980
TS100 ERZ.	98cc.	Introduced UK only November 1982
TS125 ERN.	123cc.	Introduced UK only March 1979
TS125 ERT.	123cc.	Introduced UK November 1979, USA October 1979
TS125 ERX.	123cc.	Introduced UK October 1980, USA September 1980
TS125 ERZ.	123cc.	Introduced UK only September 1982
TS185 ERN	183cc.	Introduced UK only July 1979
TS185 ERT.	183cc.	Introduced UK November 1979, USA October 1979
TS185 ERX.	183cc.	Introduced UK October 1980, USA September 1980
TS250 ERN.	246cc.	Introduced UK only August 1979
TS250 ERT.	246cc.	Introduced UK November 1979, USA October 1979
TS250 ERX.	246cc.	Introduced UK October 1980, USA September 1980

Note: US models do not take the ER suffix

ISBN 978 1 85010 260 1

© Haynes Publishing Group 1989

All rights reserved. No part of this book may be reproduced or transmitted in
any form or by any means, electronic or mechanical, including photocopying,
recording or by any information storage or retrieval system, without permission
in writing from the copyright holder.

ABCDE
FGHIJ
KL

2

Printed in the UK   (797-1004)

Haynes Publishing Group
Sparkford Nr Yeovil
Somerset BA22 7JJ England

Haynes Publications, Inc
859 Lawrence Drive
Newbury Park
California 91320 USA

British Library Cataloguing in Publication Data

Rogers, Chris
  Suzuki 100, 125, 185 and 250 trail bikes 1979 to 1987
  (Owners Workshop Manual)
  1. Suzuki motorcycle
  I. Title   II. Shoemark, Pete
  III. Series
  629.28'775   TL448.S8
  ISBN 1-85010-260-0

Library of Congress Catalog Card Number
86-80180

Acknowledgements

Thanks are due to Chris Branson of Paul Branson Motorcycles of Yeovil for the supply of technical information and for the loan of the Suzuki TS125 ER featured in the photographs throughout this Manual and the TS100 ER featured on the front cover.

Johnny Walker and Peter Hodgson of Huxhams Motorcycles Ltd of Poole, Dorset, gave invaluable assistance in supplying technical information and in organising the availability of the Suzuki TS185 ER featured in the photographs accompanying the text.

Special thanks are due to Hugh Brooker of our production department here at Sparkford, who gave his permission for the author to examine and partially dismantle his Suzuki TS250.

We should also like to thank Heron Suzuki (GB) Ltd. for permission to use their line illustrations, and members of the Technical Service Department of that company for their help and advice.

Finally, we would like to thank the Avon Rubber Company, who kindly supplied information and technical assistance on tyre fitting; NGK Spark Plugs (UK) Ltd. for information on spark plug maintenance and electrode conditions, and Renold Ltd. for advice on chain care and renewal.

About this manual

The purpose of this manual is to present the owner with a concise and graphic guide which will enable him to tackle any operation from basic routine maintenance to a major overhaul. It has been assumed that any work would be undertaken without the luxury of a well-equipped workshop and a range of manufacturer's service tools.

To this end, the machine featured in the manual was stripped and rebuilt in our own workshop, by a team comprising a mechanic, a photographer and the author. The resulting photographic sequence depicts events as they took place, the hands shown being those of the author and the mechanic.

The use of specialised, and expensive, service tools was avoided unless their use was considered to be essential due to risk of breakage or injury. There is usually some way of improvising a method of removing a stubborn component, providing that a suitable degree of care is exercised.

The author learnt his motorcycle mechanics over a number of years, faced with the same difficulties and using similar facilities to those encountered by most owners. It is hoped that this practical experience can be passed on through the pages of this manual.

Where possible, a well-used example of the machine is chosen for the workshop project, as this highlights any areas which might be particularly prone to giving rise to problems. In this way, any such difficulties are encountered and resolved before the text is written, and the techniques used to deal with them can be incorporated in the relevant section. Armed with a working knowledge of the machine, the author undertakes a considerable amount of research in order that the maximum amount of data can be included in the manual.

A comprehensive section, preceding the main part of the manual, describes procedures for carrying out the routine maintenance of the machine at intervals of time and mileage. This section is included particularly for those owners who wish to ensure the efficient day-to-day running of their motorcycle, but who choose not to undertake overhaul or renovation work.

Each Chapter is divided into numbered sections. Within these sections are numbered paragraphs. Cross reference throughout the manual is quite straightforward and logical. When reference is made 'See Section 6.10' it means Section 6, paragraph 10 in the same Chapter. If another Chapter were intended, the reference would read, for example, 'See Chapter 2, Section 6.10'. All the photographs are captioned with a section/paragraph number to which they refer and are relevant to the Chapter text adjacent.

Figures (usually line illustrations) appear in a logical but numerical order, within a given Chapter. Fig. 1.1 therefore refers to the first figure in Chapter 1.

Left-hand and right-hand descriptions of the machines and their components refer to the left and right of a given machine when the rider is seated normally.

Motorcycle manufacturers continually make changes to specifications and recommendations, and these, when notified, are incorporated into our manuals at the earliest opportunity.

Whilst every care is taken to ensure that the information in this manual is correct no liability can be accepted by the author or publishers for loss, damage or injury caused by any errors in or omissions from the information given.

Contents

	Page
Acknowledgements	2
About this manual	2
Introduction to the Suzuki TS100, 125, 185 and 250 models	6
Model dimensions and weights	6
Ordering spare parts	7
Safety first!	8
Tools and working facilities	9
Choosing and fitting accessories	12
Fault diagnosis	15
Routine maintenance	25
Chapter 1 Engine, clutch and gearbox	44
Chapter 2 Fuel system and lubrication	114
Chapter 3 Ignition system	151
Chapter 4 Frame and forks	164
Chapter 5 Wheels, brakes and tyres	186
Chapter 6 Electrical system	203
Chapter 7 The UK TS100 and 125 ERZ models	217
Wiring diagrams	230
Conversion factors	234
English/American terminology	235
Index	236

The Suzuki TS 185 ERX

The engine and gearbox unit of the Suzuki TS185 ERX

Introduction to the Suzuki TS100, 125, 185 and 250 models

Although the Suzuki Motor Company Limited commenced manufacturing motorcycles as early as 1936, it was not until the early 1960s that their machines were first imported into the US and European markets. During the years that followed, Suzuki achieved great success with numerous types of road machines and eventually, in the late 1960s, turned their attention to the market which was forming around the rapidly growing number of riders who wished to spend their leisure hours riding both on and off the open road. To this end, Suzuki produced their first trail bike, the TC120 Trail Cat.

The TC120 incorporated several unique technical innovations, one of which was the method of gearbox construction and gear selection. Suzuki's approach was to incorporate two sets of three-speed gears, the selection of which was actuated by a short lever on the crankcase casting. This made it possible to select normal gear ratios for road use, or by bringing in a reduction gear, special low gear ratios for off road use over especially rough terrain. Although this particular innovation was not continued in the following TS series models which were introduced in the early 1970s, many of the basic design features, such as the upswept exhaust system, high ground clearance and the general slimming down of component parts to reduce weight, were carried over. This resulted in a motorcycle which formed the basis for all Suzuki's future trail bike models up to the present day; that is, of a strong but lightweight frame housing a mechanically simple but extremely efficient engine and incorporating cycle parts which have been developed from Suzuki's now appreciable experience of off road sports. The TC series models ended their production run towards the end of 1977 having eventually been ousted by the now established range of TS models.

The present TS models in the UK, that is the range covered in this Manual, carry the suffix letters ER. They also carry a third suffix letter to denote the year of introduction; these letters run in the order N, T and X. US models do not have the ER suffix but use the third letter only, although the letter N should be disregarded with reference to the US model types covered in this Manual. This is because the N model of all engine capacities remained in the earlier style of Suzuki trail bikes until its development into the T model. In the US, the suffix T indicates that the machine was introduced into the market of that country in October of 1979. This was superseded by the model with the suffix letter X in September of 1980. Un-

fortunately, the introduction of the model range into the UK does not follow such a clear and concise pattern as that given for the US market and indeed, in some capacity classes, would seem to omit one suffix classification altogether. To present an accurate indication of the model types which have been introduced into the UK market and their dates of introduction, the following table has been prepared.

Model type	Introduction date
TS100 ERN	May 1979
TS100 ERT	December 1980
TS100 ERX	August 1980
TS125 ERN	March 1979
TS125 ERT	November 1979
TS125 ERX	October 1980
TS185 ERN	July 1979
TS185 ERT	November 1979
TS185 ERX	October 1980
TS250 ERN	August 1979
TS250 ERT	November 1979
TS250 ERX	October 1980

All model types throughout the capacity range feature a single cylinder two-stroke engine which is housed in a conventional tubular steel frame, although there are some design variations in these components. Styling is similar to all models and features the increased use of plastic mouldings for such items as the mudguards and side panels. Conventional steel rimmed and wire spoked wheels are fitted. The suspension is also conventional, not having followed the present day trend towards the monoshock type of rear suspension favoured by some rival manufacturers. Suzuki's experience in off road competition is however reflected in the modification or redesign of these components to give them added strength and efficiency; the most obvious change from earlier TS models being the introduction of a swinging arm fork constructed from box section material instead of the tubular steel previously used. All machines are fully equipped for road use and incorporate electrical systems which are equal to those fitted to many pukka road machines. Ignition systems vary between the model types; the conventional type of contact breaker system, Suzuki's own PEI system and their adaptation of the CDI type of system all feature throughout the model range.

Model dimensions and weights

	TS100	TS125 N	TS125 T and X	TS185	TS250
Overall length	2065 mm (81.3 in)	2075 mm (81.7 in)	2120 mm (83.5 in)	2160 mm (85.0 in)	2220 mm (87.4 in)
Overall width	800 mm (31.5 in)	850 mm (33.5 in)	800 mm (31.5 in)	835 mm (32.9 in)	850 mm (33.5 in)
Overall height	1075 mm (42.3 in)	1080 mm (42.5 in)	1135 mm (44.7 in)	1125 mm (44.3 in)	1180 mm (46.5 in)
Wheelbase	1325 mm (52.2 in)	1335 mm (52.6 in)	1350 mm (53.1 in)	1375 mm (54.1 in)	1410 mm (55.5 in)
Ground clearance	205 mm (8.1 in)	230 mm (9.1 in)	250 mm (9.8 in)	255 mm (10.0 in)	270 mm (10.6 in)
Dry weight	91 kg (201 lb)	94 kg (207 lb)	97 kg (214 lb)	102 kg (225 lb)	121 kg (267 lb)

Ordering spare parts

When ordering spare parts for any Suzuki, it is advisable to deal direct with an official Suzuki agent who should be able to supply most of the parts ex-stock. Parts cannot be obtained from Suzuki direct even if the parts required are not held in stock. Always, quote the engine and frame numbers in full, especially if parts are required for earlier models.

On TS250 models, the engine number is stamped on the top surface of the right-hand crankcase half, to the rear of the cylinder barrel. On all other models, the engine number is stamped on the top surface of the left-hand crankcase half, at the side of the cylinder barrel. The frame number of all models is stamped on the right-hand side of the steering head.

Use only genuine Suzuki spares. Some pattern parts are available that are made in Japan and may be packed in similar looking packages. They should only be used if genuine parts are hard to obtain or in an emergency, for they do not normally last as long as genuine parts, even although there may be a price advantage.

Some of the more expendable parts such as spark plugs, bulbs, oils and greases etc, can be obtained from accessory shops and motor factors, who have convenient opening hours, and can be found not far from home. It is also possible to obtain parts on a Mail Order basis from a number of specialists who advertise regularly in the motorcycle magazines.

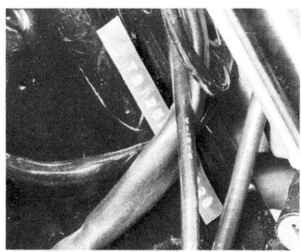

Location of frame number

Location of engine number (except 250)

Safety first!

Professional motor mechanics are trained in safe working procedures. However enthusiastic you may be about getting on with the job in hand, do take the time to ensure that your safety is not put at risk. A moment's lack of attention can result in an accident, as can failure to observe certain elementary precautions.

There will always be new ways of having accidents, and the following points do not pretend to be a comprehensive list of all dangers; they are intended rather to make you aware of the risks and to encourage a safety-conscious approach to all work you carry out on your vehicle.

Essential DOs and DON'Ts

DON'T start the engine without first ascertaining that the transmission is in neutral.

DON'T suddenly remove the filler cap from a hot cooling system – cover it with a cloth and release the pressure gradually first, or you may get scalded by escaping coolant.

DON'T attempt to drain oil until you are sure it has cooled sufficiently to avoid scalding you.

DON'T grasp any part of the engine, exhaust or silencer without first ascertaining that it is sufficiently cool to avoid burning you.

DON'T allow brake fluid or antifreeze to contact the machine's paintwork or plastic components.

DON'T syphon toxic liquids such as fuel, brake fluid or antifreeze by mouth, or allow them to remain on your skin.

DON'T inhale dust – it may be injurious to health (see *Asbestos* heading).

DON'T allow any spilt oil or grease to remain on the floor – wipe it up straight away, before someone slips on it.

DON'T use ill-fitting spanners or other tools which may slip and cause injury.

DON'T attempt to lift a heavy component which may be beyond your capability – get assistance.

DON'T rush to finish a job, or take unverified short cuts.

DON'T allow children or animals in or around an unattended vehicle.

DON'T inflate a tyre to a pressure above the recommended maximum. Apart from overstressing the carcase and wheel rim, in extreme cases the tyre may blow off forcibly.

DO ensure that the machine is supported securely at all times. This is especially important when the machine is blocked up to aid wheel or fork removal.

DO take care when attempting to slacken a stubborn nut or bolt. It is generally better to pull on a spanner, rather than push, so that if slippage occurs you fall away from the machine rather than on to it.

DO wear eye protection when using power tools such as drill, sander, bench grinder etc.

DO use a barrier cream on your hands prior to undertaking dirty jobs – it will protect your skin from infection as well as making the dirt easier to remove afterwards; but make sure your hands aren't left slippery. Note that long-term contact with used engine oil can be a health hazard.

DO keep loose clothing (cuffs, tie etc) and long hair well out of the way of moving mechanical parts.

DO remove rings, wristwatch etc, before working on the vehicle – especially the electrical system.

DO keep your work area tidy – it is only too easy to fall over articles left lying around.

DO exercise caution when compressing springs for removal or installation. Ensure that the tension is applied and released in a controlled manner, using suitable tools which preclude the possibility of the spring escaping violently.

DO ensure that any lifting tackle used has a safe working load rating adequate for the job.

DO get someone to check periodically that all is well, when working alone on the vehicle.

DO carry out work in a logical sequence and check that everything is correctly assembled and tightened afterwards.

DO remember that your vehicle's safety affects that of yourself and others. If in doubt on any point, get specialist advice.

IF, in spite of following these precautions, you are unfortunate enough to injure yourself, seek medical attention as soon as possible.

Asbestos

Certain friction, insulating, sealing, and other products – such as brake linings, clutch linings, gaskets, etc – contain asbestos. *Extreme care must be taken to avoid inhalation of dust from such products since it is hazardous to health.* If in doubt, assume that they *do* contain asbestos.

Fire

Remember at all times that petrol (gasoline) is highly flammable. Never smoke, or have any kind of naked flame around, when working on the vehicle. But the risk does not end there – a spark caused by an electrical short-circuit, by two metal surfaces contacting each other, by careless use of tools, or even by static electricity built up in your body under certain conditions, can ignite petrol vapour, which in a confined space is highly explosive.

Always disconnect the battery earth (ground) terminal before working on any part of the fuel or electrical system, and never risk spilling fuel on to a hot engine or exhaust.

It is recommended that a fire extinguisher of a type suitable for fuel and electrical fires is kept handy in the garage or workplace at all times. Never try to extinguish a fuel or electrical fire with water.

Note: *Any reference to a 'torch' appearing in this manual should always be taken to mean a hand-held battery-operated electric lamp or flashlight. It does **not** mean a welding/gas torch or blowlamp.*

Fumes

Certain fumes are highly toxic and can quickly cause unconsciousness and even death if inhaled to any extent. Petrol (gasoline) vapour comes into this category, as do the vapours from certain solvents such as trichloroethylene. Any draining or pouring of such volatile fluids should be done in a well ventilated area.

When using cleaning fluids and solvents, read the instructions carefully. Never use materials from unmarked containers – they may give off poisonous vapours.

Never run the engine of a motor vehicle in an enclosed space such as a garage. Exhaust fumes contain carbon monoxide which is extremely poisonous; if you need to run the engine, always do so in the open air or at least have the rear of the vehicle outside the workplace.

The battery

Never cause a spark, or allow a naked light, near the vehicle's battery. It will normally be giving off a certain amount of hydrogen gas, which is highly explosive.

Always disconnect the battery earth (ground) terminal before working on the fuel or electrical systems.

If possible, loosen the filler plugs or cover when charging the battery from an external source. Do not charge at an excessive rate or the battery may burst.

Take care when topping up and when carrying the battery. The acid electrolyte, even when diluted, is very corrosive and should not be allowed to contact the eyes or skin.

If you ever need to prepare electrolyte yourself, always add the acid slowly to the water, and never the other way round. Protect against splashes by wearing rubber gloves and goggles.

Mains electricity and electrical equipment

When using an electric power tool, inspection light etc, always ensure that the appliance is correctly connected to its plug and that, where necessary, it is properly earthed (grounded). Do not use such appliances in damp conditions and, again, beware of creating a spark or applying excessive heat in the vicinity of fuel or fuel vapour. Also ensure that the appliances meet the relevant national safety standards.

Ignition HT voltage

A severe electric shock can result from touching certain parts of the ignition system, such as the HT leads, when the engine is running or being cranked, particularly if components are damp or the insulation is defective. Where an electronic ignition system is fitted, the HT voltage is much higher and could prove fatal.

Tools and working facilities

The first priority when undertaking maintenance or repair work of any sort on a motorcycle is to have a clean, dry, well-lit working area. Work carried out in peace and quiet in the well-ordered atmosphere of a good workshop will give more satisfaction and much better results than can usually be achieved in poor working conditions. A good workshop must have a clean flat workbench or a solidly constructed table of convenient working height. The workbench or table should be equipped with a vice which has a jaw opening of at least 4 in (100 mm). A set of jaw covers should be made from soft metal such as aluminium alloy or copper, or from wood. These covers will minimise the marking or damaging of soft or delicate components which may be clamped in the vice. Some clean, dry, storage space will be required for tools, lubricants and dismantled components. It will be necessary during a major overhaul to lay out engine/gearbox components for examination and to keep them where they will remain undisturbed for as long as is necessary. To this end it is recommended that a supply of metal or plastic containers of suitable size is collected. A supply of clean, lint-free, rags for cleaning purposes and some newspapers, other rags, or paper towels for mopping up spillages should also be kept. If working on a hard concrete floor note that both the floor and one's knees can be protected from oil spillages and wear by cutting open a large cardboard box and spreading it flat on the floor under the machine or workbench. This also helps to provide some warmth in winter and to prevent the loss of nuts, washers, and other tiny components which have a tendency to disappear when dropped on anything other than a perfectly clean, flat, surface.

Unfortunately, such working conditions are not always available to the home mechanic. When working in poor conditions it is essential to take extra time and care to ensure that the components being worked on are kept scrupulously clean and to ensure that no components or tools are lost or damaged.

A selection of good tools is a fundamental requirement for anyone contemplating the maintenance and repair of a motor vehicle. For the owner who does not possess any, their purchase will prove a considerable expense, offsetting some of the savings made by doing-it-yourself. However, provided that the tools purchased are of good quality, they will last for many years and prove an extremely worthwhile investment.

To help the average owner to decide which tools are needed to carry out the various tasks detailed in this manual, we have compiled three lists of tools under the following headings: *Maintenance and minor repair*, *Repair and overhaul*, and *Specialized*. The newcomer to practical mechanics should start off with the simpler jobs around the vehicle. Then, as his confidence and experience grow, he can undertake more difficult tasks, buying extra tools as and when they are needed.

In this way, a *Maintenance and minor repair* tool kit can be built-up into a *Repair and overhaul* tool kit over a considerable period of time without any major cash outlays. The experienced home mechanic will have a tool kit good enough for most repair and overhaul procedures and will add tools from the specialized category when he feels the expense is justified by the amount of use these tools will be put to.

It is obviously not possible to cover the subject of tools fully here. For those who wish to learn more about tools and their use there is a book entitled *Motorcycle Workshop Practice Manual* (Book no 1454) available from the publishers of this manual.

As a general rule, it is better to buy the more expensive, good quality tools. Given reasonable use, such tools will last for a very long time, whereas the cheaper, poor quality, item will wear out faster and need to be renewed more often, thus nullifying the original saving. There is also the risk of a poor quality tool breaking while in use, causing personal injury or expensive damage to the component being worked on. It should be noted, however, that many car accessory shops and the large department stores sell tools of reasonable quality at competitive prices. The best example of this is found with socket sets, where a medium-priced socket set will be quite adequate for the home owner and yet prove less expensive than a selection of individual sockets and accessories. This is because individual pieces are usually only available from expensive, top quality, ranges and whilst they are undeniably good, it should be remembered that they are intended for professional use.

The basis of any toolkit is a set of spanners. While open-ended spanners with their slim jaws, are useful for working on awkwardly-positioned nuts, ring spanners have advantages in that they grip the nut far more positively. There is less risk of the spanner slipping off the nut and damaging it, for this reason alone ring spanners are to be preferred. Ideally, the home mechanic should acquire a set of each, but if expense rules this out a set of combination spanners (open-ended at one end and with a ring of the same size at the other) will provide a good compromise. Another item which is so useful it should be considered an essential requirement for any home mechanic is a set of socket spanners. These are available in a variety of drive sizes. It is recommended that the $\frac{1}{2}$-inch drive type is purchased to begin with as although bulkier and more expensive than the $\frac{3}{8}$-inch type, the larger size is far more common and will accept a greater variety of torque wrenches, extension pieces and socket sizes. The socket set should comprise sockets of sizes between 8 and 24 mm, a reversible ratchet drive, an extension bar of about 10 inches in length, a spark plug socket with a rubber insert, and a universal joint. Other attachments can be added to the set at a later date.

Maintenance and minor repair tool kit

Set of spanners 8 – 24 mm
Set of sockets and attachments
Spark plug spanner with rubber insert – 10, 12, or 14 mm as appropriate
Adjustable spanner
C-spanner/pin spanner
Torque wrench (same size drive as sockets)
Set of screwdrivers (flat blade)
Set of screwdrivers (cross-head)
Set of Allen keys 4 – 10 mm
Impact screwdriver and bits
Ball pein hammer – 2 lb
Hacksaw (junior)
Self-locking pliers – Mole grips or vice grips
Pliers – combination
Pliers – needle nose
Wire brush (small)
Soft-bristled brush
Tyre pump
Tyre pressure gauge
Tyre tread depth gauge
Oil can
Fine emery cloth
Funnel (medium size)
Drip tray
Grease gun
Set of feeler gauges
Brake bleeding kit
Strobe timing light
Continuity tester (dry battery and bulb)
Soldering iron and solder
Wire stripper or craft knife
PVC insulating tape
Assortment of split pins, nuts, bolts, and washers

Repair and overhaul toolkit

The tools in this list are virtually essential for anyone undertaking major repairs to a motorcycle and are additional to the tools listed above. Concerning Torx driver bits, Torx screws are encountered on some of the more modern machines where their use is restricted to fastening certain components inside the engine/gearbox unit. It is therefore recommended that if Torx bits cannot be borrowed from a local dealer, they are purchased individually as the need arises. They are not in regular use in the motor trade and will therefore only be available in specialist tool shops.

Plastic or rubber soft-faced mallet
Torx driver bits
Pliers – electrician's side cutters
Circlip pliers – internal (straight or right-angled tips are available)
Circlip pliers – external
Cold chisel
Centre punch
Pin punch
Scriber
Scraper (made from soft metal such as aluminium or copper)
Soft metal drift
Steel rule/straight edge
Assortment of files
Electric drill and bits
Wire brush (large)
Soft wire brush (similar to those used for cleaning suede shoes)
Sheet of plate glass
Hacksaw (large)
Valve grinding tool
Valve grinding compound (coarse and fine)
Stud extractor set (E-Z out)

Specialized tools

This is not a list of the tools made by the machine's manufacturer to carry out a specific task on a limited range of models. Occasional references are made to such tools in the text of this manual and, in general, an alternative method of carrying out the task without the manufacturer's tool is given where possible. The tools mentioned in this list are those which are not used regularly and are expensive to buy in view of their infrequent use. Where this is the case it may be possible to hire or borrow the tools against a deposit from a local dealer or tool hire shop. An alternative is for a group of friends or a motorcycle club to join in the purchase.

Valve spring compressor
Piston ring compressor
Universal bearing puller
Cylinder bore honing attachment (for electric drill)
Micrometer set
Vernier calipers
Dial gauge set
Cylinder compression gauge
Vacuum gauge set
Multimeter
Dwell meter/tachometer

Care and maintenance of tools

Whatever the quality of the tools purchased, they will last much longer if cared for. This means in practice ensuring that a tool is used for its intended purpose; for example screwdrivers should not be used as a substitute for a centre punch, or as chisels. Always remove dirt or grease and any metal particles but remember that a light film of oil will prevent rusting if the tools are infrequently used. The common tools can be kept together in a large box or tray but the more delicate, and more expensive, items should be stored separately where they cannot be damaged. When a tool is damaged or worn out, be sure to renew it immediately. It is false economy to continue to use a worn spanner or screwdriver which may slip and cause expensive damage to the component being worked on.

Fastening systems

Fasteners, basically, are nuts, bolts and screws used to hold two or more parts together. There are a few things to keep in mind when working with fasteners. Almost all of them use a locking device of some type; either a lock washer, lock nut, locking tab or thread adhesive. All threaded fasteners should be clean, straight, have undamaged threads and undamaged corners on the hexagon head where the spanner fits. Develop the habit of replacing all damaged nuts and bolts with new ones.

Rusted nuts and bolts should be treated with a rust penetrating fluid to ease removal and prevent breakage. After applying the rust penetrant, let it 'work' for a few minutes before trying to loosen the nut or bolt. Badly rusted fasteners may have to be chiseled off or removed with a special nut breaker, available at tool shops.

Flat washers and lock washers, when removed from an assembly should always be replaced exactly as removed. Replace any damaged washers with new ones. Always use a flat washer between a lock washer and any soft metal surface (such as aluminium), thin sheet metal or plastic. Special lock nuts can only be used once or twice before they lose their locking ability and must be renewed.

If a bolt or stud breaks off in an assembly, it can be drilled out and removed with a special tool called an E-Z out. Most dealer service departments and motorcycle repair shops can perform this task, as well as others (such as the repair of threaded holes that have been stripped out).

Spanner size comparison

Jaw gap (in)	Spanner size	Jaw gap (in)	Spanner size
0.250	$\frac{1}{4}$ in AF	0.945	24 mm
0.276	7 mm	1.000	1 in AF
0.313	$\frac{5}{16}$ in AF	1.010	$\frac{9}{16}$ in Whitworth; $\frac{5}{8}$ in BSF
0.315	8 mm	1.024	26 mm
0.344	$\frac{11}{32}$ in AF; $\frac{1}{8}$ in Whitworth	1.063	$1\frac{1}{16}$ in AF; 27 mm
0.354	9 mm	1.100	$\frac{5}{8}$ in Whitworth; $\frac{11}{16}$ in BSF
0.375	$\frac{3}{8}$ in AF	1.125	$1\frac{1}{8}$ in AF
0.394	10 mm	1.181	30 mm
0.433	11 mm	1.200	$\frac{11}{16}$ in Whitworth; $\frac{3}{4}$ in BSF
0.438	$\frac{7}{16}$ in AF	1.250	$1\frac{1}{4}$ in AF
0.445	$\frac{3}{16}$ in Whitworth; $\frac{1}{4}$ in BSF	1.260	32 mm
0.472	12 mm	1.300	$\frac{3}{4}$ in Whitworth; $\frac{7}{8}$ in BSF
0.500	$\frac{1}{2}$ in AF	1.313	$1\frac{5}{16}$ in AF
0.512	13 mm	1.390	$\frac{13}{16}$ in Whitworth; $\frac{15}{16}$ in BSF
0.525	$\frac{1}{4}$ in Whitworth; $\frac{5}{16}$ in BSF	1.417	36 mm
0.551	14 mm	1.438	$1\frac{7}{16}$ in AF
0.563	$\frac{9}{16}$ in AF	1.480	$\frac{7}{8}$ in Whitworth; 1 in BSF
0.591	15 mm	1.500	$1\frac{1}{2}$ in AF
0.600	$\frac{5}{16}$ in Whitworth; $\frac{3}{8}$ in BSF	1.575	40 mm; $\frac{15}{16}$ in Whitworth
0.625	$\frac{5}{8}$ in AF	1.614	41 mm
0.630	16 mm	1.625	$1\frac{5}{8}$ in AF
0.669	17 mm	1.670	1 in Whitworth; $1\frac{1}{8}$ in BSF
0.686	$\frac{11}{16}$ in AF	1.688	$1\frac{11}{16}$ in AF
0.709	18 mm	1.811	46 mm
0.710	$\frac{3}{8}$ in Whitworth; $\frac{7}{16}$ in BSF	1.813	$1\frac{13}{16}$ in AF
0.748	19 mm	1.860	$1\frac{1}{8}$ in Whitworth; $1\frac{1}{4}$ in BSF
0.750	$\frac{3}{4}$ in AF	1.875	$1\frac{7}{8}$ in AF
0.813	$\frac{13}{16}$ in AF	1.969	50 mm
0.820	$\frac{7}{16}$ in Whitworth; $\frac{1}{2}$ in BSF	2.000	2 in AF
0.866	22 mm	2.050	$1\frac{1}{4}$ in Whitworth; $1\frac{3}{8}$ in BSF
0.875	$\frac{7}{8}$ in AF	2.165	55 mm
0.920	$\frac{1}{2}$ in Whitworth; $\frac{9}{16}$ in BSF	2.362	60 mm
0.938	$\frac{15}{16}$ in AF		

Standard torque settings

Specific torque settings will be found at the end of the specifications section of each chapter. Where no figure is given, bolts should be secured according to the table below.

Fastener type (thread diameter)	kgf m	lbf ft
5mm bolt or nut	0.45 – 0.6	3.5 – 4.5
6 mm bolt or nut	0.8 – 1.2	6 – 9
8 mm bolt or nut	1.8 – 2.5	13 – 18
10 mm bolt or nut	3.0 – 4.0	22 – 29
12 mm bolt or nut	5.0 – 6.0	36 – 43
5 mm screw	0.35 – 0.5	2.5 – 3.6
6 mm screw	0.7 – 1.1	5 – 8
6 mm flange bolt	1.0 – 1.4	7 – 10
8 mm flange bolt	2.4 – 3.0	17 – 22
10 mm flange bolt	3.0 – 4.0	22 – 29

Choosing and fitting accessories

The range of accessories available to the modern motorcyclist is almost as varied and bewildering as the range of motorcycles. This Section is intended to help the owner in choosing the correct equipment for his needs and to avoid some of the mistakes made by many riders when adding accessories to their machines. It will be evident that the Section can only cover the subject in the most general terms and so it is recommended that the owner, having decided that he wants to fit, for example, a luggage rack or carrier, seeks the advice of several local dealers and the owners of similar machines. This will give a good idea of what makes of carrier are easily available, and at what price. Talking to other owners will give some insight into the drawbacks or good points of any one make. A walk round the motorcycles in car parks or outside a dealer will often reveal the same sort of information.

The first priority when choosing accessories is to assess exactly what one needs. It is, for example, pointless to buy a large heavy-duty carrier which is designed to take the weight of fully laden panniers and topbox when all you need is a place to strap on a set of waterproofs and a lunchbox when going to work. Many accessory manufacturers have ranges of equipment to cater for the individual needs of different riders and this point should be borne in mind when looking through a dealer's catalogues. Having decided exactly what is required and the use to which the accessories are going to be put, the owner will need a few hints on what to look for when making the final choice. To this end the Section is now sub-divided to cover the more popular accessories fitted. Note that it is in no way a customizing guide, but merely seeks to outline the practical considerations to be taken into account when adding aftermarket equipment to a motorcycle.

Fairings and windscreens

A fairing is possibly the single, most expensive, aftermarket item to be fitted to any motorcycle and, therefore, requires the most thought before purchase. Fairings can be divided into two main groups: front fork mounted handlebar fairings and windscreens, and frame mounted fairings.

The first group. the front fork mounted fairings, are becoming far more popular than was once the case, as they offer several advantages over the second group. Front fork mounted fairings generally are much easier and quicker to fit, involve less modification to the motorcycle, do not as a rule restrict the steering lock, permit a wider selection of handlebar styles to be used, and offer adequate protection for much less money than the frame mounted type. They are also lighter, can be swapped easily between different motorcycles, and are available in a much greater variety of styles. Their main disadvantages are that they do not offer as much weather protection as the frame mounted types, rarely offer any storage space, and, if poorly fitted or naturally incompatible, can have an adverse effect on the stability of the motorcycle.

The second group, the frame mounted fairings, are secured so rigidly to the main frame of the motorcycle that they can offer a substantial amount of protection to motorcycle and rider in the event of a crash. They offer almost complete protection from the weather and, if double-skinned in construction, can provide a great deal of useful storage space. The feeling of peace, quiet and complete relaxation encountered when riding behind a good full fairing has to be experienced to be believed. For this reason full fairings are considered essential by most touring motorcyclists and by many people who ride all year round. The main disadvantages of this type are that fitting can take a long time, often involving removal or modification of standard motorcycle components, they restrict the steering lock and they can add up to about 40 lb to the weight of the machine. They do not usually affect the stability of the machine to any great extent once the front tyre pressure and suspension have been adjusted to compensate for the extra weight, but can be affected by sidewinds.

The first thing to look for when purchasing a fairing is the quality of the fittings. A good fairing will have strong, substantial brackets constructed from heavy-gauge tubing; the brackets must be shaped to fit the frame or forks evenly so that the minimum of stress is imposed on the assembly when it is bolted down. The brackets should be properly painted or finished — a nylon coating being the favourite of the better manufacturers — the nuts and bolts provided should be of the same thread and size standard as is used on the motorcycle and be properly plated. Look also for shakeproof locking nuts or locking washers to ensure that everything remains securely tightened down. The fairing shell is generally made from one of two materials: fibreglass or ABS plastic. Both have their advantages and disadvantages, but the main consideration for the owner is that fibreglass is much easier to repair in the event of damage occurring to the fairing. Whichever material is used, check that it is properly finished inside as well as out, that the edges are protected by beading and that the fairing shell is insulated from vibration by the use of rubber grommets at all mounting points. Also be careful to check that the windscreen is retained by plastic bolts which will snap on impact so that the windscreen will break away and not cause personal injury in the event of an accident.

Having purchased your fairing or windscreen, read the manufacturer's fitting instructions very carefully and check that you have all the necessary brackets and fittings. Ensure that the mounting brackets are located correctly and bolted down securely. Note that some manufacturers use hose clamps to retain the mounting brackets; these should be discarded as they are convenient to use but not strong enough for the task. Stronger clamps should be substituted; car exhaust pipe clamps of suitable size would be a good alternative. Ensure that the front forks can turn through the full steering lock available without fouling the fairing. With many types of frame-mounted fairing the handlebars will have to be altered or a different type fitted and the steering lock will be restricted by stops provided with the fittings. Also check that the fairing does not foul the front wheel or mudguard, in any steering position, under full fork compression. Re-route any cables, brake pipes or electrical wiring which may snag on the fairing and take great care to protect all electrical connections, using insulating tape. If the manufacturer's instructions are followed carefully at every stage no serious problems should be encountered. Remember that hydraulic pipes that have been disconnected must be carefully re-tightened and the hydraulic system purged of air bubbles by bleeding.

Two things will become immediately apparent when taking a motorcycle on the road for the first time with a fairing – the first is the tendency to underestimate the road speed because of the lack of wind pressure on the body. This must be very carefully watched until one has grown accustomed to riding behind the fairing. The second thing is the alarming increase in engine noise which is an unfortunate but inevitable by-product of fitting any type of fairing or windscreen, and is caused by normal engine noise being reflected, and in some cases amplified, by the flat surface of the fairing.

Luggage racks or carriers

Carriers are possibly the commonest item to be fitted to modern motorcycles. They vary enormously in size, carrying capacity, and durability. When selecting a carrier, always look for one which is made specifically for your machine and which is bolted on with as few separate brackets as possible. The universal-type carrier, with its mass of brackets and adaptor pieces, will generally prove too weak to be of any real use. A good carrier should bolt to the main frame, generally using the two suspension unit top mountings and a mudguard mounting bolt as attachment points, and have its luggage platform as low and as far forward as possible to minimise the effect of any load on the machine's stability. Look for good quality, heavy gauge tubing, good welding and good finish. Also ensure that the carrier does not prevent opening of the seat, sidepanels or tail compartment, as appropriate. When using a carrier, be very careful not to overload it. Excessive weight placed so high and so far to the rear of any motorcycle will have an adverse effect on the machine's steering and stability.

Luggage

Motorcycle luggage can be grouped under two headings: soft and hard. Both types are available in many sizes and styles and have advantages and disadvantages in use.

Soft luggage is now becoming very popular because of its lower cost and its versatility. Whether in the form of tankbags, panniers, or strap-on bags, soft luggage requires in general no brackets and no modification to the motorcycle. Equipment can be swapped easily from one motorcycle to another and can be fitted and removed in seconds. Awkwardly shaped loads can easily be carried. The disadvantages of soft luggage are that the contents cannot be secure against the casual thief, very little protection is afforded in the event of a crash, and waterproofing is generally poor. Also, in the case of panniers, carrying capacity is restricted to approximately 10 lb, although this amount will vary considerably depending on the manufacturer's recommendation. When purchasing soft luggage, look for good quality material, generally vinyl or nylon, with strong, well-stitched attachment points. It is always useful to have separate pockets, especially on tank bags, for items which will be needed on the journey. When purchasing a tank bag, look for one which has a separate, well-padded, base. This will protect the tank's paintwork and permit easy access to the filler cap at petrol stations.

Hard luggage is confined to two types: panniers, and top boxes or tail trunks. Most hard luggage manufacturers produce matching sets of these items, the basis of which is generally that manufacturer's own heavy-duty luggage rack. Variations on this theme occur in the form of separate frames for the better quality panniers, fixed or quickly-detachable luggage, and in size and carrying capacity. Hard luggage offers a reasonable degree of security against theft and good protection against weather and accident damage. Carrying capacity is greater than that of soft luggage, around 15 – 20 lb in the case of panniers, although top boxes should never be loaded as much as their

apparent capacity might imply. A top box should only be used for lightweight items, because one that is heavily laden can have a serious effect on the stability of the machine. When purchasing hard luggage look for the same good points as mentioned under fairings and windscreens, ie good quality mounting brackets and fittings, and well-finished fibreglass or ABS plastic cases. Again as with fairings, always purchase luggage made specifically for your motorcycle, using as few separate brackets as possible, to ensure that everything remains securely bolted in place. When fitting hard luggage, be careful to check that the rear suspension and brake operation will not be impaired in any way and remember that many pannier kits require re-siting of the indicators. Remember also that a non-standard exhaust system may make fitting extremely difficult.

Handlebars

The occupation of fitting alternative types of handlebar is extremely popular with modern motorcyclists, whose motives may vary from the purely practical, wishing to improve the comfort of their machines, to the purely aesthetic, where form is more important than function. Whatever the reason, there are several considerations to be borne in mind when changing the handlebars of your machine. If fitting lower bars, check carefully that the switches and cables do not foul the petrol tank on full lock and that the surplus length of cable, brake pipe, and electrical wiring are smoothly and tidily disposed of. Avoid tight kinks in cable or brake pipes which will produce stiff controls or the premature and disastrous failure of an overstressed component. If necessary, remove the petrol tank and re-route the cable from the engine/gearbox unit upwards, ensuring smooth gentle curves are produced. In extreme cases, it will be necessary to purchase a shorter brake pipe to overcome this problem. In the case of higher handlebars than standard it will almost certainly be necessary to purchase extended cables and brake pipes. Fortunately, many standard motorcycles have a custom version which will be equipped with higher handlebars and, therefore, factory-built extended components will be available from your local dealer. It is not usually necessary to extend electrical wiring, as switch clusters may be used on several different motorcycles, some being custom versions. This point should be borne in mind however when fitting extremely high or wide handlebars.

When fitting different types of handlebar, ensure that the mounting clamps are correctly tightened to the manufacturer's specifications and that cables and wiring, as previously mentioned, have smooth easy runs and do not snag on any part of the motorcycle throughout the full steering lock. Ensure that the fluid level in the front brake master cylinder remains level to avoid any chance of air entering the hydraulic system. Also check that the cables are adjusted correctly and that all handlebar controls operate correctly and can be easily reached when riding.

Crashbars

Crashbars, also known as engine protector bars, engine guards, or case savers, are extremely useful items of equipment which can contribute protection to the machine's structure if a crash occurs. They do not, as has been inferred in the US, prevent the rider from crashing, or necessarily prevent rider injury should a crash occur.

It is recommended that only the smaller, neater, engine protector type of crashbar is considered. This type will offer protection while restricting, as little as is possible, access to the engine and the machine's ground clearance. The crashbars should be designed for use specifically on your machine, and should be constructed of heavy-gauge tubing with strong, integral mounting brackets. Where possible, they should bolt to a strong lug on the frame, usually at the engine mounting bolts.

The alternative type of crashbar is the larger cage type. This type is not recommended in spite of their appearance which promises some protection to the rider as well as to the machine. The larger amount of leverage imposed by the size of this type of crashbar increases the risk of severe frame damage in the event of an accident. This type also decreases the machine's ground clearance and restricts access to the engine. The amount of protection afforded the rider is open to some doubt as the design is based on the premise that the rider will stay in the normally seated position during an accident, and the crash bar structure will not itself fail. Neither result can in any way be guaranteed.

As a general rule, always purchase the best, ie usually the most expensive, set of crashbars you an afford. The investment will be repaid by minimising the amount of damage incurred, should the machine be involved in an accident. Finally, avoid the universal type of crashbar. This should be regarded only as a last resort to be used if no alternative exists. With its usual multitude of separate brackets and spacers, the universal crashbar is far too weak in design and construction to be of any practical value.

Exhaust systems

The fitting of aftermarket exhaust systems is another extremely popular pastime amongst motorcyclists. The usual motive is to gain more performance from the engine but other considerations are to gain more ground clearance, to lose weight from the motorcycle, to obtain a more distinctive exhaust note or to find a cheaper alternative to the manufacturer's original equipment exhaust system. Original equipment exhaust systems often cost more and may well have a relatively short life. It should be noted that it is rare for an aftermarket exhaust system alone to give a noticeable increase in the engine's power output. Modern motorcycles are designed to give the highest power output possible allowing for factors such as quietness, fuel economy, spread of power, and long-term reliability. If there were a magic formula which allowed the exhaust system to produce more power without affecting these other considerations you can be sure that the manufacturers, with their large research and development facilities, would have found it and made use of it. Performance increases of a worthwhile and noticeable nature only come from well-tried and properly matched modifications to the entire engine, from the air filter, through the carburettors, port timing or camshaft and valve design, combustion chamber shape, compression ratio, and the exhaust system. Such modifications are well outside the scope of this manual but interested owners might refer to the 'Piper Tuning Manual' produced by the publisher of this manual; this book goes into the whole subject in great detail.

Whatever your motive for wishing to fit an alternative exhaust system, be sure to seek expert advice before doing so. Changes to the carburettor jetting will almost certainly be required for which you must consult the exhaust system manufacturer. If he cannot supply adequately specific information it is reasonable to assume that insufficient development work has been carried out, and that particular make should be avoided. Other factors to be borne in mind are whether the exhaust system allows the use of both centre and side stands, whether it allows sufficient access to permit oil and filter changing and whether modifications are necessary to the standard exhaust system. Many two-stroke expansion chamber systems require the use of the standard exhaust pipe; this is all very well if the standard exhaust pipe and silencer are separate units but can cause problems if the two, as with so many modern two-strokes, are a one-piece unit. While the exhaust pipe can be removed easily by means of a hacksaw it is not so easy to refit the original silencer should you at any time wish to return the machine to standard trim. The same applies to several four-stroke systems.

On the subject of the finish of aftermarket exhausts, avoid black-painted systems unless you enjoy painting. As any trail-bike owner will tell you, rust has a great affinity for black exhausts and re-painting or rust removal becomes a task which must be carried out with monotonous regularity. A bright chrome finish is, as a general rule, a far better proposition as it is much easier to keep clean and to prevent rusting. Although the general finish of aftermarket exhaust systems is not always

up to the standard of the original equipment the lower cost of such systems does at least reflect this fact.

When fitting an alternative system always purchase a full set of new exhaust gaskets, to prevent leaks. Fit the exhaust first to the cylinder head or barrel, as appropriate, tightening the retaining nuts or bolts by hand only and then line up the exhaust rear mountings. If the new system is a one-piece unit and the rear mountings do not line up exactly, spacers must be fabricated to take up the difference. Do not force the system into place as the stress thus imposed will rapidly cause cracks and splits to appear. Once all the mountings are loosely fixed, tighten the retaining nuts or bolts securely, being careful not to overtighten them. Where the motorcycle manufacturer's torque settings are available, these should be used. Do not forget to carry out any carburation changes recommended by the exhaust system's manufacturer.

Electrical equipment

The vast range of electrical equipment available to motorcyclists is so large and so diverse that only the most general outline can be given here. Electrical accessories vary from electric ignition kits fitted to replace contact breaker points, to additional lighting at the front and rear, more powerful horns, various instruments and gauges, clocks, anti-theft systems, heated clothing, CB radios, radio-cassette players, and intercom systems, to name but a few of the more popular items of equipment.

As will be evident, it would require a separate manual to cover this subject alone and this section is therefore restricted to outlining a few basic rules which must be borne in mind when fitting electrical equipment. The first consideration is whether your machine's electrical system has enough reserve capacity to cope with the added demand of the accessories you wish to fit. The motorcycle's manufacturer or importer should be able to furnish this sort of information and may also be able to offer advice on uprating the electrical system. Failing this, a good dealer or the accessory manufacturer may be able to help. In some cases, more powerful generator components may be available, perhaps from another motorcycle in the manufacturer's range. The second consideration is the legal requirements in force in your area. The local police may be prepared to help with this point. In the UK for example, there are strict regulations governing the position and use of auxiliary riding lamps and fog lamps.

When fitting electrical equipment always disconnect the battery first to prevent the risk of a short-circuit, and be careful to ensure that all connections are properly made and that they are waterproof. Remember that many electrical accessories are designed primarily for use in cars and that they cannot easily withstand the exposure to vibration and to the weather. Delicate components must be rubber-mounted to insulate them from vibration, and sealed carefully to prevent the entry of rainwater and dirt. Be careful to follow exactly the accessory manufacturer's instructions in conjunction with the wiring diagram at the back of this manual.

Accessories – general

Accessories fitted to your motorcycle will rapidly deteriorate if not cared for. Regular washing and polishing will maintain the finish and will provide an opportunity to check that all mounting bolts and nuts are securely fastened. Any signs of chafing or wear should be watched for, and the cause cured as soon as possible before serious damage occurs.

As a general rule, do not expect the re-sale value of your motorcycle to increase by an amount proportional to the amount of money and effort put into fitting accessories. It is usually the case that an absolutely standard motorcycle will sell more easily at a better price than one that has been modified. If you are in the habit of exchanging your machine for another at frequent intervals, this factor should be borne in mind to avoid loss of money.

Fault diagnosis

Contents

Introduction 1

Engine does not start when turned over
No fuel flow to carburettor 2
Fuel not reaching cylinder 3
Engine flooding 4
No spark at plug 5
Weak spark at plug 6
Compression low 7

Engine stalls after starting
General causes 8

Poor running at idle and low speed
Weak spark at plug or erratic firing 9
Fuel/air mixture incorrect 10
Compression low 11

Acceleration poor
General causes 12

Poor running or lack of power at high speeds
Weak spark at plug or erratic firing 13
Fuel/air mixture incorrect 14
Compression low 15

Knocking or pinking
General causes 16

Overheating
Firing incorrect 17
Fuel/air mixture incorrect 18
Lubrication inadequate 19
Miscellaneous causes 20

Clutch operating problems
Clutch slip 21
Clutch drag 22

Gear selection problems
Gear lever does not return 23
Gear selection difficult or impossible 24
Jumping out of gear 25
Overselection 26

Abnormal engine noise
Knocking or pinking 27
Piston slap or rattling from cylinder 28
Other noises 29

Abnormal transmission noise
Clutch noise 30
Transmission noise 31

Exhaust smokes excessively
White/blue smoke (caused by oil burning) 32
Black smoke (caused by over-rich mixture) 33

Poor handling or roadholding
Directional instability 34
Steering bias to left or right 35
Handlebar vibrates or oscillates 36
Poor front fork performance 37
Front fork judder when braking 38
Poor rear suspension performance 39

Abnormal frame and suspension noise
Front end noise 40
Rear suspension noise 41

Brake problems
Brakes are spongy or ineffective – disc brakes 42
Brakes drag – disc brakes 43
Brake lever or pedal pulsates in operation – disc brakes ... 44
Disc brake noise 45
Brakes are spongy or ineffective – drum brakes 46
Brake drag – drum brakes 47
Brake lever or pedal pulsates in operation – drum brakes . 48
Drum brake noise 49
Brake induced fork judder 50

Electrical problems
Battery dead or weak 51
Battery overcharged 52
Total electrical failure 53
Circuit failure 54
Bulbs blowing repeatedly 55

1 Introduction

This Section provides an easy reference-guide to the more common ailments that are likely to afflict your machine. Obviously, the opportunities are almost limitless for faults to occur as a result of obscure failures, and to try and cover all eventualities would require a book. Indeed, a number have been written on the subject.

Successful fault diagnosis is not a mysterious 'black art' but the application of a bit of knowledge combined with a systematic and logical approach to the problem. Approach any fault diagnosis by first accurately identifying the symptom and then checking through the list of possible causes, starting with the simplest or most obvious and progressing in stages to the most complex. Take nothing for granted, but above all apply liberal quantities of common sense.

The main symptom of a fault is given in the text as a major heading below which are listed, as Sections headings, the various systems or areas which may contain the fault. Details of each possible cause for a fault and the remedial action to be taken are given, in brief, in the paragraphs below each Section heading. Further information should be sought in the relevant Chapter.

Engine does not start when turned over

2 No fuel flow to carburettor

● Fuel tank empty or level too low. Check that the tap is turned to 'On' or 'Reserve' position as required. If in doubt, prise off the fuel feed pipe at the carburettor end and check that fuel runs from pipe when the tap is turned on.

● Tank filler cap vent obstructed. This can prevent fuel from flowing into the carburettor float bowl bcause air cannot enter the fuel tank to replace it. The problem is more likely to appear when the machine is being ridden. Check by listening close to the filler cap and releasing it. A hissing noise indicates that a blockage is present. Remove the cap and clear the vent hole with wire or by using an air line from the inside of the cap.

● Fuel tap or filter blocked. Blockage may be due to accumulation of rust or paint flakes from the tank's inner surface or of foreign matter from contaminated fuel. Remove the tap and clean it and the filter. Look also for water droplets in the fuel.

● Fuel line blocked. Blockage of the fuel line is more likely to result from a kink in the line rather than the accumulation of debris.

3 Fuel not reaching cylinder

● Float chamber not filling. Caused by float needle or floats sticking in up position. This may occur after the machine has been left standing for an extended length of time allowing the fuel to evaporate. When this occurs a gummy residue is often left which hardens to a varnish-like substance. This condition may be worsened by corrosion and crystalline deposits produced prior to the total evaporation of contaminated fuel. Sticking of the float needle may also be caused by wear. In any case removal of the float chamber will be necessary for inspection and cleaning.

● Blockage in starting circuit, slow running circuit or jets. Blockage of these items may be attributable to debris from the fuel tank by-passing the filter system or to gumming up as described in paragraph 1. Water droplets in the fuel will also block jets and passages. The carburettor should be dismantled for cleaning.

● Fuel level too low. The fuel level in the float chamber is controlled by float height. The fuel level may increase with wear or damage but will never reduce, thus a low fuel level is an inherent rather than developing condition. Check the float height, adjustng the float tang if required.

4 Engine flooding

● Float valve needle worn or stuck open. A piece of rust or other debris can prevent correct seating of the needle against the valve seat thereby permitting an uncontrolled flow of fuel. Similarly, a worn needle or needle seat will prevent valve closure. Dismantle the carburettor float bowl for cleaning and, if necessary, renewal of the worn components.

● Fuel level too high. The fuel level is controlled by the float height which may increase due to wear of the float needle, pivot pin or operating tang. Check the float height, and make any necessary adjustments. A leaking float will cause an increase in fuel level, and thus should be renewed.

● Cold starting mechanism. Check the choke (starter mechanism) for correct operation. If the mechanism jams in the 'On' position subsequent starting of a hot engine will be difficult.

● Blocked air filter. A badly restricted air filter will cause flooding. Check the filter and clean or renew as required. A collapsed inlet hose will have a similar effect. Check that the air filter inlet has not become blocked by a rag or similar item.

5 No spark at plug

● Ignition switch not on.
● Engine stop switch off.
● Spark plug dirty, oiled or 'whiskered'. Because the induction mixture of a two-stroke engine is inclined to be of a rather oily nature it is comparatively easy to foul the plug electrodes, especially where there have been repeated attempts to start the engine. A machine used for short journeys will be more prone to fouling because the engine may never reach full operating temperature, and the deposits will not burn off. On rare occasions a change of plug grade may be required but the advice of a dealer should be sought before making such a change. 'Whiskering' is a comparatively rare occurrence on modern machines but may be encountered where pre-mixed petrol and oil (petroil) lubrication is employed. An electrode deposit in the form of a barely visible filament across the plug electrodes can short circuit the plug and prevent its sparking. On all two-stroke machines it is a sound precaution to carry a new spare spark plug for substitution in the event of fouling problems.

● Spark plug failure. Clean the spark plug thoroughly and reset the electrode gap. Refer to the spark plug section and the colour condition guide in Chapter 3. If the spark plug shorts internally or has sustained visible damage to the electrodes, core or ceramic insulator it should be renewed. On rare occasions a plug that appears to spark vigorously will fail to do so when refitted to the engine and subjected to the compression pressure in the cylinder.

● Spark plug cap or high tension (HT) lead faulty. Check condition and security. Replace if deterioration is evident. Most spark plugs have an internal resistor designed to inhibit electrical interference with radio and television sets. On rare occasions the resistor may break down, thus preventing sparking. If this is suspected, fit a new cap as a precaution.

● Spark plug cap loose. Check that the spark plug cap fits securely over the plug and, where fitted, the screwed terminal on the plug end is secure.

● Shorting due to moisture. Certain parts of the ignition system are susceptible to shorting when the machine is ridden or parked in wet weather. Check particularly the area from the spark plug cap back to the ignition coil. A water dispersant spray may be used to dry out waterlogged components.

Recurrence of the problem can be prevented by using an ignition sealant spray after drying out and cleaning.
● Ignition or stop switch shorted. May be caused by water corrosion or wear. Water dispersant and contact cleaning sprays may be used. If this fails to overcome the problem dismantling and visual inspection of the switches will be required.
● Shorting or open circuit in wiring. Failure in any wire connecting any of the ignition components will cause ignition malfunction. Check also that all connections are clean, dry and tight.
● Ignition coil failure. Check the coil, referring to Chapter 3.
● Capacitor (condenser) failure (where fitted). The capacitor may be checked most easily by substitution with a replacement item. Blackened contact breaker points indicate capacitor malfunction but this may not always occur.
● Contact breaker points (where fitted) pitted, burned or closed up. Check the contact breaker points, referring to Chapter 3. Check also that the low tension leads at the contact breaker are secure and not shorting out.
● Ignition source coil (contact breaker models) or pulser coil (PEI models) faulty. See Chapter 3.
● CDI unit faulty (PEI models). See Chapter 3.

6 Weak spark at plug

● Feeble sparking at the plug may be caused by any of the faults mentioned in the preceding Section other than those items in the first two paragraphs. Check first the contact breaker assembly and the spark plug, these being the most likely culprits.

7 Compression low

● Spark plug loose. This will be self-evident on inspection, and may be accompanied by a hissing noise when the engine is turned over. Remove the plug and check that the threads in the cylinder head are not damaged. Check also that the plug sealing washer is in good condition.
● Cylinder head gasket leaking. This condition is often accompanied by a high pitched squeak from around the cylinder head and oil loss, and may be caused by insufficiently tightened cylinder head fasteners, a warped cylinder head or mechanical failure of the gasket material. Re-torqueing the fasteners to the correct specification may seal the leak in some instances but if damage has occurred this course of action will provide, at best, only a temporary cure.
● Low crankcase compression. This can be caused by worn main bearings and seals and will upset the incoming fuel/air mixture. A good seal in these areas is essential on any two-stroke engine.
● Piston rings sticking or broken. Sticking of the piston rings may be caused by seizure due to lack of lubrication or overheating as a result of poor carburation or incorrect fuel type. Gumming of the rings may result from lack of use, or carbon deposits in the ring grooves. Broken rings result from over-revving, over-heating or general wear. In either case a top-end overhaul will be required.

Engine stalls after starting

8 General causes

● Improper cold start mechanism operation. Check that the operating controls function smoothly and, where applicable, are correctly adjusted. A cold engine may not require application of an enriched mixture to start initially but may baulk without choke once firing. Likewise a hot engine may start with an enriched mixture but will stop almost immediately if the choke is inadvertently in operation.
● Ignition malfunction. See Section 9. Weak spark at plug.

● Carburettor incorrectly adjusted. Maladjustment of the mixture strength or idle speed may cause the engine to stop immediately after starting. See Chapter 2.
● Fuel contamination. Check for filter blockage by debris or water which reduces, but does not completely stop, fuel flow, or blockage of the slow speed circuit in the carburettor by the same agents. If water is present it can often be seen as droplets in the bottom of the float bowl. Clean the filter and, where water is in evidence, drain and flush the fuel tank and float bowl.
● Intake air leak. Check for security of the carburettor mounting and hose connections, and for cracks or splits in the hoses. Check also that the carburettor top is secure and that the vacuum gauge adaptor plug (where fitted) is tight.
● Air filter blocked or omitted. A blocked filter will cause an over-rich mixture; the omission of a filter will cause an excessively weak mixture. Both conditions will have a detrimental effect on carburation. Clean or renew the filter as necessary.
● Fuel filler cap air vent blocked. Usually caused by dirt or water. Clean the vent orifice.
● Choked exhaust system. Caused by excessive carbon build-up in the system, particularly around the silencer baffles. In many cases these can be detached for cleaning, though mopeds have one-piece systems which require a rather different approach. Refer to Chapter 2 for further information.
● Excessive carbon build-up in the engine. This can result from failure to decarbonise the engine at the specified interval or through excessive oil consumption. On pump-fed engines check pump adjustment. On pre-mix (petroil) systems check that oil is mixed in the recommended ratio.

Poor running at idle and low speed

9 Weak spark at plug or erratic firing

● Battery voltage low. In certain conditions low battery charge, especially when coupled with a badly sulphated battery, may result in misifirng. If the battery is in good general condition it should be recharged; an old battery suffering from sulphated plates should be renewed.
● Spark plug fouled, faulty or incorrectly adjusted. See Section 4 or refer to Chapter 3.
● Spark plug cap or high tension lead shorting. Check the condition of both these items ensuring that they are in good condition and dry and that the cap is fitted correctly.
● Spark plug type incorrect. Fit plug of correct type and heat range as given in Specifications. In certain conditions a plug of hotter or colder type may be required for normal running.
● Contact breaker points (where fitted) pitted, burned or closed-up. Check the contact breaker assembly, referring to Chapter 3.
● Ignition timing incorrect. Check the ignition timing statically or dynamically, ensuring that the advance (PEI models) is functioning correctly.
● Faulty ignition coil. Partial failure of the coil internal insulation will diminish the performance of the coil. No repair is possible, a new component must be fitted.
● Faulty capacitor (condenser) (where fitted). A failure of the capacitor will cause blackening of the contact breaker point faces and will allow excessive sparking at the points. A faulty capacitor may best be checked by substitution of a serviceable replacement item.
● Defective flywheel generator ignition source. Refer to Chapter 3 for further details on test procedures.

10 Fuel/air mixture incorrect

● Intake air leak. Check carburettor mountings and air cleaner hoses for security and signs of splitting. Ensure that carburettor top is tight and that the vacuum gauge take-off plug (where fitted) is tight.

● Mixture strength incorrect. Adjust slow running mixture strength using pilot adjustment screw.
● Carburettor synchronisation.
● Pilot jet or slow running circuit blocked. The carburettor should be removed and dismantled for thorough cleaning. Blow through all jets and air passages with compressed air to clear obstructions.
● Air cleaner clogged or omitted. Clean or fit air cleaner element as necessary. Check also that the element and air filter cover are correctly seated.
● Cold start mechanism in operation. Check that the choke has not been left on inadvertently and the operation is correct. Where applicable check the operating cable free play.
● Fuel level too high or too low. Check the float height, adjusting the float tang if required. See Section 3 or 4.
● Fuel tank air vent obstructed. Obstructions usually caused by 4dirt or water. Clean vent orifice.

11 Compression low

● See Section 7. *Acceleration poor*

12 General causes

● All items as for previous Section.
● Choked air filter. Failure to keep the air filter element clean will allow the build-up of dirt with proportional loss of performance. In extreme cases of neglect acceleration will suffer.
● Choked exhaust system. This can result from failure to remove accumulations of carbon from the silencer baffles at the prescribed intervals. The increased back pressure will make the machine noticeably sluggish. Refer to Chapter 2 for further information on decarbonisation.
● Excessive carbon build-up in the engine. This can result from failure to decarbonise the engine at the specified interval or through excessive oil consumption. Check pump adjustment.
● Ignition timing incorrect. Check the timing statically or dynamically.
● Carburation fault. See Section 10.
● Mechanical resistance. Check that the brakes are not binding. On small machines in particular note that the increased rolling resistance caused by under-inflated tyres may impede acceleration.

Poor running or lack of power at high speeds

13 Weak spark at plug or erratic firing

● All items as for Section 9.
● HT lead insulation failure. Insulation failure of the HT lead and spark plug cap due to old age or damage can cause shorting when the engine is driven hard. This condition may be less noticeable, or not noticeable at all at lower engine speeds.

14 Fuel/air mixture incorrect

● All items as for Section 10, with the exception of items relative exclusively to low speed running.
● Main jet blocked. Debris from contaminated fuel, or from the fuel tank, and water in the fuel can block the main jet. Clean the fuel filter, the float bowl area, and if water is present, flush and refill the fuel tank.
● Main jet is the wrong size. The standard carburettor jetting is for sea level atmospheric pressure. For high altitudes, usually above 5000 ft, a smaller main jet will be required.
● Jet needle and needle jet worn. These can be renewed

individually but should be renewed as a pair. Renewal of both items requires partial dismantling of the carburettor.
● Air bleed holes blocked. Dismantle carburettor and use compressed air to blow out all air passages.
● Reduced fuel flow. A reduction in the maximum fuel flow from the fuel tank to the carburettor will cause fuel starvation, proportionate to the engine speed. Check for blockages through debris or a kinked fuel line.

15 Compression low

● See Section 7.

Knocking or pinking

16 General causes

● Carbon build-up in combustion chamber. After high mileages have been covered large accumulations of carbon may occur. This may glow red hot and cause premature ignition of the fuel/air mixture, in advance of normal firing by the spark plug. Cylinder head removal will be required to allow inspection and cleaning.
● Fuel incorrect. A low grade fuel, or one of poor quality may result in compression induced detonation of the fuel resulting in knocking and pinking noises. Old fuel can cause similar problems. A too highly leaded fuel will reduce detonation but will accelerate deposit formation in the combustion chamber and may lead to early pre-ignition as described in item 1.
● Spark plug heat range incorrect. Uncontrolled pre-ignition can result from the use of a spark plug the heat range of which is too hot.
● Weak mixture. Overheating of the engine due to a weak mixture can result in pre-ignition occurring where it would not occur when engine temperature was within normal limits. Maladjustment, blocked jets or passages and air leaks can cause this condition.

Overheating

17 Firing incorrect

● Spark plug fouled, defective or maladjusted. See Section 5.
● Spark plug type incorrect. Refer to the Specifications and ensure that the correct plug type is fitted.
● Incorrect ignition timing. Timing that is far too much advanced or far too much retarded will cause overheating. Check the ignition timing is correct.

18 Fuel/air mixture incorrect

● Slow speed mixture strength incorrect. Adjust pilot air screw.
● Main jet wrong size. The carburettor is jetted for sea level atmospheric conditions. For high altitudes, usually above 5000 ft, a smaller main jet will be required.
● Air filter badly fitted or omitted. Check that the filter element is in place and that it and the air filter box cover are sealing correctly. Any leaks will cause a weak mixture.
● Induction air leaks. Check the security of the carburettor mountings and hose connections, and for cracks and splits in the hoses. Check also that the carburettor top is secure and that the vacuum gauge adaptor plug (where fitted) is tight.
● Fuel level too low. See Section 3.
● Fuel tank filler cap air vent obstructed. Clear blockage.

19 Lubrication inadequate

● Oil pump settings incorrect. The oil pump settings are of great importance since the quantities of oil being injected are

very small. Any variation in oil delivery will have a significant effect on the engine. Refer to Chapter 3 for further information.
● Oil tank empty or low. This will have disastrous consequences if left unnoticed. Check and replenish tank regularly.
● Transmission oil low or worn out. Check the level regularly and investigate any loss of oil. If the oil level drops with no sign of external leakage it is likely that the crankshaft main bearing oil seals are worn, allowing transmission oil to be drawn into the crankcase during induction.

20 Miscellaneous causes

● Engine fins clogged. A build-up of mud in the cylinder head and cylinder barrel cooling fins will decrease the cooling capabilities of the fins. Clean the fins as required.

Clutch operating problems

21 Clutch slip

● No clutch lever play. Adjust clutch lever end play according to the procedure in Chapter 1.
● Friction plates worn or warped. Overhaul clutch assembly, replacing plates out of specification.
● Steel plates worn or warped. Overhaul clutch assembly, replacing plates out of specification.
● Clutch spring broken or worn. Old or heat-damaged (from slipping clutch) springs should be replaced with new ones.
● Clutch release not adjusted properly. See the adjustments section of Chapter 1.
● Clutch inner cable snagging. Caused by a frayed cable or kinked outer cable. Replace the cable with a new one. Repair of a frayed cable is not advised.
● Clutch release mechanism defective. Worn or damaged parts in the clutch release mechanism could include the shaft, cam, actuating arm or pivot. Replace parts as necessary.
● Clutch hub and outer drum worn. Severe indentation by the clutch plate tangs of the channels in the hub and drum will cause snagging of the plates preventing correct engagement. If this damage occurs, renewal of the worn components is required.
● Lubricant incorrect. Use of a transmission lubricant other than that specified may allow the plates to slip.

22 Clutch drag

● Clutch lever play excessive. Adjust lever at bars or at cable end if necessary.
● Clutch plates warped or damaged. This will cause a drag on the clutch, causing the machine to creep. Overhaul clutch assembly.
● Clutch spring tension uneven. Usually caused by a sagged or broken spring. Check and replace springs.
● Transmission oil deteriorated. Badly contaminated transmission oil and a heavy deposit of oil sludge on the plates will cause plate sticking. The oil recommended for this machine is of the detergent type, therefore it is unlikely that this problem will arise unless regular oil changes are neglected.
● Transmission oil viscosity too high. Drag in the plates will result from the use of an oil with too high a viscosity. In very cold weather clutch drag may occur until the engine has reached operating temperature.
● Clutch hub and outer drum worn. Indentation by the clutch plate tangs of the channels in the hub and drum will prevent easy plate disengagement. If the damage is light the affected areas may be dressed with a fine file. More pronounced damage will necessitate renewal of the components.

● Clutch housing seized to shaft. Lack of lubrication, severe wear or damage can cause the housing to seize to the shaft. Overhaul of the clutch, and perhaps the transmission, may be necessary to repair damage.
● Clutch release mechanism defective. Worn or damaged release mechanism parts can stick and fail to provide leverage. Overhaul clutch cover components.
● Loose clutch hub nut. Causes drum and hub misalignment, putting a drag on the engine. Engagement adjustment continually varies. Overhaul clutch assembly.

Gear selection problems

23 Gear lever does not return

● Weak or broken centraliser spring. Renew the spring.
● Gearchange shaft bent or seized. Distortion of the gearchange shaft often occurs if the machine is dropped heavily on the gear lever. Provided that damage is not severe straightening of the shaft is permissible.

24 Gear selection difficult or impossible

● Clutch not disengaging fully. See Section 22.
● Gearchange shaft bent. This often occurs if the machine is dropped heavily on the gear lever. Straightening of the shaft is permissible if the damage is not too great.
● Gearchange arms, pawls or pins worn or damaged. Wear or breakage of any of these items may cause difficulty in selecting one or more gears. Overhaul the selector mechanism.
● Gearchange drum stopper cam or detent plunger damaged. Failure, rather than wear of these items may jam the drum thereby preventing gearchanging or causing false selection at high speed.
● Selector forks bent or seized. This can be caused by dropping the machine heavily on the gearchange lever or as a result of lack of lubrication. Though rare, bending of a shaft can result from a missed gearchange or false selection at high speed.
● Selector fork end and pin wear. Pronounced wear of these items and the grooves in the gearchange drum can lead to imprecise selection and, eventually, no selection. Renewal of the worn components will be required.
● Structural failure. Failure of any one component of the selector rod and change mechanism will result in improper or fouled gear selection.

25 Jumping out of gear

● Detent plunger assembly worn or damaged. Wear of the plunger and the cam with which it locates and breakage of the detent spring can cause imprecise gear selection resulting in jumping out of gear. Renew the damaged components.
● Gear pinion dogs worn or damaged. Rounding off the dog edges and the mating recesses in adjacent pinion can lead to jumping out of gear when under load. The gears should be inspected and renewed. Attempting to reprofile the dogs is not recommended.
● Selector forks, gearchange drum and pinion grooves worn. Extreme wear of these interconnected items can occur after high mileages especially when lubrication has been neglected. The worn components must be renewed.
● Gear pinions, bushes and shafts worn. Renew the worn components.
● Bent gearchange shaft. Often caused by dropping the machine on the gear lever.
● Gear pinion tooth broken. Chipped teeth are unlikely to cause jumping out of gear once the gear has been selected fully; a tooth which is completely broken off, however, may cause

problems in this respect and in any event will cause transmission noise.

26 Overselection

● Pawl spring weak or broken. Renew the spring.
● Detent plunger worn or broken. Renew the damaged items.
● Stopper arm spring worn or broken. Renew the spring.
● Gearchange arm stop pads worn. Repairs can be made by welding and reprofiling with a file.
● Selector limiter claw components (where fitted) worn or damaged. Renew the damaged items.

Abnormal engine noise

27 Knocking or pinking

● See Section 16.

28 Piston slap or rattling from cylinder

● Cylinder bore/piston clearance excessive. Resulting from wear, or partial seizure. This condition can often be heard as a high, rapid tapping noise when the engine is under little or no load, particularly when power is just beginning to be applied. Reboring to the next correct oversize should be carried out and a new oversize piston fitted.
● Connecting rod bent. This can be caused by over-revving, trying to start a very badly flooded engine (resulting in a hydraulic lock in the cylinder) or by earlier mechanical failure. Attempts at straightening a bent connecting rod from a high performance engine are not recommended. Careful inspection of the crankshaft should be made before renewing the damaged connecting rod.
● Gudgeon pin, piston boss bore or small-end bearing wear or seizure. Excess clearance or partial seizure between normal moving parts of these items can cause continuous or intermittent tapping noises. Rapid wear or seizure is caused by lubrication starvation.
● Piston rings worn, broken or sticking. Renew the rings after careful inspection of the piston and bore.

29 Other noises

● Big-end bearing wear. A pronounced knock from within the crankcase which worsens rapidly is indicative of big-end bearing failure as a result of extreme normal wear or lubrication failure. Remedial action in the form of a bottom end overhaul should be taken; continuing to run the engine will lead to further damage including the possibility of connecting rod breakage.
● Main bearing failure. Extreme normal wear or failure of the main bearings is characteristically accompanied by a rumble from the crankcase and vibration felt through the frame and footrests. Renew the worn bearings and carry out a very careful examination of the crankshaft.
● Crankshaft excessively out of true. A bent crank may result from over-revving or damage from an upper cylinder component or gearbox failure. Damage can also result from dropping the machine on either crankshaft end. Straightening of the crankshaft is not be possible in normal circumstances; a replacement item should be fitted.
● Engine mounting loose. Tighten all the engine mounting nuts and bolts.
● Cylinder head gasket leaking. The noise most often associated with a leaking head gasket is a high pitched squeaking, although any other noise consistent with gas being forced out

under pressure from a small orifice can also be emitted. Gasket leakage is often accompanied by oil seepage from around the mating joint or from the cylinder head holding down bolts and nuts. Leakage results from insufficient or uneven tightening of the cylinder head fasteners, or from random mechanical failure. Retightening to the correct torque figure will, at best, only provide a temporary cure. The gasket should be renewed at the earliest opportunity.
● Exhaust system leakage. Popping or crackling in the exhaust system, particularly when it occurs with the engine on the overrun, indicates a poor joint either at the cylinder port or at the exhaust pipe/silencer connection. Failure of the gasket or looseness of the clamp should be looked for.

Abnormal transmission noise

30 Clutch noise

● Clutch outer drum/friction plate tang clearance excessive.
● Clutch outer drum/spacer clearance excessive.
● Clutch outer drum/thrust washer clearance excessive.
● Primary drive gear teeth worn or damaged.
● Clutch shock absorber assembly worn or damaged.

31 Transmission noise

● Bearing or bushes worn or damaged. Renew the affected components.
● Gear pinions worn or chipped. Renew the gear pinions.
● Metal chips jammed in gear teeth.This can occur when pieces of metal from any failed component are picked up by a meshing pinion. The condition will lead to rapid bearing wear or early gear failure.
● Engine/transmission oil level too low. Top up immediately to prevent damage to gearbox and engine.
● Gearchange mechanism worn or damaged. Wear or failure of certain items in the selection and change components can induce mis-selection of gears (see Section 24) where incipient engagement of more than one gear set is promoted. Remedial action, by the overhaul of the gearbox, should be taken without delay.
● Chain snagging on cases or cycle parts. A badly worn chain or one that is excessively loose may snag or smack against adjacent components.

Exhaust smokes excessively

32 White/blue smoke (caused by oil burning)

● Piston rings worn or broken. Breakage or wear of any ring, but particularly the oil control ring, will allow engine oil past the piston into the combustion chamber. Examine and renew, where necessary, the cylinder barrel and piston.
● Cylinder cracked, worn or scored. These conditions may be caused by overheating, lack of lubrication, component failure or advanced normal wear. The cylinder barrel should be renewed and, if necessary, a new piston fitted.
● Petrol/oil ratio incorrect. Ensure that oil is mixed with the petrol in the correct ratio. The manufacturer's recommendation must be adhered to if excessive smoking or under-lubrication is to be avoided.
● Oil pump settings incorrect. Check and reset the oil pump as described in Chapter 2.
● Crankshaft main bearing oil seals worn. Wear in the main bearing oil seals, often in conjunction with wear in the bearings themselves, can allow transmission oil to find its way into the crankcase and thence to the combustion chamber. This

condition is often indicated by a mysterious drop in the transmission oil level with no sign of external leakage.

● Accumulated oil deposits in exhaust system. If the machine is used for short journeys only it is possible for the oil residue in the exhaust gases to condense in the relatively cool silencer. If the machine is then taken for a longer run in hot weather, the accumulated oil will burn off producing ominous smoke from the exhaust.

33 Black smoke (caused by over-rich mixture)

● Air filter element clogged. Clean or renew the element.
● Main jet loose or too large. Remove the float chamber to check for tightness of the jet. If the machine is used at high altitudes rejetting will be required to compensate for the lower atmospheric pressure.
● Cold start mechanism jammed on. Check that the mechanism works smoothly and correctly and that, where fitted, the operating cable is lubricated and not snagged.
● Fuel level too high. The fuel level is controlled by the float height which can increase as a result of wear or damage. Remove the float bowl and check the float height. Check also that floats have not punctured; a punctured float will lose buoyancy and allow an increased fuel level.
● Float valve needle stuck open. Caused by dirt or a worn valve. Clean the float chamber or renew the needle and, if necessary, the valve seat.

Poor handling or roadholding

34 Directional instability

● Steering head bearing adjustment too tight. This will cause rolling or weaving at low speeds. Re-adjust the bearings.
● Steering head bearing worn or damaged. Correct adjustment of the bearing will prove impossible to achieve if wear or damage has occurred. Inconsistent handling will occur including rolling or weaving at low speed and poor directional control at indeterminate higher speeds. The steering head bearing should be dismantled for inspection and renewed if required. Lubrication should also be carried out.
● Bearing races pitted or dented. Impact damage caused, perhaps, by an accident or riding over a pot-hole can cause indentation of the bearing, usually in one position. This should be noted as notchiness when the handlebars are turned. Renew and lubricate the bearings.
● Steering stem bent. This will occur only if the machine is subjected to a high impact such as hitting a curb or a pot-hole. The lower yoke/stem should be renewed; do not attempt to straighten the stem.
● Front or rear tyre pressures too low.
● Front or rear tyre worn. General instability, high speed wobbles and skipping over white lines indicates that tyre renewal may be required. Tyre induced problems, in some machine/tyre combinations, can occur even when the tyre in question is by no means fully worn.
● Swinging arm bearings worn. Difficulties in holding line, particularly when cornering or when changing power settings indicates wear in the swinging arm bearings. The swinging arm should be removed from the machine and the bearings renewed.
● Swinging arm flexing. The symptoms given in the preceding paragraph will also occur if the swinging arm fork flexes badly. This can be caused by structural weakness as a result of corrosion, fatigue or impact damage, or because the rear wheel spindle is slack.
● Wheel bearings worn. Renew the worn bearings.
● Loose wheel spokes. The spokes should be tightened evenly to maintain tension and trueness of the rim.

● Tyres unsuitable for machine. Not all available tyres will suit the characteristics of the frame and suspension, indeed, some tyres or tyre combinations may cause a transformation in the handling characteristics. If handling problems occur immediately after changing to a new tyre type or make, revert to the original tyres to see whether an improvement can be noted. In some instances a change to what are, in fact, suitable tyres may give rise to handling deficiences. In this case a thorough check should be made of all frame and suspension items which affect stability.

35 Steering bias to left or right

● Rear wheel out of alignment. Caused by uneven adjustment of chain tensioner adjusters allowing the wheel to be askew in the fork ends. A bent rear wheel spindle will also misalign the wheel in the swinging arm.
● Wheels out of alignment. This can be caused by impact damage to the frame, swinging arm, wheel spindles or front forks. Although occasionally a result of material failure or corrosion it is usually as a result of a crash.
● Front forks twisted in the steering yokes. A light impact, for instance with a pot-hole or low curb, can twist the fork legs in the steering yokes without causing structural damage to the fork legs or the yokes themselves. Re-alignment can be made by loosening the yoke pinch bolts, wheel spindle and mudguard bolts. Re-align the wheel with the handlebars and tighten the bolts working upwards from the wheel spindle. This action should be carried out only when there is no chance that structural damage has occurred.

36 Handlebar vibrates or oscillates

● Tyres worn or out of balance. Either condition, particularly in the front tyre, will promote shaking of the fork assembly and thus the handlebars. A sudden onset of shaking can result if a balance weight is displaced during use.
● Tyres badly positioned on the wheel rims. A moulded line on each wall of a tyre is provided to allow visual verification that the tyre is correctly positioned on the rim. A check can be made by rotating the tyre; any misalignment will be immediately obvious.
● Wheels rims warped or damaged. Inspect the wheels for runout as described in Chapter 5.
● Swinging arm bearings worn. Renew the bearings.
● Wheel bearings worn. Renew the bearings.
● Steering head bearings incorrectly adjusted. Vibration is more likely to result from bearings which are too loose rather than too tight. Re-adjust the bearings.
● Loosen fork component fasteners. Loose nuts and bolts holding the fork legs, wheel spindle, mudguards or steering stem can promote shaking at the handlebars. Fasteners on running gear such as the forks and suspension should be check tightened occasionally to prevent dangerous looseness of components occurring.
● Engine mounting bolts loose. Tighten all fasteners.

37 Poor front fork performance

● Damping fluid level incorrect. If the fluid level is too low poor suspension control will occur resulting in a general impairment of roadholding and early loss of tyre adhesion when cornering and braking. Too much oil is unlikely to change the fork characteristics unless severe overfilling occurs when the fork action will become stiffer and oil seal failure may occur.
● Damping oil viscosity incorrect. The damping action of the fork is directly related to the viscosity of the damping oil. The lighter the oil used, the less will be the damping action

imparted. For general use, use the recommended viscosity of oil, changing to a slightly higher or heavier oil only when a change in damping characteristic is required. Overworked oil, or oil contaminated with water which has found its way past the seals, should be renewed to restore the correct damping performance and to prevent bottoming of the forks.

● Damping components worn or corroded. Advanced normal wear of the fork internals is unlikely to occur until a very high mileage has been covered. Continual use of the machine with damaged oil seals which allows the ingress of water, or neglect, will lead to rapid corrosion and wear. Dismantle the forks for inspection and overhaul.

● Weak fork springs. Progressive fatigue of the fork springs, resulting in a reduced spring free length, will occur after extensive use. This condition will promote excessive fork dive under braking, and in its advanced form will reduce the at-rest extended length of the forks and thus the fork geometry. Renewal of the springs as a pair is the only satisfactory course of action.

● Bent stanchions or corroded stanchions. Both conditions will prevent correct telescoping of the fork legs, and in an advanced state can cause sticking of the fork in one position. In a mild form corrosion will cause stiction of the fork thereby increasing the time the suspension takes to react to an uneven road surface. Bent fork stanchions should be attended to immediately because they indicate that impact damage has occurred, and there is a danger that the forks will fail with disastrous consequences.

38 Front fork judder when braking (see also Section 41)

● Wear between the fork stanchions and the fork legs. Renewal of the affected components is required.
● Slack steering head bearings. Re-adjust the bearings.
● Warped brake disc or drum. If irregular braking action occurs fork judder can be induced in what are normally serviceable forks. Renew the damaged brake components.

39 Poor rear suspension performances

● Rear suspension unit damper worn out or leaking. The damping performance of most rear suspension units falls off with age. This is a gradual process, and thus may not be immediately obvious. Indications of poor damping include hopping of the rear end when cornering or braking, and a general loss of positive stability.
● Weak rear springs. If the suspension unit springs fatigue they will promote excessive pitching of the machine and reduce the ground clearance when cornering. Although replacement springs are available separately from the rear suspension damper unit it is probable that if spring fatigue has occurred the damper units will also require renewal.
● Swinging arm flexing or bearings worn. See Sections 34 and 36.
● Bent suspension unit damper rod. This is likely to occur only if the machine is dropped or if seizure of the piston occurs. If either happens the suspension units should be renewed as a pair.

Abnormal frame and suspension noise

40 Front end noise

● Oil level low or too thin. This can cause a 'spurting' sound and is usually accompanied by irregular fork action.
● Spring weak or broken. Makes a clicking or scraping sound. Fork oil will have a lot of metal particles in it.

● Steering head bearings loose or damaged. Clicks when braking. Check, adjust or replace.
● Fork clamps loose. Make sure all fork clamp pinch bolts are tight.
● Fork stanchion bent. Good possibility if machine has been dropped. Repair or replace tube.

41 Rear suspension noise

● Fluid level too low. Leakage of a suspension unit, usually evident by oil on the outer surfaces, can cause a spurting noise. The suspension units should be renewed as a pair.
● Defective rear suspension unit with internal damage. Renew the suspension units as a pair.

Brake problems

42 Brakes are spongy or ineffective – disc brakes

● Air in brake circuit. This is only likely to happen in service due to neglect in checking the fluid level or because a leak has developed. The problem should be identified and the brake system bled of air.
● Pad worn. Check the pad wear against the wear lines provided and renew the pads if necessary.
● Contaminated pads. Cleaning pads which have been contaminated with oil, grease or brake fluid is unlikely to prove successful; the pads should be renewed.
● Pads glazed. This is usually caused by overheating. The surface of the pads may be roughened using glass-paper or a fine file.
● Brake fluid deterioration. A brake which on initial operation is firm but rapidly becomes spongy in use may be failing due to water contamination of the fluid. The fluid should be drained and then the system refilled and bled.
● Master cylinder seal failure. Wear or damage of master cylinder internal parts will prevent pressurisation of the brake fluid. Overhaul the master cylinder unit.
● Caliper seal failure. This will almost certainly be obvious by loss of fluid, a lowering of fluid in the master cylinder reservoir and contamination of the brake pads and caliper. Overhaul the caliper assembly.
● Brake lever or pedal improperly adjusted. Adjust the clearance between the lever end and master cylinder plunger to take up lost motion, as recommended in Routine maintenance.

43 Brakes drag – disc brakes

● Disc warped. The disc must be renewed.
● Caliper piston, caliper or pads corroded. The brake caliper assembly is vulnerable to corrosion due to water and dirt, and unless cleaned at regular intervals and lubricated in the recommended manner, will become sticky in operation.
● Piston seal deteriorated. The seal is designed to return the piston in the caliper to the retracted position when the brake is released. Wear or old age can affect this function. The caliper should be overhauled if this occurs.
● Brake pad damaged. Pad material separating from the backing plate due to wear or faulty manufacture. Renew the pads. Faulty installation of a pad also will cause dragging.
● Wheel spindle bent. The spindle may be straightened if no structural damage has occurred.
● Brake lever or pedal not returning. Check that the lever or pedal works smoothly throughout its operating range and does not snag on any adjacent cycle parts. Lubricate the pivot if necessary.

● Twisted caliper support bracket. This is likely to occur only after impact in an accident. No attempt should be made to re-align the caliper; the bracket should be renewed.

44 Brake lever or pedal pulsates in operation – disc brakes

● Disc warped or irregularly worn. The disc must be renewed.
● Wheel spindle bent. The spindle may be straightened provided no structural damage has occurred.

45 Disc brake noise

● Brake squeal. This can be caused by the omission or incorrect installation of the anti-squeal shim fitted to the rear of one pad. The arrow on the shim should face the direction of wheel normal rotation. Squealing can also be caused by dust on the pads, usually in combination with glazed pads, or other contamination from oil, grease, brake fluid or corrosion. Persistent squealing which cannot be traced to any of the normal causes can often be cured by applying a thin layer of high temperature silicone grease to the rear of the pads. Make absolutely certain that no grease is allowed to contaminate the braking surface of the pads.
● Glazed pads. This is usually caused by high temperatures or contamination. The pad surfaces may be roughened using glass-paper or a fine file. If this approach does not effect a cure the pads should be renewed.
● Disc warped. This can cause a chattering, clicking or intermittent squeal and is usually accompanied by a pulsating brake lever or pedal or uneven braking. The disc must be renewed.
● Brake pads fitted incorrectly or undersize. Longitudinal play in the pads due to omission of the locating springs (where fitted) or because pads of the wrong size have been fitted will cause a single tapping noise every time the brake is operated. Inspect the pads for correct installation and security.

46 Brakes are spongy or ineffective – drum brakes

● Brake cable deterioration. Damage to the outer cable by stretching or being trapped will give a spongy feel to the brake lever. The cable should be renewed. A cable which has become corroded due to old age or neglect of lubrication will partially seize making operation very heavy. Lubrication at this stage may overcome the problem but the fitting of a new cable is recommended.
● Worn brake linings. Determine lining wear using the external brake wear indicator on the brake backplate, or by removing the wheel and withdrawing the brake backplate. Renew the shoe/lining units as a pair if the linings are worn below the recommended limit.
● Worn brake camshaft. Wear between the camshaft and the bearing surface will reduce brake feel and reduce operating efficiency. Renewal of one or both items will be required to rectify the fault.
● Worn brake cam and shoe ends. Renew the worn components.
● Linings contaminated with dust or grease. Any accumulations of dust should be cleaned from the brake assembly and drum using a petrol dampened cloth. Do not blow or brush off the dust because it is asbestos based and thus harmful if inhaled. Light contamination from grease can be removed from the surface of the brake linings using a solvent; attempts at removing heavier contamination are less likely to be successful because some of the lubricant will have been absorbed by the lining material which will severely reduce the braking performance.

47 Brake drag – drum brakes

● Incorrect adjustment. Re-adjust the brake operating mechanism.
● Drum warped or oval. This can result from overheating or impact or uneven tension of the wheel spokes. The condition is difficult to correct, although if slight ovality only occurs, skimming the surface of the brake drum can provide a cure. This is work for a specialist engineer. Renewal of the complete wheel hub is normally the only satisfactory solution.
● Weak brake shoe return springs. This will prevent the brake lining/shoe units from pulling away from the drum surface once the brake is released. The springs should be renewed.
● Brake camshaft, lever pivot or cable poorly lubricated. Failure to attend to regular lubrication of these areas will increase operating resistance which, when compounded, may cause tardy operation and poor release movement.

48 Brake lever or pedal pulsates in operation – drum brakes

● Drums warped or oval. This can result from overheating or impact or uneven spoke tension. This condition is difficult to correct, although if slight ovality only occurs skimming the surface of the drum can provide a cure. This is work for a specialist engineer. Renewal of the hub is normally the only satisfactory solution.

49 Drum brake noise

● Drum warped or oval. This can cause intermittent rubbing of the brake linings against the drum. See the preceding Section.
● Brake linings glazed. This condition, usually accompanied by heavy lining dust contamination, often induces brake squeal. The surface of the linings may be roughened using glass-paper or a fine file.

50 Brake induced fork judder

● Worn front fork stanchions and legs, or worn or badly adjusted steering head bearings. These conditions, combined with uneven or pulsating braking as described in Sections 44 and 48 will induce more or less judder when the brakes are applied, dependent on the degree of wear and poor brake operation. Attention should be given to both areas of malfunction. See the relevant Sections.

Electrical problems

51 Battery dead or weak

● Battery faulty. Battery life should not be expected to exceed 3 to 4 years, particularly where a starter motor is used regularly. Gradual sulphation of the plates and sediment deposits will reduce the battery performance. Plate and insulator damage can often occur as a result of vibration. Complete power failure, or intermittent failure, may be due to a broken battery terminal. Lack of electrolyte will prevent the battery maintaining charge.
● Battery leads making poor contact. Remove the battery leads and clean them and the terminals, removing all traces of corrosion and tarnish. Reconnect the leads and apply a coating of petroleum jelly to the terminals.
● Load excessive. If additional items such as spot lamps, are fitted, which increase the total electrical load above the

maximum alternator output, the battery will fail to maintain full charge. Reduce the electrical load to suit the electrical capacity.
● Rectifier or ballast resistor failure.
● Alternator generating coils open-circuit or shorted.
● Charging circuit shorting or open circuit. This may be caused by frayed or broken wiring, dirty connectors or a faulty ignition switch. The system should be tested in a logical manner. See Section 54.

52 Battery overcharged

● Regulator or ballast resistor faulty. Overcharging is indicated if the battery becomes hot or it is noticed that the electrolyte level falls repeatedly between checks. In extreme cases the battery will boil causing corrosive gases and electrolyte to be emitted through the vent pipes.
● Battery wrongly matched to the electrical circuit. Ensure that the specified battery is fitted to the machine.

53 Total electrical failure

● Fuse blown. Check the main fuse. If a fault has occurred, it must be rectified before a new fuse is fitted.
● Battery faulty. See Section 51.
● Earth failure. Check that the frame main earth strap from the battery is securely affixed to the frame and is making a good contact.
● Ignition switch or power circuit failure. Check for current flow through the battery positive lead (red) to the ignition switch. Check the ignition switch for continuity.

54 Circuit failure

● Cable failure. Refer to the machine's wiring diagram and check the circuit for continuity. Open circuits are a result of loose or corroded connections, either at terminals or in-line connectors, or because of broken wires. Occasionally, the core of a wire will break without there being any apparent damage to the outer plastic cover.
● Switch failure. All switches may be checked for continuity in each switch position, after referring to the switch position boxes incorporated in the wiring diagram for the machine. Switch failure may be a result of mechanical breakage, corrosion or water.
● Fuse blown. Refer to the wiring diagram to check whether or not a circuit fuse is fitted. Replace the fuse, if blown, only after the fault has been identified and rectified.

55 Bulbs blowing repeatedly

● Vibration failure. This is often an inherent fault related to the natural vibration characteristics of the engine and frame and is, thus, difficult to resolve. Modifications of the lamp mounting, to change the damping characteristics, may help.
● Intermittent earth. Repeated failure of one bulb, particularly where the bulb is fed directly from the generator, indicates that a poor earth exists somewhere in the circuit. Check that a good contact is available at each earthing point in the circuit.
● Reduced voltage. Where a quartz-halogen bulb is fitted the voltage to the bulb should be maintained or early failure of the bulb will occur. Do not overload the system with additional electrical equipment in excess of the system's power capacity and ensure that all circuit connections are maintained clean and tight.

Routine maintenance

Periodic routine maintenance is a continuous process which should commence immediately the machine is used. The object is to maintain all adjustments and to diagnose and rectify minor defects before they develop into more extensive, and often more expensive, problems.

It follows that if the machine is maintained properly, it will both run and perform with optimum efficiency, and be less prone to unexpected breakdowns. Regular inspection of the machine will show up any parts which are wearing, and with a little experience, it is possible to obtain the maximum life from any one component, renewing it when it becomes so worn that it is liable to fail.

Regular cleaning can be considered as important as mechanical maintenance. This will ensure that all the cycle parts are inspected regularly and are kept free from accumulations of road dirt and grime.

Cleaning is especially important during the winter months, despite its appearance of being a thankless task which very soon seems pointless. On the contrary, it is during these months that the paintwork, chromium plating, and the alloy casings suffer the ravages of abrasive grit, rain and road salt. A couple of hours spent weekly on cleaning the machine will maintain its appearance and value, and highlight small points, like chipped paint, before they become a serious problem.

The various maintenance tasks are described under their respective mileage and calendar headings, and are accompanied by diagrams and photographs where pertinent.

It should be noted that the intervals between each maintenance task serve only as a guide. As the machine gets older, or if it is used under particularly arduous conditions, it is advisable to reduce the period between each check.

For ease of reference, most service operations are described in detail under the relevant heading. However, if further general information is required, this can be found under the pertinent Section heading and Chapter in the main text.

Although no special tools are required for routine maintenance, a good selection of general workshop tools is essential. Included in the tools must be a range of metric ring or combination spanners, a selection of crosshead screwdrivers, and two pairs of circlip pliers, one external opening and the other internal opening. Additionally, owing to the extreme tightness of most casing screws on Japanese machines, an impact screwdriver, together with a choice of large or small cross-head screw bits, is absolutely indispensable. This is particularly so if the engine has not been dismantled since leaving the factory.

Weekly, or every 200 miles (300 km)

---

1 Safety inspection
Give the complete machine a close and thorough visual inspection, checking for loose nuts, bolts and fittings frayed control cables, severe oil and petrol leaks, etc.

2 Legal inspection
Check the operation of the electrical system, ensuring that the lights and horn are working properly and that the lenses are clean. Note that in the UK it is an offence to use a vehicle on which the lights are defective. This applies even when the machine is used during daylight hours. The horn is also a statutory requirement.

3 Tyre pressures
Check the tyre pressures. Always check with the tyres cold, using a pressure gauge known to be accurate. It is recommended that a pocket pressure gauge is purchased to offset any fluctuation between garage forecourt instruments. The tyre pressures should be as follows:

	TS100 models	All other models
Solo:		
Front	21 psi	21 psi
	(1.5 kg/cm²)	(1.5 kg/cm²)
Rear	28 psi	25 psi
	(2.0 kg/cm²)	(1.75 kg/cm²)
With pillion:		
Front	25 psi	21 psi
	(1.75 kg/cm²)	(1.5 kg/cm²)
Rear	31 psi	28 psi
	(2.25 kg/cm²)	(2.0 kg/cm²)

At this juncture also inspect the actual condition of the tyres, ensuring there are no splits or cracks which may develop into serious problems. Also remove any small stones or other small objects of road debris which may be lodged between the tread blocks. A small flat-bladed screwdriver will be admirable for this job. Examine the amount of tread remaining on the tyre.

Check the tyre pressures with an accurate gauge

Check the tyre tread depth

The manufacturer's recommended minimum tread depth is 4 mm (0.16 in). A tyre with a tread depth below this figure should be renewed.

4   Engine oil level

The oil tank level should be checked on a daily basis, prior to starting the engine. A small plastic window or sight glass gives an immediate visual warning of whether the oil level has dropped too low. Although it is quite safe to use the machine as long as oil is visible in the sight glass, it is recommended that the level is maintained to within about an inch of the tank filler neck, to allow a good reserve. It is advised that the oil tank level is checked whenever the machine is refuelled, and topped up as required.

5   Control cable lubrication

Apply a few drops of motor oil to the exposed inner portion of each control cable. This will prevent drying-up of the cables before the more thorough lubrication that should be carried out during the 2000 mile/3 monthly service.

6   Final drive chain lubrication

In order that the life of the final drive chain be extended as much as possible, regular lubrication is essential. Intermediate lubrication should take place with the chain in position on the machine. The chain should be lubricated by the application of one of the proprietary chain greases contained in an aerosol can. Ordinary engine oil can be used, though owing to the speed with which it is flung off the rotating chain, its effective life is limited.

Check the engine oil level through the sight glass

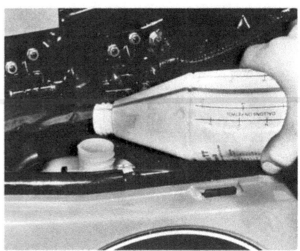

Keep the engine oil well topped up

Monthly, or every 600 miles (1000 km)

Complete the tasks listed under the weekly/200 mile heading and then carry out the following checks:

1   Final drive chain adjustment

Check the slack in the final drive chain. The correct up and down movement, as measured at the mid-point of the chain lower run, should be as listed in the following table:

Model type	Chain movement
TS100	30 – 40 mm (1.2 – 1.6 in)
TS125	40 – 50 mm (1.6 – 2.0 in)
TS185	35 – 45 mm (1.4 – 1.8 in)
UK TS250	35 – 45 mm (1.4 – 1.8 in)
US TS250	50 – 60 mm (2.0 – 2.4 in)

Note that each measurement listed in the above table must be made with the rider off the machine and with the machine supported on its prop stand. The measurement of 15 – 20 mm (0.6 – 0.8 in) given by Suzuki in some of their service literature applies only when the rider is seated and with the rear suspension compressed. Remember that the chain may not have worn evenly. Rotate the rear wheel to check along the chain length for its tight spot and make the necessary measurement with this tight spot in the centre of the lower chain run. If necessary, adjust the chain by carrying out the following procedure.

Position the machine securely on its stand, remove the split-pin from the wheel spindle and slacken the wheel nut. Loosen the locknuts on the two chain adjuster bolts and slacken off the brake torque rod nuts.

Rotation of the adjuster bolts in a clockwise direction will tighten the chain. Tighten each bolt a similar number of turns so that wheel alignment is maintained. This can be verified by checking that the mark on the outer face of each chain adjuster is aligned with the same aligning mark on each fork end. With the adjustment correct, tighten the wheel nut to its specified torque setting (see Specifications, Chapter 5) and lock it in

Use an aerosol chain lubricant at regular intervals

Check the wheel alignment marks

Lock each adjuster bolt in position with its locknut

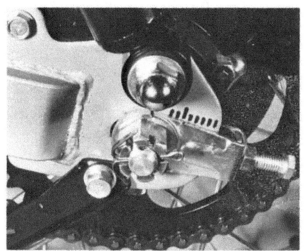

Lock the wheel spindle nut with a new split-pin

Check the battery electrolyte level

position with a new split-pin. Spin the rear wheel to ensure that it rotates freely, adjust the rear brake operating mechanism (where necessary) and retighten the torque arm retaining nuts to their specified torque setting. Finally, retighten the adjuster bolt locknuts.

Note that under no circumstances should the chain be run overtight to compensate for uneven wear. A tight chain will place excessive stresses on the gearbox and rear wheel bearings leading to their early failure. It will also absorb a surprising amount of power.

2 Battery electrolyte level

Gain access to the battery by removing the right-hand side panel from the machine. Unplug the battery terminal connections, remove the battery from its tray and place it on a flat and level work surface.

Check the level of the electrolyte through the translucent case of the battery. If the electrolyte level does not lie between the upper and lower level marks on the battery case, then replenish each cell with distilled water until the level rises to the upper mark. Take care not to overfill the battery.

Before reconnecting the battery terminal connections, make sure that they are free of contamination before protecting them with petroleum jelly (**not** grease). The battery must be

retained securely in its tray and have its vent pipe correctly connected and routed.

Three monthly, or every 2000 miles (3000 km)

Complete the checks listed under the preceding Routine Maintenance Sections and then complete the following:

1 Final drive chain lubrication

Lubrication of the final drive chain should be carried out at short intervals as described in Section 6 of the weekly/200 mile maintenance schedule. A more thorough lubrication of the chain should be carried out every three months/2000 miles by carrying out the following procedure.

Move the machine until the split link appears at the rear wheel sprocket. Support the machine securely on its prop stand. Separate the ends of the chain by removing the split link and then run the chain off the gearbox sprocket. If an old chain is available, interconnect the old and new chain, before the new chain is run off the sprockets. In this way the old chain can be pulled into place on the gearbox sprocket and then used to pull the regreased chain into place with ease. Otherwise, it will be necessary to remove the gearbox sprocket cover before refitting the chain.

Clean the chain thoroughly in a paraffin bath and then finally with petrol. The petrol will wash the paraffin out of the links and rollers which will then dry more quickly. Remember to observe the necessary fire precautions during this cleaning procedure.

Allow the chain to dry and then immerse it in a molten lubricant such as Linklyfe or Chainguard. These lubricants must be used hot and will achieve better penetration of the links and rollers. They are less likely to be thrown off by centrifugal force when the chain is in motion.

Refit the newly greased chain onto the sprocket, refitting the spring link. This is accomplished most easily when the free ends of the chain are pushed into mesh on the rear wheel sprocket. The spring link must be fitted so that the closed end faces the normal direction of chain travel. Adjust the chain to the correct tension before the machine is used.

2   Control cable lubrication

Lubricate the control cables thoroughly with motor oil or an all-purpose oil. A good method of lubricating the cables is shown in the accompanying illustration, using a plasticine funnel. This method has the disadvantage that the cables usually need removing from the machine. A hydraulic cable oiler which pressurises the lubricant overcomes this problem. Do not lubricate nylon lined cables (which may have been fitted as replacements), as the oil may cause the nylon to swell, thereby causing total cable seizure.

3   General lubrication

Work around the machine, applying grease or oil to any pivot points. These points should include the handlebar lever pivots and the prop stand pivot.

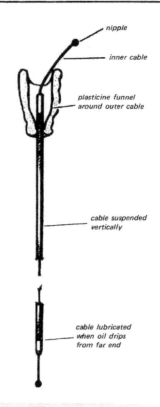

Oiling a control cable

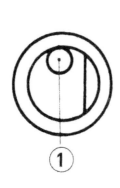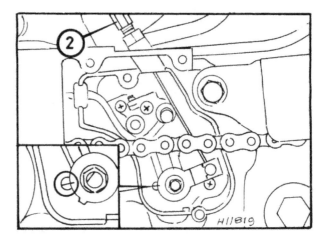

Oil pump cable adjustment marks

1   Throttle valve punch mark   2   Cable adjuster

4   Oil pump adjustment

In order to check the oil pump for correct adjustment, it is first necessary to expose the pump by removing both the gearbox sprocket cover and the pump cover plate. Continue by removing the single screw with sealing washer from the wall of the carburettor mixing chamber. Check the condition of the sealing washer and renew it if necessary. On TS185 and 250 models, this screw is located on the left-hand side of the carburettor, which makes a sighting comparison between the throttle valve and the oil pump relatively easy. Unfortunately, this is not so with TS100 and 125 models, because the screw is located on the right-hand side of the carburettor and therefore faces away from the oil pump. In this instance, it is best to recruit the help of an assistant to sight the position of

the throttle valve whilst the oil pump control cable is adjusted.

Rotate the throttle twistgrip until the circular indicator mark on the side of the throttle valve comes into alignment with the upper edge of the screw hole in the wall of the mixing chamber (see accompanying figure). With the throttle set in this position, check that the mark scribed on the pump lever boss is in exact alignment with the mark cast in the pump body. If this is not the case, then the marks should be made to align by rotating the control cable adjuster, after having released the locknut. On completion of the adjustment procedure, retighten the locknut whilst holding the cable adjuster in position and slide the rubber sealing cap down the cable to cover the adjuster. It should be noted that any adjustment of the oil pump control cable may well affect the adjustment of the throttle cable. It is, therefore,

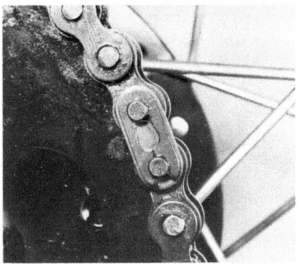

Fit the spring link correctly

Remove the blanking screw ...

... to allow sighting of the throttle valve alignment mark

Bring the pump lever mark into alignment with the mark on the pump body

On TS 100 and 125 models, examine the detent plunger assembly

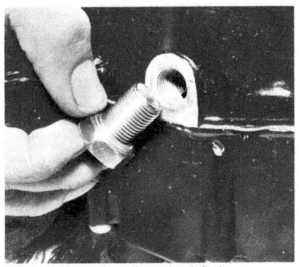

Fit a new sealing washer to the gearbox drain plug

necessary to refer to paragraph 6 of this Section and to check the throttle cable for correct adjustment before proceeding further.

Upon completion of adjustment, refit and tighten the screw, with its sealing washer, to the carburettor. Place the oil pump cover plate in position and fit and tighten its two securing screws. Wipe clean the mating surfaces of the gearbox sprocket cover and the crankcase and refit the cover. Tighten the cover retaining screws evenly and in a diagonal sequence in order to lessen the risk of the cover becoming distorted. On TS250 models, check the condition of the spring washers fitted beneath the heads of the cover retaining screws. If any one of these washers is flattened, then it should be renewed. Finally, if removed, the gearchange lever can now be slid into position on the gearchange shaft. Ensure that the lever is positioned correctly in relation to the footrest and that its bolt hole is aligned with the channel in the shaft spline before inserting and tightening its retaining bolt.

5 Changing the gearbox oil

In order to preclude any risk of the gearbox components having to run in oil that has deteriorated to the point where it is ineffective in its prime function as a lubricating medium, it is advisable to change the gearbox oil every three months/2000 miles.

Start the machine and run it until the engine reaches its normal operating temperature. Doing this will thin the oil in the gearbox thereby allowing it to drain more effectively. Stop the engine and position the machine on a flat and level piece of ground.

Position a container of at least 900 cc (1.90/1.58 US/Imp pint) capacity beneath the engine unit and then wipe clean the area of crankcase around the gearbox filler plug and its drain plug. The drain plug is located on the underside of the right-hand crankcase half on all models. Remove this plug and allow the oil to drain from the gearbox. It will be seen that on TS100 and 125 models, the drain plug is in fact the housing for the gearbox detent plunger and spring. Whilst waiting for the gearbox to drain, examine and clean the detent assembly. The spring of this assembly must be renewed if it shows signs of fatigue or failure. The sealing washer should be renewed as a matter of course; always replace the used washer with a new one of an identical type, never omit to fit the washer or replace it with any form of sealant as this will cause the plunger assembly to malfunction because of the incorrect pressure placed upon the spring. Once cleaned, the assembly should be placed to one side ready for refitting. It must be kept clean. Note that care should be taken to avoid movement of the gearchange lever whilst the detent assembly is removed.

On completion of draining, refit the drain plug with its new washer and tighten it fully. Remove both the plastic filler plug and the oil level screw, which is located just forward of the kickstart shaft. Replenish the gearbox by pouring the specified amount of SAE 20W/40 oil through the filler orifice. The oil level is correct when oil begins to appear through the oil level hole with the machine supported off its prop stand in an upright position. Finally, refit and tighten the oil level screw, after having placed a new washer beneath its head, and refit the filler plug.

6 Carburettor adjustment

Adjustment of the carburettor should be carried out with the machine supported securely on its stand and with the engine running at its normal operating temperature. Commence adjustment by setting the throttle stop screw to give the slowest possible idle speed. Turn the pilot air screw in by a fraction of a turn at a time until the engine starts to falter. Now back the screw off progressively whilst noting the number of turns required to reach the point where the engine again starts to run erratically. The correct position for the pilot air screw is mid-way between these two extremes, when it will be found that the engine is idling at its fastest. This should be close to the

Remove the gearbox filler plug

Replenish the gearbox with oil

Remove the gearbox oil level screw to determine the correct oil level

specified setting (see Specifications, Chapter 2).

At this point, the engine should be idling at the recommended speed. If the reading on the tachometer (where fitted) indicates an idle speed which is slightly outside that recommended, then the throttle stop screw should be turned until the indicated speed is correct.

Always guard against the possibility of incorrect carburettor adjustment which will result in a weak mixture. Two-stroke engines are very susceptible to this type of fault, causing rapid overheating and often subsequent engine seizure. Changes in carburation leading to a weak mixture will occur if the air cleaner is removed or disconnected, or if the silencer is tampered with in any way.

7 Throttle cable adjustment

Adjustment of the throttle cable is correct when there is 0.5 – 1.0 mm (0.02 – 0.04 in) of free movement in the cable outer when it is pulled out of its adjuster at the carburettor top. If cable adjustment is found to be incorrect, then loosen the adjuster locknut and rotate the adjuster the required amount before retightening the locknut. Initial adjustment should always be carried out at the carburettor end of the cable. Where the machine has an adjuster at the throttle twistgrip, then use this adjuster for any fine adjustment that may be necessary. Upon completion of adjustment, ensure that the throttle functions smoothly over its full operating range.

8 Cleaning the air filter element

The air filter assembly consists of a large oil-impregnated polyurethane foam filter element which is housed in a frame-mounted container located just to the rear of the carburettor. This container is connected to the carburettor mouth by means of a large-diameter rubber hose and is attached to the frame by means of a mounting bracket with two bolts and washers. TS185 and 250 models incorporate also two mounting rubbers with spacers in this assembly. The element must be removed from its container for the purposes of examination and cleaning. To remove the filter element, expose the side cover of the filter housing by detaching the left-hand sidepanel from the machine. Remove the filter element as follows.

US TS250 and all TS100, 125 and 185 models

The filter housing cover of these machines is retained in position by a single, centrally-located nut with washer. With this nut removed, the cover can be pulled clear of its mounting stud. Note the fitted position of the spacer (and seal on TS185 models) which is located between the cover and filter frame before removing both it and the frame from the mounting stud. The element can now be detached from the frame by unhooking its retaining clip and easing it from position.

UK TS250 models

The filter housing cover of these machines is retained in position by two clips which slide upwards off projections on both the cover and housing body. In practice, it was found that these clips were difficult to dislodge, the best method of removal being to push one of the clips completely off its location with the flat of a large screwdriver, thus taking the tension off the remaining clip which may then be removed without much effort. The element should now be carefully drawn clear of its mounting frame.

All models

Carry out a close inspection of the filter element. If the foam of the element shows signs of having become hardened with age, or is seen to be very badly clogged, then it must be renewed. If the element is found to be serviceable then it should be cleaned as follows.

Immerse the element in a non-flammable solvent such as white spirit (available as Stoddard solvent in the US), whilst gently squeezing it to remove any oil and dust. After cleaning, squeeze out the foam by pressing it between the palms of both

Check adjustment of the throttle cable at the carburettor top

Locate the air filter element over its mounting frame

Retain the filter element in position with its spring clip

Clean the filter housing and check each seal

Fit the filter element over its mounting stud

Do not omit the spacer piece ...

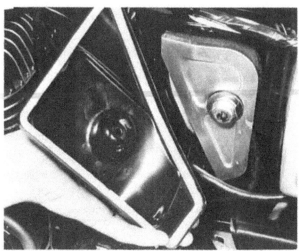

... before fitting the filter housing cover with its seal

hands and then allow a short time for any solvent remaining in the foam to evaporate. Do not wring out the foam as this will cause damage and thus lead to the need for early renewal.

Reimpregnate the foam with clean 2-stroke oil and gently squeeze out any excess. Refit the element onto its mounting frame and reassemble the component parts of the filter housing by reversing the procedure used for their removal. Check the condition of the spacer seal fitted to the TS185 models and renew it if necessary. Take great care when positioning each component, to ensure that no incoming air is allowed to bypass the element. If this is allowed to happen, it will allow any dirt or dust that it normally retained by the element to find its way into the carburettor and crankcase assembies, it will also effectively weaken the fuel/air mixture.

Finally, note that if the machine is being run in a particularly dusty or moist atmosphere, then it is advisable to increase the frequency of cleaning and reimpregnating the element. Never run the engine without the element fitted. This is because the carburettor is specially jetted to compensate for the addition of this component and the resulting weak mixture will cause overheating of the engine with the probable risk of severe engine damage.

9 Checking the spark plug

Suzuki fit NGK or ND spark plugs as standard equipment to all the model types covered in this Manual. Refer to the Specifications Section at the beginning of Chapter 3 for the exact plug type fitted to each model. The recommended gap between the plug electrodes is 0.6 – 0.8 mm (0.024 – 0.031 in). The plug should be cleaned and the gap checked and reset at regular service intervals. In addition, in the event of a roadside breakdown where the engine has mysteriously 'died', the spark plug should be the first item checked.

The plug should be cleaned thoroughly by using one of the following methods. The most efficient method of cleaning the electrodes is by using a bead blasting machine. It is quite possible that a local garage or motorcycle dealer has one of these machines installed on the premises and will be willing to clean any plugs for a nominal fee. Remember, before fitting a plug cleaned by this method, to ensure that there is none of the blasting medium left between the porcelain insulator and the plug body. An alternative method of cleaning the plug electrodes is to use a small brass-wire brush. Most motorcycle dealers sell such brushes which are designed specifically for this purpose. Any stubborn deposits of hard carbon may be removed by judicious scraping with a pocket knife. Take great care not to chip the porcelain insulator round the centre electrode whilst doing this. Ensure that the electrode faces are clean by passing a small fine file between them; alternatively, use emery paper but make sure that all traces of the abrasive material are removed from the plug on completion of cleaning.

To reset the gap between the plug electrodes, bend the outer electrode away from or closer to the central electrode and check that a feeler gauge of the correct size can be inserted between the electrodes. The gauge should be a light sliding fit.

Never bend the central electrode or the insulator will crack, causing engine damage if the particles fall in whilst the engine is running.

Always carry a spare spark plug of the correct type. The plug in a two-stroke engine leads a particularly hard life and is liable to fail more readily than when fitted to a four-stroke.

Beware of overtightening the spark plug, otherwise there is risk of stripping the threads from the aluminium alloy cylinder head. The plug should be sufficiently tight to seat firmly on its sealing washer, and no more. Use a spanner which is a good fit to prevent the spanner from slipping and breaking the insulator.

Before fitting the spark plug in the cylinder head, coat its threads sparingly with a graphited grease. This will prevent the plug from becoming seized in the head and therefore aid future removal.

When reconnecting the suppressor cap to the plug, make sure that the cap is a good, firm fit and is in good condition; renew its rubber seals if they are in any way damaged or perished. The cap contains the suppressor that eliminates both radio and TV interference.

10 Checking the contact breaker points – contact breaker models

Access to the contact breaker assembly can be gained by removing the cover from the left-hand crankcase. Avoid distorting the cover by loosening its four retaining screws evenly and in a diagonal sequence and take care to avoid tearing its gasket as the cover is lifted from position. This will reveal the flywheel generator rotor, which has four elongated holes in its outer face. The larger two of these holes are provided to permit inspection and adjustment of the contact breaker points.

Using the flat of a small screwdriver, open the contact breaker points against the pressure of the return spring so that the condition of the point contact faces may be checked. A piece of stiff card or crokus paper may be used to remove any light surface deposits, but if burnt or pitted, the complete contact breaker assembly must be removed to facilitate further examination of the point contact faces, and if necessary, renewal of the assembly. Refer to Section 5 of Chapter 3 for full details on removal, renovation and fitting of the assembly. If the points are found to be in sound condition, then proceed with adjustment as follows.

Rotate the crankshaft slowly until the contact breaker points are seen to be in their fully open position. Measure the gap between the points with a feeler gauge. If the gap is correct, a gauge of 0.35 mm (0.014 in) thickness will be a light sliding fit between the point faces. If this is not the case, slacken the single crosshead screw which serves to retain the fixed contact in position, just enough to allow movement of the contact. With the flat of a screwdriver placed in the indentation provided in the edge of the fixed contact plate, move the plate in the appropriate direction until the gap is correct. Retighten the retaining screw and recheck the gap setting; it is not unknown for this setting to alter slightly upon retightening of the screw.

Prior to refitting the crankcase cover, apply one or two drops of light machine oil to the cam lubricating wick whilst taking care not to allow excess oil to foul the point contact surfaces. Renew the cover sealing gasket if it is in any way damaged.

Inspect the suppressor cap seals

Measure the contact breaker gap

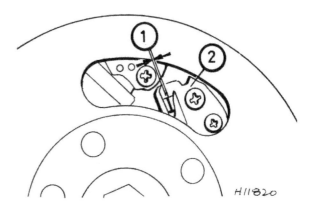

Measuring the contact breaker gap

1   0.3 – 0.4 mm (0.012 – 0.016 in)   2   Adjusting plate

11 Checking the ignition timing (static) – contact breaker models

It cannot be overstressed that optimum performance of the engine depends on the accuracy with which the ignition timing is set. Even a small error can cause a marked reduction in performance and the possibility of engine damage as the result of overheating.

Static timing of the ignition can be carried out simply by aligning the timing mark scribed on the wall of the flywheel generator rotor with the corresponding mark on the crankcase and then checking that the contact breaker points are just on the point of separation.

Before commencing a check of the ignition timing, refer to the previous Section of this Chapter and check that the contact breaker points are both clean and correctly gapped. In order to provide an accurate indication as to when the contact breaker points begin to separate, it will be necessary to obtain certain items of electrical equipment. This equipment may take the form of a multimeter or ohmmeter, or a high wattage 6 volt bulb, complete with three lengths of electrical lead, which will be used in conjunction with the battery of the motorcycle.

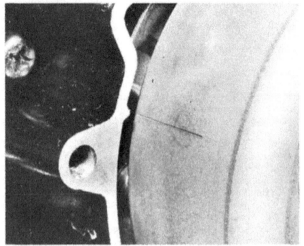
Align the timing marks

When carrying out the above described method of static timing and the method of dynamic timing described in the following Section of this Chapter, note that the accuracy of both methods of timing depends very much on whether the flywheel generator rotor is set correctly on the crankshaft. Any amount of wear between the keyways in both the crankshaft and rotor bore and the Woodruff key will cause some amount of variation between the timing marks which will, in turn, lead to inaccurate timing. Inaccuracy in the timing mark position may also be a result of manufacturing error. The only means of overcoming this is to remove any movement between the two components and then to set the piston at a certain position within the cylinder bore before checking that the timing marks have remained in correct alignment. In order to accurately position the piston, it will be necessary to remove the spark plug and replace it with a dial gauge or a slide gauge either one of which is adapted to fit into the spark plug hole of the cylinder head.

Position the piston in the cylinder bore by first rotating the crankshaft until the piston is set in the top dead centre (TDC) position. Set the gauge at zero on that position and then rotate the crankshaft backwards (clockwise) until the piston has passed down the cylinder bore a distance of at least 4 mm (0.16 in). Reverse the direction of rotation of the crankshaft until the piston is exactly 1.90 mm (0.075 in) from TDC. The timing marks should now be in exact alignment. If this is not the case, then new marks will have to be made. All subsequent adjustment of the timing may be made using these marks.

A standard reference for timing is provided

Commence a check of the ignition timing by tracing the electrical lead from the fixed contact point (colour code, Black/Yellow) to a point where it can be disconnected. To set up a multimeter, set it to its resistance function, connect one of its probes to the lead end and the other probe to a good earthing point on the crankcase. Use a similar method to set up an ohmmeter. In each case, opening of the contact breaker points will be indicated by a deflection of the instrument needle from one reading of resistance to another.

When using the battery and bulb method, remove the battery from the machine and position it at a convenient point next to the left-hand side of the engine. Connect one end of a wire to the positive (+) terminal of the battery and one end of another wire to the negative (-) terminal of the battery. The negative lead may now be earthed to a point on the engine casing. Ensure the earth point on the casing is clean and that the wire is positively connected; a crocodile clip fastened to the wire is ideal. Take the free end of the positive lead and connect it to the bulb. The third length of wire may now be connected between the bulb and the electrical lead of the fixed contact point. As the final connection is made, the bulb should light with the points closed. Opening of the contact breaker points will be indicated by a dimming of the bulb. The reason for recommending the use of a high voltage bulb is that this

Move the stator plate to alter the ignition timing

Use a 'strobe' to 'freeze' the moving marks on the rotor

dimming will be more obvious to the eye.

Ignition timing is correct when the contact breaker points are seen to open just as the timing marks come into alignment. If this is not the case, then the points will have to be adjusted by moving the stator plate clockwise or anti-clockwise, depending on whether the opening point needs to be advanced or retarded. This is accomplished by removing the flywheel generator rotor in order to gain full access to the three stator plate retaining screws, each one of which passes through an elongated slot cut in the plate.

Full details of removal and fitting of the flywheel generator rotor are contained in Section 3 of Chapter 3. With the rotor removed, it will be appreciated that resetting of the timing is a repetitive process of loosening the three stator plate retaining screws, rotating the stator plate a small amount in the required direction, retightening the screws, pushing the rotor back onto the crankshaft taper and then rechecking the timing. With the ignition timing correctly set, check tighten the stator plate retaining screws and then refit the rotor.

Finally, on completion of the timing procedure, remove all test equipment from the machine, reconnect any disturbed electrical connections and then refit the rotor cover.

12 Checking the ignition timing (dynamic)

On those models equipped with a contact breaker assembly, checking the ignition timing by the following method provides an alternative to the method of static timing described in the previous Section. On those models equipped with a PEI system, this method provides the only means of checking the ignition timing.

Ignition timing by the dynamic method is carried out with the engine running whilst using a stroboscopic lamp. This will entail gaining access to the wall of the flywheel generator rotor by removing its cover. When the light from the lamp is aimed at the timing marks on the crankcase and rotor wall, it has the effect of 'freezing' the moving mark on the rotor in one position and thus the accuracy of the timing can be seen. It cannot be overstressed that optimum performance of the engine depends on the accuracy with which the ignition timing is set. Even a small error can cause a marked reduction in performance and the possibility of engine damage as the result of overheating.

Prepare the timing marks by degreasing them and then coating each one with a trace of white paint. This is not absolutely necessary but will make the position of each mark far easier to observe if the light from the 'strobe' is weak or if the timing operation is carried out in bright conditions.

Two basic types of stroboscopic lamp are available, namely the neon and xenon tube types. Of the two, the neon type is much cheaper and will usually suffice if used in a shaded position, its light output being rather limited. The brighter but more expensive xenon types are preferable, if funds permit, because they produce a much clearer image.

Connect the 'strobe' to the HT lead, following the maker's instructions. If an external 6 volt power source is required, use the battery from the machine but make sure that when the leads from the machine's electrical system are disconnected from the terminals of the battery, they have their ends properly insulated with tape in order to prevent them from shorting on the cycle components.

Start the engine and aim the 'strobe' at the timing marks. The models equipped with a PEI system incorporate a means of automatically advancing and retarding the ignition timing throughout the engine speed range. Because of this, it is necessary to bring the ignition timing to its fully advanced position before noting the position of the timing marks. The ignition timing is fully advanced with the engine speed set at 6000 rpm. Models equipped with a contact breaker system have fixed ignition timing which means that the position of the timing marks can be noted at any given engine speed. On all models, ignition timing is correct when the mark on the crankcase will be seen to be in exact alignment with the mark on the rotor wall (the centre mark where three marks are present).

If it is found necessary to adjust the ignition timing, then it is first necessary to remove the flywheel generator rotor in order to gain full access to the three stator plate retaining screws, each one of which passes through an elongated slot cut in the plate. Before commencing this task, take careful note of the information given in paragraphs 4 and 5 of Section 3, Chapter 3. Full information for rotor removal on each model type is contained in Section 4 of Chapter 3.

With the rotor removed, it will be appreciated that resetting of the timing is a repetitive process of loosening the three stator plate retaining screws, rotating the stator plate a small amount in the required direction to advance or retard the timing as required, retightening the screws, setting the rotor back onto the crankshaft taper, starting the engine and rechecking the timing. With the ignition timing correctly set, check tighten the stator plate retaining screws and then refit the rotor by following the instructions given in Section 4 of Chapter 3.

Finally, on completion of the timing procedure, refit the rotor cover, reconnect any disturbed electrical connections and remove all test equipment from the machine.

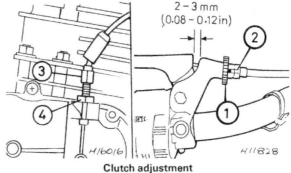

Clutch adjustment

1	Locking ring	3	Casing adjuster
2	Handlebar adjuster	4	Locknut

13 Clutch adjustment

The clutch is in correct adjustment when there is 2 – 3 mm (0.08 – 0.12 in) of free play in the cable inner. The amount of free play is measured between the pivot end of the handlebar lever and its retaining clamp.

If necessary, carry out adjustment of the clutch cable by rotating the adjuster at the engine end of the cable through its crankcase location until the amount of free play measured is correct. Hold the adjuster in position and tighten its locknut. Any subsequent fine adjustment to the amount of free play in the cable inner may now be achieved by rotating the knurled adjuster at the handlebar lever end. This adjuster is locked in position by its knurled lock ring. On completion of adjustment, relocate any displaced sealing caps over the cable ends and check the clutch operation.

14 Check tightening the cylinder head nuts

Suzuki recommend that the nuts which retain the cylinder head in position and the two bolts which retain the exhaust pipe to the cylinder barrel be checked for security at regular service intervals. Any loosening of the cylinder head retaining nuts whilst the machine is in use will result in leakage around the head to cylinder barrel mating face, with the subsequent risk of distortion of the cylinder head casting. Leakage of exhaust gases from the pipe to cylinder barrel seal will affect the performance of the engine.

Check tighten the cylinder head retaining nuts by tightening them to the torque setting given in the Specifications Section of Chapter 1. Tighten in even increments whilst working in a diagonal sequence. If the nuts are found to be loose and the cylinder head to barrel joint leaking, then the cylinder head gasket must be renewed.

If the exhaust pipe retaining bolts are found to be loose, then remove them and check the condition of the spring washers fitted beneath their heads. If these washers have become flattened then they must be renewed before the bolts are refitted and tightened. It is a good idea to renew the sealing ring between the barrel and pipe to prevent leakage.

15 Adjusting the front brake

Adjustment of the front brake is correct when there is 20 – 30 mm (0.8 – 1.2 in) of clearance between the end of the handlebar lever and the throttle twistgrip with the lever fully applied.

To adjust the clearance between the lever and twistgrip, simply loosen the locknut(s) at the cam shaft operating arm end of the cable and turn the cable adjuster the required amount in the appropriate direction before retightening the locknut(s). Any minor adjustments necessary may then be made with the cable adjuster at the handlebar lever bracket. To use this adjuster, simply loosen the lock ring, turn the knurled adjuster the required amount and then retighten the lock ring.

On completion of adjustment, check the brake for correct operation by spinning the wheel and applying the brake lever. There should be no indication of the brake binding as the wheel is spun. If the brake shoes are heard to be brushing against the surface of the wheel drum back off on the cable adjuster slightly until all indication of binding disappears. The brake may be readjusted after a period of bedding-in has been allowed for the brake shoes.

16 Adjusting the rear brake

Adjustment of the rear brake is correct when there is 20 – 30 mm (0.8 – 1.2 in) of movement, measured at the forward point of the brake pedal, between the point at which the brake pedal is depressed and the point where it abuts against its return stop.

The range of pedal movement may be adjusted simply by turning the nut on the wheel end of the brake operating rod in the required direction.

On completion of brake adjustment, check that the stop lamp switch operates the stop lamp as soon as the brake pedal is depressed. If necessary, adjust the height setting of the switch in accordance with the instructions given in Chapter 6.

17 Checking brake shoe wear

An indication of brake shoe lining wear is provided by an indicator line which is cast into the brake backplate. If, when the brake is correctly adjusted and applied fully, the line on the end of the brake cam spindle is seen to align with a point outside the arc shown by the indicator line, then the lining on the brake shoes can be assumed to have worn beyond limits and should be renewed at the earliest possible opportunity.

Commence adjustment of the front brake by turning the coarse adjuster

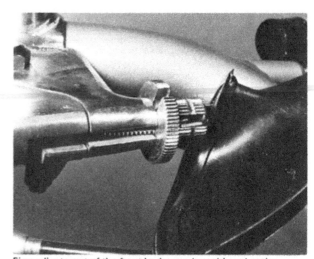

Fine adjustment of the front brake can be achieved at the handlebar end of the cable

Rotate the adjuster nut on the brake rod to alter the rear brake setting

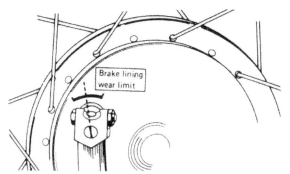

The extension line of the index mark is within the range.

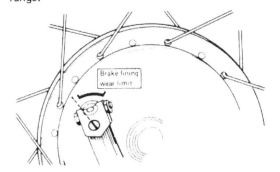

The extension line of the index mark is beyond the range.

Brake shoe lining wear indicator mark

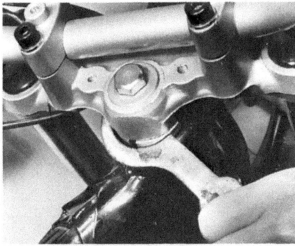

Do not overtighten the steering head bearing adjuster ring

18 Checking the steering head bearings

Support the machine in a stable position with the front wheel clear of the ground. This is best accomplished by positioning a stout wooden crate or blocks beneath the engine crankcase guard. Make sure that there is no danger of the machine toppling whilst it is being worked on.

Grasp the front fork legs near the wheel spindle and push and pull firmly in a fore and aft direction. If play is evident between the upper and lower steering yokes and the head lug casting, then the steering head bearings are in need of adjustment. Imprecise handling or a tendency for the front forks to judder may be caused by this fault.

Bearing adjustment is correct when the adjuster ring is tightened until resistance to movement is felt and then loosened $\frac{1}{8}$ to $\frac{1}{4}$ of a turn. The adjuster ring should be rotated by means of a C-spanner.

Take great care not to overtighten the nut. It is possible to place a pressure of several tons on the head bearings by overtightening even though the handlebars may seem to turn quite freely. Overtight bearings will cause the machine to roll at low speeds and give imprecise steering. Adjustment is correct if there is no play in the bearings and the handlebars swing freely to full lock in either direction. Only a light tap on each end of the handlebars should cause them to swing.

Six monthly, or every 4000 miles (6000 km)

Complete the checks listed under the preceding Routine Maintenance Sections and then complete the following:

1 Spark plug renewal

Remove and discard the existing spark plug, regardless of its condition. Although the plug may give acceptable performance after it has reached this mileage, it is unlikely that it is still working at peak efficiency.

Suzuki fit NGK or ND spark plugs as standard equipment to all the model types covered in this Manual. Refer to the Specifications Section at the beginning of Chapter 3 for the exact plug type fitted to each model.

Before fitting the new plug, adjust the gap between the electrodes to 0.6 – 0.8 mm (0.024 – 0.031 in), coat its threads sparingly with graphite grease and check that the aluminium crush washer is in place on the plug.

2 Cleaning the fuel tap collector bowl

The type of tap fitted to the machines covered in this Manual incorporates a detachable plastic base which acts as a collecting bowl for any heavy sediment or any moisture that may find its way past the fuel tap filter stack. This bowl must be removed at frequent intervals for cleaning and examination. Its contents will provide a good guide to the cleanliness of the inside of the fuel tank.

Some machines will be found to incorporate a filter element within the collector bowl. The purpose of this element is to provide a back-up filter to the fuel tap filter stack in preventing any contamination in the fuel from being passed directly from the tank to the carburettor, thus precluding the likelihood of any one of the jets within the carburettor becoming blocked. Contamination caught by the filter is deposited in the bowl.

It should be realised that any loss in engine performance or a refusal of the engine to run for any more than a short period of time, might be attributable to fuel starvation caused by a blocked or partially blocked filter element. Any suspected contamination of the fuel will therefore lead to the need to clean the element and bowl at more frequent intervals than that recommended. It may well also be necessary to remove and clean out the fuel tank.

To remove the collector bowl (and filter, where fitted) turn the tap lever to the 'Off' position, place an open-ended spanner over the squared end of the bowl and turn it anti-clockwise to unscrew the bowl. It was found in practice that the O-ring between the bowl and the tap casing had formed a semi-permanent seal between the two components and some effort was required to effect an initial release of the bowl.

Where a filter element is fitted, the element may be gently eased out its location within the tap whilst taking care to note the positioning of the hole within the filter element in relation to the corresponding fuel line within the tap. Clean the filter element by rinsing it in clean fuel. Any stubborn traces of contamination may be removed from the element by gently brushing it with a small soft-bristled brush soaked in fuel; a

The fuel tap incorporates a collector bowl

used toothbrush is ideal. Remember to take the necessary fire precautions when cleaning the element and always wear eye protection against any fuel that may spray back from the brush. Once it is cleaned, closely inspect the gauze of the element for any splits or holes that will allow the passage of sediment through it and onto the carburettor. Renew the element if it is in any way defective.

The collector bowl and the inside of the base of the fuel tap should be cleaned by using a procedure similar to that described in the proceeding paragraph. Note and adhere to the safety precautions that accompany this procedure.

Inspect the condition of the O-ring; if it is flattened, perished, or in any way damaged, then it must be renewed. Position the filter element (where fitted) at the base of the fuel tap. Remember to align the hole in the filter with the corresponding fuel line within the tap. Fit the serviceable O-ring and fit and tighten the collector bowl. Take care not to overtighten the bowl, it need only be nipped tight. Finally, turn the tap lever to the 'On' position and carry out a check for any leakage of fuel around the bowl to tap joint. Cure any leak found by nipping the bowl a little tighter. If this fails, remove the bowl and check that the O-ring has seated correctly.

3   Decarbonising the cylinder head and barrel

Decarbonising of the cylinder head and cylinder barrel can be undertaken with a minimal amount of dismantling, namely removal of the cylinder head, cylinder barrel and, as an additional requirement, the exhaust silencer baffle (except TS250 models).

Carbon build up in a two-stroke engine is more rapid than that of its four-stroke counterpart, due to the oily nature of the combustion mixture. It is however, rather softer and is therefore more easily removed.

The object of the exercise is to remove all traces of carbon whilst avoiding the removal of the metal surface on which it is deposited. It follows that care must be taken when dealing with the relatively soft alloy cylinder head and piston. Never use a steel scraper or screwdriver for carbon removal. A hardwood, brass or aluminium scraper is the ideal tool as these are harder than the carbon, but no harder than the underlying metal. Once the bulk of the carbon has been removed, a brass wire brush of the type used to clean suede shoes can be used to good effect.

The whole of the combustion chamber should be cleaned, as should the piston crown. It is recommended that as smooth a finish as possible is obtained, as this will slow the subsequent build up of carbon. If desired, metal polish can be used to obtain a smooth surface. The exhaust port must also be cleaned out, as a build up of carbon in this area will restrict the flow of exhaust gases from the cylinder. Take care to remove all traces of debris

from the cylinder and ports, prior to reassembly, by washing the components thoroughly in fuel or paraffin whilst taking care to observe the necessary fire precautions.

Full details of decarbonising the silencer assembly are given in Section 18 of Chapter 2.

4   Speedometer and tachometer cable lubrication

To grease either the speedometer or tachometer cable, uncouple both ends and withdraw the inner cable. (On some model types this may not be possible in which case a badly seized cable will have to be renewed as a complete assembly). After removing any old grease, clean the inner cable with a petrol soaked rag and examine the cable for broken strands or other damage. Do not check the cable for broken strands by passing it through the fingers or palm of the hand, this may well cause a painful injury if a broken strand snags the skin. It is best to wrap a piece of rag around the cable and pull the cable through it, any broken strands will snag on the rag.

Regrease the cable with high melting point grease, taking care not to grease the last six inches closest to the instrument head. If this precaution is not observed, grease will work into the instrument and immobilise the sensitive movement.

5   Brake cam shaft lubrication

In order to gain access to the brake cam shaft of each drum brake assembly it is necessary to remove the wheel and withdraw the brake backplate from the wheel hub. It is a good idea to combine this operation with the examination sequence given in Section 13 of Chapter 5. Note that failure to lubricate the cam shaft could well result in seizure of the shaft during operation of the brake with disastrous consequences. After displacing the brake shoes the cam shaft can be removed by withdrawing the retaining bolt on the operating arm and pulling the arm off the shaft. Before removing the arm, it is advisable to mark its position in relation to the shaft, so that it can be relocated correctly.

Remove any deposits of hardened grease or corrosion from the bearing surface of the brake cam shaft by rubbing it lightly with a strip of fine emery paper or by applying solvent with a piece of rag. Lightly grease the length of the shaft and the face of the operating cam prior to reassembly. Clean and grease the pivot stub which is set in the backplate.

Check the condition of the O-ring which prevents the escape of grease from the end of the cam shaft. If it is in any way damaged or perished, then it must be renewed before the shaft is relocated in the backplate. Relocate the cam shaft and align and fit the operating arm with the O-ring and plain washer.

The bolt which retains the arm in position on the shaft should be tightened to the torque loading specified in Chapter 5.

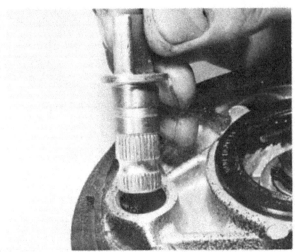

The brake cam shaft must be clean and lubricated

The brake cam shaft must be fitted with a serviceable O-ring

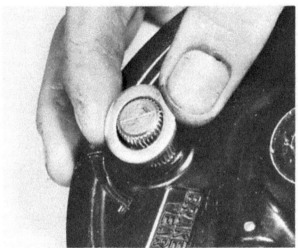

Cover the shaft O-ring with a plain washer

Align the brake operating arm with the cam shaft

Annually, or every 8000 miles (12 000 km)

Complete the checks listed under the preceding Routine Maintenance Sections and then complete the following:

1 Cleaning the carburettor

Suzuki recommend that the carburettor is removed and dismantled for cleaning at every 8000 mile service interval. This should only be necessary if there is real evidence of fuel contamination or if the performance of the machine is thought to be affected by a fault in carburation.

Evidence of fuel contamination will be found in the fuel tap collector bowl. Full details of servicing the carburettor may be found in Chapter 2 of this Manual.

2 Renewing the contact breaker points

Suzuki recommend that the contact breaker point assembly is renewed at every 8000 mile service interval. This is, of course, only necessary if the points have a substantial amount of material missing from their faces, due to routine dressing or spark erosion.

The contact breaker assembly forms part of the flywheel generator stator plate assembly. To gain full access to this assembly, it is necessary to remove the crankcase covers surrounding the flywheel generator and then to remove the rotor. Full instructions for the removal and fitting of these components are given in the relevant Sections of Chapter 1.

The contact breaker assembly is secured to the stator plate by means of a single crosshead screw with a spring washer beneath its head. Before the assembly can be removed, it is first necessary to detach both the condenser and coil leads from the spring blade of the moving point. They are retained by a single crosshead screw with spring washer, which should be relocated immediately after the leads have been detached in order to avoid loss.

Using the flat of a small electrical screwdriver, prise off the clip which retains the moving contact assembly to its pivot pin. Remove the plain washer, followed by the moving contact complete with insulating washers. Make a note of the order in which components are removed, as they are easily assembled incorrectly. Release the fixed contact by unscrewing the single retaining screw which passes through its backplate. Do not, under any circumstances, loosen the three stator mounting screws, otherwise the ignition timing will be lost.

To fit the new points assembly, reverse the above dismantling sequence whilst taking care to ensure that the insulating washers are replaced correctly. If this precaution is not

The contact breaker assembly forms part of the stator assembly

observed, it is easy to inadvertently earth the assembly thus rendering it inoperative. The pivot pin should be greased sparingly, and a few drops of oil applied to the cam lubricating wick.

On completion of reassembling and fitting the renewed contact breaker assembly, and after having refitted the generator rotor, adjust the contact breaker gap and check the ignition timing, as detailed in the relevant Sections of this Chapter.

3 Changing the fork oil

Because of the likely deterioration of the oil contained in each front fork leg due to the constant working of the suspension whilst the machine is in motion and to the possible ingress of moisture and build up of contamination in each leg, Suzuki recommend that each fork leg should be drained of oil and replenished with new oil of the specified grade and quantity at each 8000 mile service interval.

Where the front fork legs have a drain plug fitted, draining of the fork oil can be accomplished simply by placing a container beneath the leg to be drained, removing the drain plug and then pumping the leg up and down, whilst applying the front brake, to eject the oil from the leg. With all the fork oil thus drained, check that the sealing washer fitted beneath the head of each drain plug is serviceable before refitting and tightening each plug.

Refer to the Specifications Section of Chapter 4 and determine the correct grade and quantity of oil required for the fork legs of the type of machine being worked on. Support the machine so that both fork legs are fully extended. This can be accomplished by positioning a stout wooden crate or blocks beneath the engine crankcase guard. Ensure that there is no danger of the machine toppling whilst it is being worked on.

Each fork leg can now be replenished with oil through the top of the fork stanchion. This will necessitate removal of the fork spring retaining plug from each fork leg. To achieve this, proceed as follows. Cover the forward part of the fuel tank with an old blanket, or similar, to protect its painted surface from damage.

Remove each of the two rubber plugs from the upper face of the warning light console. This will expose the heads of the two console securing screws which should now be removed to free the console from its upper yoke attachment points. Using a C-spanner or a small soft-metal drift and hammer, release the ignition switch retaining ring. With this ring removed, the console can be manoeuvred rearwards to clear the handlebar securing clamps. Note that the procedure for removing the handlebar cowl from machines which have their warning lights contained within what is normally the tachometer instrument head casing is identical to that described above. Slacken and remove the four handlebar clamp retaining bolts, lift off the top halves of the clamps and ease the handlebars rearwards to rest on the protected area of tank. It is now possible to remove each fork spring retaining plug. Full details for removal and fitting of the different types of plug are contained in Section 3 of Chapter 4.

With each fork replenished with oil and its fork spring retaining plug refitted, refit the handlebar assembly by reversing its removal procedure.

When refitting the handlebars, note that they must be fitted so that the punch mark on the handlebar is directly in line with the rear mating faces of the handlebar clamps. Ensure that the clamp retaining bolts are fitted with serviceable spring washers beneath their heads and are tightened evenly so that the gap between the forward mating surfaces of the clamps is equal to that between the rear mating surfaces.

Note that where the front fork legs have no drain plugs fitted, each leg will have to be removed from the machine to allow the oil contained within it to be drained through the top of the fork stanchion, once the fork spring retaining plug is removed. This process will of course negate the need to remove the handlebar assembly from the fork upper yoke. Full instruc-

Check that each fork leg has a drain plug fitted

Replenish each fork leg with the correct amount of oil

tions for removal and refitting of the fork legs will be found in Section 2 of Chapter 4.

Eighteen monthly, or every 12 000 miles (20 000 km)

---

1 Lubricating the steering head and swinging arm fork bearings

Because neither the steering head nor the swinging arm fork pivot assemblies are equipped with grease nipples, Suzuki recommend that each assembly is fully dismantled for the purposes of cleaning and lubrication at every 12 000 mile service interval.

Full details of dismantling each pivot assembly are contained in the relevant Sections of Chapter 4 of this Manual. It is well worth taking note of the examination and renovation procedures also contained in Chapter 4 and renewing any defective component parts as required.

Two yearly, or every 16 000 miles (24 000 km)

---

1 Renewing the fuel feed pipe

Because the fuel feed pipe from the fuel tank to the carburettor is constructed of thin-walled synthetic rubber which will be affected by heat and the elements over a period of time, Suzuki recommend that, to anticipate any risk of fuel leakage, the pipe is renewed at every 16 000 mile service interval.

Additional maintenance items

Cleaning the machine

Keeping the motorcycle clean should be considered as an important part of the routine maintenance, to be carried out whenever the need arises. A machine cleaned regularly will not only succumb less speedily to the inevitable corrosion of external surfaces and hence maintain its market value, but will be far more approachable when the time comes for maintenance or service work. Furthermore, loose or failing components are more readily spotted when not partially obscured by a mantle of road grime and oil.

Metal components

Surface dirt on the metal cycle parts of the machine should be removed using a sponge and warm, soapy water, the latter being applied copiously to remove the particles of grit which might otherwise cause damage to the paintwork and polished surfaces.

Oil and grease is removed most easily by the application of a cleaning solvent such as 'Gunk' or 'Jizer'. The solvent should be applied when the parts are still dry and worked in with a stiff brush. Large quantities of water should be used when rinsing off, taking care that water does not enter the carburettors, air cleaners or electrics.

If desired a polish such as Solvol Autosol can be applied to the aluminium alloy parts to restore the original lustre. This does not apply in instances, much favoured by Japanese manufacturers, where the components are lacquered. Application of a wax polish to the cycle parts and a good chrome cleaner to the chrome parts will also give a good finish. Always wipe the machine down if used in the wet, and make sure the chain is well oiled. There is less chance of water getting into control cables if they are regularly lubricated, which will prevent stiffness of action.

Plastic components

The moulded plastic cycle parts, which include the headlamp nacelle, the side panels and the mudguards, need to be treated in a different manner from any metal cycle parts when it comes to cleaning. The plastic from which the parts are made will be adversely affected by traditional cleaning and polishing techniques, and lead as a result, to the surface finish deteriorating. Avoid the use of strong detergents which contain bleaching additives, scouring powders or other abrasive cleaning agents, including all but the finest aerosol polishes. Cleaning agents with an abrasive additive will score the surface of the parts thereby making them more receptive to dirt and permanently damaging the surface finish. The most satisfactory method of cleaning moulded plastic parts is to 'float' off any dirt from their surface by washing them thoroughly with a mild solution of soapy water and then wiping them dry with a clean chamois leather. A light coat of polish may then be applied to each item as necessary if it is thought that it is beginning to lose its original shine.

SUZUKI TRAIL BIKES

Check list

Weekly or every 200 miles (300 km)

1. Inspect the machine for loose fittings, frayed cables and leaks
2. Check the correct operation of the electrical system
3. Check the tyre pressures (cold)
4. Check the level of oil in the engine oil tank
5. Lubricate the exposed portions of each control cable
6. Lubricate the final drive chain

Monthly or every 600 miles (1000 km)

1. Check and adjust the final drive chain tension
2. Check the level of electrolyte in the battery

Three monthly or every 2000 miles (3000 km)

1. Remove and clean the final drive chain
2. Lubricate the control cables
3. Lubricate all pivot points on the machine
4. Adjust the oil pump setting
5. Change the gearbox oil
6. Adjust the carburettor
7. Adjust the throttle cable
8. Remove and clean the air filter element
9. Clean and adjust the spark plug
10. Adjust the contact breaker points – contact breaker models
11. Check the ignition timing
12. Adjust the clutch cable
13. Check that the cylinder head nuts are tightened to the correct torque setting
14. Adjust the front brake
15. Adjust the rear brake
16. Check the degree of brake shoe wear
17. Check for play in the steering head bearings

Six monthly or every 4000 miles (6000 km)

1. Renew the spark plug
2. Clean the fuel tap collector bowl
3. Decarbonize the cylinder head and barrel
4. Lubricate the speedometer and tachometer drive cables
5. Remove and lubricate each brake camshaft

Annually or every 8000 miles (12 000 km)

1. Dismantle and clean the carburettor
2. Renew the contact breaker points – contact breaker models
3. Change the front fork oil

Eighteen monthly or every 12 000 miles (20 000 km)

1. Lubricate the steering head and swinging arm bearings

Two yearly or every 16 000 miles (24 000 km)

1. Renew the fuel feed pipe

Additional maintenance items

1. Clean the machine

Adjustment data

Tyre pressures	Solo	With pillion
TS100 model		
Front	21 psi (1.5 kg/cm²)	25 psi (1.75 kg/cm²)
Rear	28 psi (2.0 kg/cm²)	31 psi (2.25 kg/cm²)
All other models		
Front	21 psi (1.5 kg/cm²)	21 psi (1.5 kg/cm²)
Rear	25 psi (1.75 kg/cm²)	28 psi (2.0 kg/cm²)

Spark plug type
TS125 N UK — NGK BP8ES or ND W24EP
TS185 and 250 — NGK BP7ES or ND W22EP
All other models — NGK B8ES or ND W24ES

Spark plug gap 0.6 – 0.8 mm (0.024 – 0.031 in)

Contact breaker er gap 0.3 – 0.4 mm (0.012 – 0.016 in)

Ignition timing
TS100 and 125 UK — 20° ± 2° BTDC (1.90 mm/0.075 in piston position)

TS100 and 125 US — 20° BTDC @ 6000 rpm
TS185 — 21.5° BTDC @ 6000 rpm
TS250 — 24° BTDC @ 6000 rpm

Idle speed 1300 ± 100 rpm

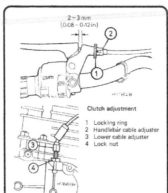

2 – 3 mm (0.08 – 0.12 in)

Clutch adjustment

1. Locking ring
2. Handlebar cable adjuster
3. Lower cable adjuster
4. Lock nut

Recommended lubricants

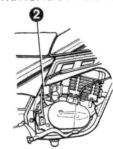

Component	Quantity	Type/viscosity
1 Engine oil	1.2 lit (2.8/2.3 US/Imp pt)	Suzuki CCI, CCI Super or equivalent 2-stroke oil
		SAE 20W/40
2 Transmission oil:- At oil change:		
TS100	800cc (1.69/1.41 US/Imp pt)	
TS125	700cc (1.48/1.24 US/Imp pt)	
TS185	700cc (1.48/1.24 US/Imp pt)	
TS250	850cc (1.80/1.50 US/Imp pt)	
3 Front forks:		
TS100	130cc (4.39/4.58 US/Imp fl oz)	
TS125 N	182cc (6.15/6.40 US/Imp fl oz)	SAE 5W/20 mineral oil or ATF
TS125 T and X	177cc (5.98/6.23 US/Imp fl oz)	
TS185 UK	166cc (5.61/5.84 US/Imp fl oz)	50/50 mixture of SAE 10W/30 motor oil and ATF
TS250 UK	242cc (8.17/8.52 US/Imp fl oz)	
TS185 US	166cc (5.61/5.84 US/Imp fl oz)	SAE 10 fork oil
TS250 US	242cc (8.17/8.52 US/Imp fl oz)	SAE 10 fork oil
4 Final drive chain	As required	Aerosol chain lubricant
5 Wheel bearings	As required	High melting point grease
6 Steering head bearings	As required	High melting point grease
7 Swinging arm bearings	As required	High melting point grease
8 Pivot points	As required	High melting point grease
9 Control cables	As required	Light machine oil

ROUTINE MAINTENANCE GUIDE

Refer To Chapter 7 for information relating to UK TS100 and 125 ERZ models

Castrol Lubricants

Castrol Engine Oils

Castrol Grand Prix

Castrol Grand Prix 10W/40 four stroke motorcycle oil is a superior quality lubricant designed for air or water cooled four stroke motorcycle engines, operating under all conditions.

Castrol Super TT Two Stroke Oil

Castrol Super TT Two Stroke Oil is a superior quality lubricant specially formulated for high powered Two Stroke engines. It is readily miscible with fuel and contains selective modern additives to provide excellent protection against deposit induced pre-ignition, high temperature ring sticking and scuffing, wear and corrosion.
Castrol Super TT Two Stroke Oil is recommended for use at petrol mixture ratios of up to 50:1.

Castrol R40

Castrol R40 is a castor-based lubricant specially designed for racing and high speed rallying, providing the ultimate in lubrication. Castrol R40 should never be mixed with mineral-based oils, and further additives are unnecessary and undesirable. A specialist oil for limited applications.

Castrol Gear Oils

Castrol Hypoy EP90

An SAE 90 mineral-based extreme pressure multi-purpose gear oil, primarily recommended for the lubrication of conventional hypoid differential units operating under moderate service conditions. Suitable also for some gearbox applications.

Castrol Hypoy Light EP 80W

A mineral-based extreme pressure multi-purpose gear oil with similar applications to Castrol Hypoy but an SAE rating of 80W and suitable where the average ambient temperatures are between 32°F and 10°F. Also recommended for manual transmissions where manufacturers specify an extreme pressure SAE 80 gear oil.

Castrol Hypoy B EP80 and B EP90

Are mineral-based extreme pressure multi-purpose gear oils with similar applications to Castrol Hypoy, operating in average ambient temperatures between 90°F and 32°F. The Castrol Hypoy B range provides added protection for gears operating under very stringent service conditions.

Castrol Greases

Castrol LM Grease

A multi-purpose high melting point lithium-based grease suitable for most automotive applications, including chassis and wheel bearing lubrication.

Castrol MS3 Grease

A high melting point lithium-based grease containing molybdenum disulphide. Suitable for heavy duty chassis application and some CV joints where a lithium-based grease is specified.

Castrol BNS Grease

A bentone-based non melting high temperature grease for ultra severe applications such as race and rally car front wheel bearings.

Other Castrol Products

Castrol Girling Universal Brake and Clutch Fluid

A special high performance brake and clutch fluid with an advanced vapour lock performance. It is the only fluid recommended by Girling Limited and surpasses the performance requirements of the current SAE J1703 Specification and the United States Federal Motor Vehicle Safety Standard No. 116 DOT 3 Specification.
In addition, Castrol Girling Universal Brake and Clutch fluid fully meets the requirements of the major vehicle manufacturers.

Castrol Fork Oil

A specially formulated fluid for the front forks of motorcycles, providing excellent damping and load carrying properties.

Castrol Chain Lubricant

A specially developed motorcycle chain lubricant containing non-drip, anti corrosion and water resistant additives which afford excellent penetration, lubrication and protection of exposed chains.

Castrol Everyman Oil

A light-bodied machine oil containing anti-corrosion additives for both household use and cycle lubrication.

Castrol DWF

A de-watering fluid which displaces moisture, lubricates and protects against corrosion of all metals. Innumerable uses in both car and home. Available in 400gm and 200gm aerosol cans.

Castrol Easing Fluid

A rust releasing fluid for corroded nuts, locks, hinges and all mechanical joints. Also available in 250ml tins.

Castrol Antifreeze

Contains anti-corrosion additives with ethylene glycol. Recommended for the cooling system of all petrol and diesel engines.

Chapter 1 Engine, clutch and gearbox

For modifications and information relating to later models, see Chapter 7

Contents

General description 1
Operations with the engine/gearbox unit in the frame 2
Operations with the engine/gearbox unit removed from
the frame 3
Removing the engine/gearbox unit from the frame 4
Dismantling the engine/gearbox unit: preliminaries 5
Dismantling the engine/gearbox unit: removing the
cylinder head 6
Dismantling the engine/gearbox unit: removing the
cylinder barrel, piston and small-end bearing 7
Dismantling the engine/gearbox unit: removing the
flywheel generator assembly 8
Dismantling the engine/gearbox unit: removing the
neutral indicator switch assembly and the gearbox
sprocket 9
Dismantling the engine/gearbox unit: removing the oil
pump assembly 10
Dismantling the engine/gearbox unit: removing the
right-hand crankcase cover, primary drive pinion and
clutch assembly 11
Dismantling the engine/gearbox unit: removing the
kickstart drive and idler pinions and the oil pump drive
– TS100, 125 and 250 models 12
Dismantling the engine/gearbox unit: removing the
kickstart return spring assembly – TS185 models 13
Dismantling the engine/gearbox unit: removing the
gearchange shaft and pawl mechanism – TS100, 125
and 250 models 14
Dismantling the engine/gearbox unit: removing the
gearchange mechanism – TS185 models 15
Dismantling the engine/gearbox unit: separating the
crankcase halves 16
Dismantling the engine/gearbox unit: removing the
kickstart shaft assembly, gearbox components, tachometer
drive gear and crankshaft 17
Examination and renovation: general 18
Examination and renewal: crankcase and gearbox oil
seals 19
Examination and renovation: crankshaft main bearings and
gearbox bearings 20
Examination and renovation: crankshaft assembly 21
Examination and renovation: decarbonising 22

Examination and renovation: cylinder head 23
Examination and renovation: cylinder barrel 24
Examination and renovation: piston and piston rings 25
Examination and renovation: gearbox components 26
Examination and renovation: gearchange shaft and pawl
mechanisms – TS100, 125 and 250 models 27
Examination and renovation: gearchange mechanism –
TS185 models 28
Examination and renovation: clutch assembly and primary
drive 29
Examination and renovation: kickstart assembly 30
Examination and renovation: oil pump assembly 31
Examination and renovation: tachometer drive assembly ... 32
Examination and renovation: reed valve assembly 33
Engine reassembly: general 34
Reassembling the engine/gearbox unit: reassembling
the gearbox components 35
Reassembling the engine/gearbox unit: fitting the
crankcase components and joining the crankcase halves .. 36
Reassembling the engine/gearbox unit: fitting the
gearchange shaft and pawl mechanism – TS100, 125
and 250 models 37
Reassembling the engine/gearbox unit: fitting the
gearchange mechanism – TS185 models 38
Reassembling the engine/gearbox unit: fitting the kickstart
drive and idler pinions and the oil pump drive – TS100,
125 and 250 models 39
Reassembling the engine/gearbox unit: fitting the primary
drive pinion, clutch assembly and right-hand crankcase
cover 40
Reassembling the engine/gearbox unit: fitting the oil pump
assembly 41
Reassembling the engine/gearbox unit: fitting the neutral
indicator switch assembly and gearbox sprocket 42
Reassembling the engine/gearbox unit: fitting the flywheel
generator assembly 43
Reassembling the engine/gearbox unit: fitting the
small-end bearing, piston, cylinder barrel and cylinder
head 44
Fitting the engine/gearbox unit into the frame 45
Starting and running the rebuilt engine 46
Taking the rebuilt machine on the road 47

Specifications

	TS100	TS125
Engine		
Type	Air cooled, single cylinder, two-stroke	
Bore	50.0 mm (1.969 in)	56.0 mm (2.205 in)
Stroke	50.0 mm (1.969 in)	50.0 mm (1.969 in)
Capacity	98 cc (6.0 cu in)	123 cc (7.5 cu in)
Compression ratio	6.4 : 1	6.6 : 1
Cylinder barrel		
Standard bore	50.0 - 50.15 mm (1.9685 - 1.9691 in)	56.0 - 56.015 mm (2.2047 - 2.2053 in)
Service limit	50.090 mm (1.9720 in)	56.085 mm (2.2080 in)
Barrel to head face distortion limit	0.05 mm (0.002 in)	0.05 mm (0.002 in)
Cylinder bore to piston clearance	0.040 - 0.050 mm (0.0016 - 0.0020 in)	0.045 - 0.055 mm (0.0018 - 0.0021 in)
Service limit	0.120 mm (0.0047 in)	0.120 mm (0.0047 in)
Cylinder head		
Head to barrel face distortion limit	0.05 mm (0.002 in)	0.05 mm (0.002 in)
Piston		
Outside diameter	49.955 - 49.970 mm (1.9667 - 1.9673 in)	55.950 - 55.965 mm (2.2028 - 2.2033 in)
Service limit	49.880 mm (1.9638 in)	55.880 mm (2.2000 in)
Gudgeon pin OD	13.995 - 14.0 mm (0.5510 - 0.5512 in)	
Service limit	13.980 mm (0.5504 in)	13.980 mm (0.5504 in)
Gudgeon pin hole bore diameter	13.998 - 14.006 mm (0.5511 - 0.5514 in)	
Service limit	14.030 mm (0.5524 in)	14.030 mm (0.5524 in)
Gudgeon pin to piston clearance	0.002 - 0.011 mm (0.0001 - 0.0004 in)	
Service limit	0.080 mm (0.0031 in)	
Piston rings (top and second)		
Ring to groove clearance	–	0.01 - 0.05 mm (0.0004 - 0.0020 in)
End gap:		
Fitted	0.15 - 0.35 mm (0.006 - 0.013 in)	
Service limit	0.80 mm (0.031 in)	0.80 mm (0.031 in)
Free	5.0 mm (0.20 in) – Nippon (N) 4.5 mm (0.18 in) – Teikoku (T)	7.5 mm (0.29 in)
Service limit	4.0 mm (0.16 in) – N 3.6 mm (0.14 in) – T	6.0 mm (0.24 in)
Crankshaft assembly	**TS100 and 125 models**	
Maximum runout	0.05 mm (0.002 in)	
Total web width	56.0 ± 0.1 mm (2.205 ± 0.004 in)	
Maximum connecting rod deflection	3.0 mm (0.12 in)	
Maximum big-end radial play	0.08 mm (0.003 in)	
Small-end bore	18.003 - 18.011 mm (0.7088 - 0.7091 in)	
Service limit	18.040 mm (0.7102 in)	
Clutch	**TS100 and TS125**	
Type	Wet, multiplate	
Spring free length	31.1 mm (1.224 in)	
Service limit	32.6 mm (1.283 in)	
Friction plate:		
Thickness	2.9 - 3.1 mm (0.114 - 0.122 in)	
Service limit	2.6 mm (0.102 in)	
Tang width	11.8 - 12.0 mm (0.465 - 0.472 in)	
Service limit	11.0 mm (0.33 in)	
Maximum warpage	0.4 mm (0.016 in)	

Plain plate:

Thickness; type A .... 1.60 ± 0.06 mm
(0.063 ± 0.002 in)

Thickness; type B .... 2.00 ± 0.06 mm
(0.079 ± 0.002 in) - TS100 only

Maximum warpage .... 0.1 mm (0.004 in)

Gearbox

	TS100	TS125
Type	5-speed, constant mesh	6-speed, constant mesh
Gear ratios (no of teeth):		
1st	3.090:1 (34/11)	3.090:1 (34/11)
2nd	2.000:1 (30/15)	2.000:1 (30/15)
3rd	1.368:1 (26/19)	1.368:1 (26/19)
4th	1.095:1 (23/21)	1.095:1 (23/21)
5th	0.875:1 (21/24)	0.956:1 (22/23)
6th	–	0.840:1 (21/25)
Backlash in all gears	0.05 - 0.10 mm (0.002 - 0.004 in)	
Service limit	0.15 mm (0.006 in)	0.15 mm (0.006 in)
Selector fork to pinion groove clearance	0.05 - 0.25 mm (0.002 - 0.010 in)	
Service limit	0.45 mm (0.018 in)	0.45 mm (0.018 in)
Selector fork claw end thickness:		
Nos 1 and 2	4.30 - 4.40 mm (0.169 - 0.173 in)	
No 3	5.30 - 5.40 mm (0.209 - 0.213 in)	
Pinion groove width:		
Nos 1 and 2	4.45 - 4.55 mm (0.175 - 0.179 in)	
No 3	5.45 - 5.55 mm (0.215 - 0.218 in)	
Primary reduction ratio	3.562:1 (57/16)	3.562:1 (57/16)
Primary drive gear backlash	0.02 - 0.07 mm (0.0008 - 0.0027 in)	
Service limit	0.10 mm (0.004 in)	0.10 mm (0.004 in)
Final reduction ratio	3.285:1 (46/14)	3.066:1 (46/15)

Torque wrench settings

	kgf m	lbf ft
Cylinder head retaining nuts:		
US TS100	2.0 - 2.5	14.5 - 18.0
All other TS100 and TS125	1.5 - 2.0	11.0 - 14.5
Flywheel generator rotor retaining nut	3.0 - 4.0	21.5 - 29.0
Clutch hub retaining nut:		
US TS125	2.0 - 3.0	14.5 - 21.5
All other TS125 and TS100	3.0 - 5.0	21.5 - 36.0
Primary drive pinion:		
US TS125	4.0 - 5.5	29.0 - 39.5
All other TS125 and TS100	4.0 - 6.0	29.0 - 43.5
Gearbox sprocket retaining nut	4.0 - 6.0	29.0 - 43.5
Gearchange shaft stop stud	1.5 - 2.3	11.0 - 16.5
Gearbox detent plunger housing	1.8 - 2.8	13.0 - 20.0
Reed valve securing screws	0.07 - 0.09	0.5 - 0.7
Exhaust pipe to cylinder securing bolts	0.4 - 0.7	3.0 - 5.0
Engine mounting bolts and nuts:		
8 mm	1.3 - 2.3	9.5 - 16.5
10 mm	2.5 - 4.0	18.0 - 29.0

Engine

	TS185	TS250
Type	Air cooled, single cylinder, two-stroke	
Bore	64.0 mm (2.520 in)	70.0 mm (2.756 in)
Stroke	57.0 mm (2.244 in)	64.0 mm (2.520 in)
Capacity	183 cc (11.2 cu in)	246 cc (15.0 cu in)
Compression ratio	6.2:1	5.9:1

Cylinder barrel

Standard bore	64.0 - 64.015 mm (2.5197 - 2.5203 in)	70.0 - 70.015 mm (2.7559 - 2.7565 in)
Service limit	64.080 mm (2.5228 in)	70.070 mm (2.7587 in)
Barrel to head face distortion limit	0.05 mm (0.002 in)	0.05 mm (0.002 in)
Cylinder bore to piston clearance	0.050 - 0.060 mm (0.0020 - 0.0024 in)	0.060 - 0.070 mm (0.0024 - 0.0028 in)
Service limit	0.120 mm (0.0047 in)	0.120 mm (0.0047 in)

Cylinder head

Head to barrel face distortion limit .... 0.05 mm (0.002 in)   0.05 mm (0.002 in)

Piston

Outside diameter	63.945 - 63.960 mm (2.5175 - 2.5181 in)	69.935 - 69.950 mm (2.7533 - 2.7539 in)
Service limit	63.880 mm (2.5150 in)	69.880 mm (2.7512 in)
Gudgeon pin OD	15.995 - 16.0 mm (0.6297 - 0.6299 in)	17.995 - 18.0 mm (0.7085 - 0.7087 in)
Service limit	15.980 mm (0.6291 in)	17.980 mm (0.7079 in)
Gudgeon pin hole bore diameter	15.998 - 16.006 mm (0.6298 - 0.6302 in)	17.998 - 18.006 mm (0.7086 - 0.7089 in)
Service limit	16.030 mm (0.6311 in)	18.030 mm (0.7098 in)

Piston rings (top and second)

Ring to groove clearance	0.02 - 0.06 mm (0.001 - 0.002 in)	0.03 - 0.07 mm (0.001 - 0.003 in)
End gap:		
Fitted	0.15 - 0.35 mm (0.006 - 0.014 in)	0.2 - 0.4 mm (0.008 - 0.016 in)
Service limit	0.80 mm (0.031 in)	0.85 mm (0.033 in)
Free	8.0 mm (0.31 in)	7.5 mm (0.30 in)
Service limit	6.4 mm (0.25 in)	6.0 mm (0.24 in)

Crankshaft assembly

Maximum runout	0.05 mm (0.002 in)	0.05 mm (0.002 in)
Total web width	56.0 ± 0.1 mm (2.205 ± 0.004 in)	60.0 ± 0.1 mm (2.36 ± 0.004 in)
Maximum connecting rod deflection	3.0 mm (0.12 in)	3.0 mm (0.12 in)
Small-end bore	21.003 - 21.011 mm (0.8269 - 0.8272 in)	23.003 - 23.011 mm (0.9056 - 0.9059 in)
Service limit	21.040 mm (0.8283 in)	23.040 mm (0.9071 in)

Clutch

Type	Wet, multiplate	
Spring free length service limit	32.6 mm (1.28 in)	36.5 mm (1.44 in)
Friction plate:		
Thickness	2.9 - 3.1 mm (0.11 - 0.12 in)	3.4 - 3.6 mm (0.13 - 0.14 in)
Service limit	2.6 mm (0.10 in)	3.1 mm (0.12 in)
Tang width	11.8 - 12.0 mm (0.46 - 0.47 in)	15.8 - 16.0 mm (0.62 - 0.63 in)
Service limit	11.3 mm (0.44 in)	15.3 mm (0.60 in)
Maximum warpage	0.4 mm (0.016 in)	0.4 mm (0.016 in)
Plain plate:		
Thickness	1.6 ± 0.06 mm (0.06 ± 0.002 in)	2.0 ± 0.06 mm (0.078 ± 0.002 in)
Maximum warpage	0.1 mm (0.004 in)	0.1 mm (0.004 in)

Gearbox

Type	5-speed, constant mesh	
Gear ratios (no of teeth):		
1st	2.750:1 (33/12)	2.727:1 (30/11)
2nd	1.812:1 (19/16)	1.800:1 (27/15)
3rd	1.250:1 (25/20)	1.277:1 (23/18)
4th	1.000:1 (23/23)	1.000:1 (21/21)
5th	0.800:1 (20/25)	0.826:1 (19/23)
Selector fork to pinion groove clearance	0.20 - 0.40 mm (0.008 - 0.016 in)	0.40 - 0.60 mm (0.016 - 0.024 in)
Service limit	0.60 mm (0.024 in)	0.80 mm (0.031 in)
Selector fork claw end thickness	5.30 - 5.40 mm (0.209 - 0.213 in)	3.45 - 3.55 mm (0.136 - 0.140 in)
Pinion groove width	5.60 - 5.70 mm (0.220 - 0.224 in)	3.95 - 4.05 mm (0.156 - 0.159 in)
Primary reduction ratio	3.210:1 (61/19)	3.190:1 (67/21)
Final reduction ratio	3.250:1 (39/12)	2.666:1 (40/15)

Torque wrench settings

	kgf m	lbf ft
Cylinder head retaining nuts	1.3 - 2.3	9.5 - 16.5
Cylinder barrel retaining nuts:		
8 mm	1.3 - 2.3	9.5 - 16.5

10 mm	2.5 - 4.0	18.0 - 29.0
Flywheel generator rotor retaining nut	5.5 - 6.5	40.0 - 47.0
Clutch hub retaining nut:		
TS185	3.0 - 5.0	21.5 - 36.0
TS250	4.0 - 6.0	29.0 - 43.5
Primary drive pinion retaining nut:		
TS185	4.0 - 6.0	29.0 - 43.5
TS250	5.0 - 7.0	36.0 - 50.5
Gearbox sprocket retaining nut	4.0 - 6.0	29.0 - 43.5
Engine mounting bolts and nuts:		
8 mm	1.8 - 2.8	13.0 - 20.0
10 mm	4.0 - 6.0	29.0 - 43.5

1 General description

The Suzuki range of trail bikes in the 100 cc to 250 cc capacity class employs two basic designs of single cylinder, air cooled, 2-stroke engine. Both represent a familiar design amongst small Suzuki motorcycles and have proved extremely reliable and robust over several years of manufacture.

Both types of unit employ vertically split crankcases which house both the crankshaft assembly and the gear clusters. The built-up crankshaft has full flywheels and runs on two journal ball bearings which are housed in the crankcase, one each side of the flywheels. Both small-end and big-end bearings are of the caged needle roller type. As indicated in this Chapter, there are several variations of gearbox design incorporated within the two basic designs of engine unit; the most distinct difference being between the gear selector mechanism fitted to the 185 cc models and that fitted to other models in the range.

The induction system employed on all engine types is of the piston port and reed valve design; where the induction of fuel/air mixture into the crankcase is timed by the reciprocating piston skirt and where an even flow of this mixture is maintained by the reed valve. This valve also serves to reduce the possibility of any blow-back of the combustible gases. This type of system contributes towards an economical and powerful engine unit.

Engine lubrication is effected by means of the Suzuki CCI system, which takes the form of a gear driven oil pump drawing oil from a separate frame-mounted oil tank and distributing it to the various working parts of the engine. The pump is interconnected to the throttle so that optimum lubrication is achieved at all times, thereby corresponding to the requirements of both engine speed and throttle opening. This system completely obviates the need to pre-mix fuel and oil and the problems that more often than not occur whilst doing so. Lubrication for the gearbox and primary transmission is provided by an oil reservoir shared between the two interconnected assemblies.

A flywheel generator is mounted on the left-hand end of the crankshaft and is fully enclosed behind a detachable crankcase side cover. The clutch is mounted on the right-hand end of the gearbox input shaft and is also fully enclosed. Engine starting is by kickstart; drive being passed from the kickstart shaft pinion to the crankshaft via an idler pinion to the clutch drum and then from the clutch drum onto the crankshaft-mounted primary drive pinion.

2 Operations with the engine/gearbox unit in the frame

1 It is not necessary to remove the engine/gearbox unit from the frame in order to carry out the following service operations:

1 Removal and fitting of the cylinder head
2 Removal and fitting of the cylinder barrel
3 Removal and fitting of the piston assembly
4 Removal and fitting of the carburettor assembly
5 Removal and fitting of the reed valve
6 Removal and fitting of the oil pump unit
7 Removal and fitting of the clutch assembly and primary drive pinion
8 Removal and fitting of the gearchange shaft and pawl or pin mechanism
9 Removal and fitting of the kickstart drive gear assembly
10 Removal and fitting of the flywheel generator and contact breaker assembly (where fitted)
11 Removal and fitting of the gearbox sprocket
12 Removal and fitting of the neutral indicator switch

2 If it is found necessary to carry out several of these operations at the same time, it may be considered advantageous to remove the complete engine/gearbox unit from the frame in order to gain better access and more working space. This operation is comparatively simple and should take no more than an hour, whilst working at a leisurely pace without any assistance.

3 Operations with the engine/gearbox unit removed from the frame

1 Certain operations can be accomplished only if the complete engine unit is removed from the frame. This is because it is necessary to separate the crankcase to gain access to the parts concerned. These operations include:

1 Removal and fitting of the crankshaft assembly
2 Removal and fitting of the main bearings
3 Removal and fitting of the gearbox shaft and pinion assemblies, the gearbox bearings and the gear selector components
4 Removal and fitting of the kickstart shaft
5 Removal and fitting of the kickstart spring and guide (all models except TS185)

4 Removing the engine/gearbox unit from the frame

1 Removal of the engine/gearbox unit will be made much easier by raising the machine to an acceptable working height and thus preventing the discomfort of squatting or kneeling down to work on the various component parts. Raising the machine may be achieved by using either a purpose built lift or a stout table or by building a platform from substantial planks and concrete blocks.

2 With the machine positioned on the chosen work surface and supported properly on its prop stand, apply the front brake and lock the handlebar lever in the on position by wrapping a large rubber band or similar item around it and the throttle twistgrip. Doing this will prevent the machine from rolling forward off its stand during engine removal and fitting.

3 Position a container of at least 900 cc (1.90/1.58 US/Imp pint) capacity beneath the engine unit and proceed to remove the gearbox oil drain plug. This plug is located on the underside of the right-hand crankcase half on all models and has a sealing

washer fitted beneath its head. The washer should be renewed as a matter of course. It will be seen that on TS100 and 125 models, the plug is in fact the housing for the gearbox detent plunger and spring and as the plug is withdrawn, the plunger and spring will be seen to follow it.

4 If working on a TS100 or 125 model, use the time it takes the oil to drain from the gearbox to examine and clean the detent plunger assembly. The spring must be renewed if it shows signs of fatigue or failure. The sealing washer should be renewed as a matter of course; always replace the used washer with a new one of an identical type, never omit to fit the washer or replace it with any form of sealant as this will cause the plunger assembly to malfunction because of the incorrect pressure placed upon the spring. Once cleaned, the assembly may be stored in a safe place ready for refitting. Note that if it is intended not to separate the crankcase halves then great care should be taken to avoid movement of the gearchange components whilst the detent plunger assembly is removed.

5 Detach each sidepanel from the machine by first removing its securing screw(s) and then lifting it clear of its two upper frame mounting hooks. Lift the seat and remove the split-pin from each of the two seat pivot pins. With the split-pins thus removed, draw each pivot pin from position and lift the seat clear of the machine.

6 Remove the fuel tank by first turning the lever of the fuel tap to its 'Off' position and then releasing the fuel pipe retaining clip. This will allow the pipe to be pulled off the stub at the rear of the tap. Careful use of a small screwdriver may be necessary to help ease the pipe off the stub. Once the pipe is detached, allow any fuel in the pipe to drain into a small clean container. The tank may now be detached from the frame by unscrewing the single retaining bolt at the rear of the tank, unclipping the tank cap breather pipe end attachment (where fitted) from the handlebar cross-member and then pulling the tank up and rearwards off its front mounting rubbers. In practice, it was found that some degree of force was required to free the tank from these rubbers and it helped to have an assistant pushing the tank rearwards as it was moved from side to side. Once removed, place the tank retaining bolt, together with any associated mounting components, in a safe place ready for refitting. Inspect the mounting rubbers for signs of damage or deterioration and if necessary, renew them before refitting of the tank is due to take place. Carefully place the fuel tank, together with the seat and sidepanels, in a storage space where they are likely to be safe from damage. In the case of the fuel tank, the space must also be well ventilated and free from any source of naked flame or sparks.

7 Move to the right-hand side of the machine and isolate the battery from the electrical system by disconnecting the lead to the battery negative (–) terminal. This simple precaution will ensure that no shorting of exposed contacts occurs whilst disconnecting electrical components during removal of the engine unit. It is advisable at this stage to remove completely the battery from the machine so that it may be properly serviced in accordance with the instructions given in Chapter 6 of this Manual.

8 Reference to the figures contained in Chapter 2 of this manual will show that the design of the exhaust system fitted to the TS100 and 125 models differs from that of the system fitted to the TS185 and 250 models. Nevertheless, the same basic procedure for removal, and the precautions involved, apply equally to each system type. Commence removal of the exhaust system by unscrewing the two bolts which serve to retain the exhaust pipe to the cylinder head. If working on a TS185 or 250 model, unhook the pipe retaining spring by using a pair of pliers. Check that the spring washer located beneath each bolt head has not become flattened; if this is the case, the washers should be renewed before reassembly takes place. Where necessary, unclip the clutch cable from its guide on the exhaust system. It is now necessary to move along the complete system, undoing each mounting bolt and noting the fitted position of each washer, spacer and mounting rubber. With all the mounting bolts removed, the system can be lifted

away from the machine and placed in safe storage along with its mounting assemblies. Note that under no circumstances should the system be allowed to hang unsupported from the cylinder barrel mounting as this will impose an unacceptable strain on the threads of the two mounting bolts.

9 Where fitted, disconnect the tachometer drive cable from the top of the gearbox housing by unscrewing its knurled retaining ring and pulling the cable from position.

10 Remove both crosshead screws which retain the cover of the clutch operating mechanism to the front of the engine crankcase cover. Lift the cover from position and mark the operating lever in relation to the splined shaft so that it may be fitted in the correct position during engine reassembly. Slide the rubber boot up to clear the cable adjuster, loosen the adjuster locknut and wind the adjuster fully down into the crankcase cover; doing this will serve to give enough slack in the cable inner to allow easy removal of the lever from its shaft. Remove the lever retaining bolt, pull the lever off its shaft, thread the cable adjuster up out of the crankcase cover and detach the cable, with lever, from the engine. Attach the cable end to a point on the frame which is well clear of the engine.

11 Move to the left-hand side of the machine and unscrew the top of the mixing chamber from the carburettor body. Carefully lift the chamber top away from the carburettor whilst easing the throttle valve and jet needle out of their respective locations. Manoeuvre the assembly clear of the engine and attach it to a point on the frame. It is a good idea to prevent contamination of the throttle valve assembly by wrapping it in a polythene bag. Unscrew the clamp which retains the air inlet hose to the mouth of the carburettor and ease the hose rearwards to clear its attachment point. Leave the clamp pushed up over the hose to avoid loss. Carry out a quick inspection of the inlet hose, it must be renewed if split or perished.

12 Detach the suppressor cap from the spark plug and tuck both it and the HT lead clear of the cylinder head at a suitable point on the frame top tube. Loosen the spark plug now, as it is easier to free a really tight plug with the engine supported in the frame rather than with it loose on a work surface.

13 Trace the electrical leads from the flywheel generator stator assembly and disconnect them at their nearest push connectors. All of these wires should be colour coded to avoid confusion whilst reconnecting; if in any doubt, clearly label each wire before disconnecting. Note on TS185 and 250 models that there are two wires which are remote from the main connecting block and these are both situated alongside the frame tube which supports the fuel tank. Unclip the electrical leads from their frame attachment points and lay them neatly on the engine crankcase.

14 Unscrew the gearchange lever retaining bolt and pull the lever off its shaft end. Remove the gearbox sprocket cover. On all but the TS250 models, this cover is combined with the flywheel generator rotor cover and is held in position by four retaining screws. On the TS250 models, this cover is separate and is retained by three screws.

15 It is now advisable at this stage to loosen the gearbox sprocket retaining nut as this will save work during the engine dismantling sequence. Knock back the tab washer from the flat of the sprocket retaining nut and place a close-fitting socket or box spanner over the nut; an open-ended or ring spanner can be used, but if the nut is really tight, then it may not be possible to obtain enough leverage to loosen it. Use as long an extension as is possible on the socket or spanner. Instruct an assistant to apply the rear brake whilst bearing down on the rear wheel and then attempt to loosen the sprocket retaining nut. If the nut proves to be stubborn, try striking the side of the extension bar at a point near the socket or spanner whilst the nut is under load. If all attempts to loosen the nut result in failure, then leave it tight and deal with it during the engine dismantling sequence as described in Section 9.

16 With the gearbox sprocket still in position on the gearbox output shaft, locate the split link in the final drive chain and remove it. This can be achieved by removing the spring clip of the split link with a pair of flat-nose pliers and withdrawing the

4.4 Examine the detent plunger assembly (100 and 125)

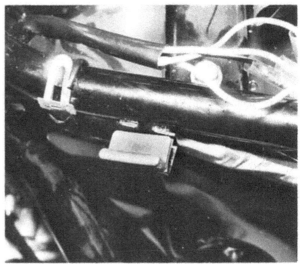

4.5 Do not displace the rubber shield from each sidepanel mounting hook

4.10 Expose the clutch operating mechanism by removing its cover

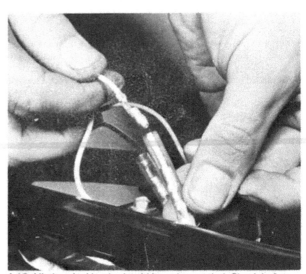

4.13 All electrical leads should be colour coded. Check before disconnecting

4.18 Remove the oil pump cover plate

4.20a Remove the two forward bolts of the footrest mounting bracket ...

4.20b ... to allow removal of the engine crankcase guard (100 and 125)

link to allow the ends of the chain to separate. Clear the chain from the gearbox sprocket and allow its ends to rest on a piece of clean rag or paper placed beneath the machine. It may be considered beneficial at this stage, to remove completely the chain from the machine so that it may be properly cleaned, inspected for wear and relubricated.

17 On TS250 models, remove the four screws that retain the cover of the flywheel generator rotor in position and detach the cover.

18 Remove the two screws that hold the oil pump cover plate in position and manoeuvre the cover clear of the machine. Disconnect the pump control cable from the pump lever by pushing the end of the lever up so that tension is taken off the cable inner and then detaching the cable nipple from the operating arm. With the rubber sealing cap detached from the cable adjuster, the cable may now be pulled through the adjuster and clear of the machine.

19 Prepare to drain the oil from the frame mounted oil tank by obtaining a clean container of at least 1.2 litre (2.6/2.3 US/Imp pint) capacity. Trace the oil feed pipe from the base of the oil tank to its retaining stub on the oil pump. Pinch the end of this pipe to prevent oil loss and detach it from the oil pump by releasing its retaining clip (where fitted) and then pulling it off its retaining stub. Pull the pipe, with its rubber grommet, clear of the oil pump housing and place its end in the container to allow the oil to drain. On completion seal the container to prevent any ingress of contamination before storing it in a safe place.

20 On TS100 and 125 models only, it is now necessary to remove the engine crankcase guard from its frame mounting attachments so that the engine unit can be removed from the machine. Commence removal of this guard by unscrewing the two bolts which pass through the two forward supports of the footrest mounting bracket. Remove these bolts together with the one bolt which secures the guard to the front mounting for the engine unit. Take note of the fitted position of each washer whilst doing this. If the guard will not pull free of its rear attachment points, then release it by loosening each of the two bolts which pass through the two rear supports of the footrest mounting bracket.

21 Move around the engine to frame attachment points, removing the nut and washers from each mounting bolt and noting the fitted position of each washer. Position a strip of wood between the frame bottom tubes and the forward section of the underneath of the crankcase. Make sure the wood is a tight fit so that when the mounting bolts are withdrawn, the engine is prevented from falling forward and damaging the painted surface of the frame tubes. It is a good idea to wrap all

potential frame 'contact' points in rag to prevent their surfaces from coming into contact with the engine unit as it is lifted from position.

22 The engine is now ready for removal from the frame. The unit itself is light and can easily be lifted away by one person. It is helpful, however, to have an assistant present who can help to steady the machine and withdraw the mounting bolts. Carry out a final check around the engine unit to ensure that no control cables, etc, remain connected or will impede the progress of the unit out of the frame.

23 Withdraw each of the engine mounting bolts whilst working from the upper rear bolt in a clockwise direction (viewed from the left-hand side of the machine) to the front mounting with its detachable plate(s). Allow the engine unit to drop onto the wood strip placed beneath it and using this strip as a support, manoeuvre the unit out away from the machine towards the left-hand side. Once removed, carefully position the engine unit on a solid, prepared work surface.

5 Dismantling the engine/gearbox unit: preliminaries

1 Before any dismantling work is undertaken, the external surfaces of the unit should be thoroughly cleaned and degreased. This will prevent the contamination of the engine internals, and will also make working a lot easier and cleaner. A high flash point solvent, such as paraffin (kerosene) can be used, or better still, a proprietary engine degreaser such as Gunk. Use old paintbrushes and toothbrushes to work the solvent into the various recesses of the engine castings. Take care to exclude solvent or water from the electrical components and inlet and exhaust ports. The use of petrol (gasoline) as a cleaning medium should be avoided, because the vapour is explosive and can be toxic if used in a confined space.

2 When clean and dry, arrange the unit on the workbench, leaving a suitable clear area for working. Gather a selection of small containers and plastic bags so that parts can be grouped together in an easily identifiable manner. Some paper and a pen should be on hand to permit notes to be made and labels attached where necessary. A supply of clean rag is also required.

3 Before commencing work, read through the appropriate section so that some idea of the necessary procedure can be gained. When removing the various engine components it should be noted that great force is seldom required, unless specified. In many cases, a component's reluctance to be removed is indicative of an incorrect approach or removal method. If in any doubt, re-check with the text.

4 Mention has already been made of the benefits of owning an impact driver. Most of these tools are equipped with a standard $\frac{1}{2}$ inch drive and an adaptor which can take a variety of screwdriver bits. It will be found that most engine casing screws will need jarring free due both to the effects of assembly by power tools and an inherent tendency for screws to become pinched in alloy castings. If an impact screwdriver is not available, it is often possible to use a crosshead screwdriver fitted with a T-handle as a substitute.

5 A cursory glance over many machines of only a few years' use, will almost invariably reveal an array of well-chewed screw heads. Not only is this unsightly, it can also make emergency repairs impossible. It should also be borne in mind that there are a number of types of crosshead screwdrivers which differ in the angle and design of the driving tangs. To this end, it is always advisable to ensure that the correct tool is available to suit a particular screw.

6 Dismantling the engine/gearbox unit: removing the cylinder head

1 Prior to removing the cylinder head, remove the spark plug and inspect it in accordance with the instructions given in

Chapter 3 of this Manual. On TS100 and 125 models, the cylinder head will be retained by four flange nuts. The TS185 and 250 models have six of these nuts retaining the cylinder head. The cylinder head may be removed easily, irrespective of whether or not the engine unit is fitted in the frame.

2   To remove the cylinder head, slacken its retaining nuts evenly and in a diagonal sequence; this will avoid any risk of distortion. Remove the nuts and lift the head off its retaining studs. If the head appears to be stuck to the cylinder barrel, tap it lightly around its base with a soft-faced mallet; this should be sufficient to break any seal between the two components. Under no circumstances attempt to lever the head from position as this will only crack or distort the head casting.

3   With the head removed, discard the gasket. A new item should be fitted on reassembly.

7   Dismantling the engine/gearbox unit: removing the cylinder barrel, piston and small-end bearing

1   Commence removal of the cylinder barrel by first detaching the carburettor. On TS100 and 125 models the carburettor has a mounting flange which is part of the carburettor body. Remove the two nuts with spring washers that hold this flange against the barrel and detach the carburettor together with its spacer plate and sealing gasket. Ensure that the O-ring fitted in the flange of the carburettor is not misplaced. Carburettors fitted to the TS185 and 250 models are retained to a separate inlet stub by means of a clamp. Release this clamp and pull the carburettor away from the stub. There is no need to remove this inlet stub unless it is intended to carry out a full decarbonisation process or to renew either the barrel or stub, in which case it should be detached by removing its retaining bolts with spring washers.

2   The TS100 and 125 models have one oil feed pipe running from the oil pump to the base of the cylinder barrel, whereas the TS185 and 250 models have two. Carefully pull each pipe off its retaining stub at the cylinder barrel and tape the open end of both pipe and union to prevent any ingress of contamination.

3   To free the cylinder barrel from the crankcase, slacken its six retaining nuts in a diagonal sequence; this will avoid any risk of distortion. Remove each nut and attempt to ease the barrel upwards clear of the crankcase. In practice, it was found that the barrel was stuck firmly to the crankcase and had to be sharply tapped at the strengthened areas around the inlet and exhaust ports with a length of wood and a mallet before it could be lifted clear. Under no circumstances should the areas of unsupported finning around the barrel be struck as this will only serve to break off the fins resulting in an area of barrel which will be inadequately cooled.

4   With the barrel freed from the crankcase, ease it gently upwards off its retaining studs. Take care to support the piston and rings as it emerges from the cylinder bore, otherwise there is risk of damage or ring breakage. If the crankcases are not to be separated, it is advisable to pack the crankcase mouth with clean rag before the piston is withdrawn from the bore, in case the piston rings have broken. This will prevent sections of broken ring from falling into the crankcase.

5   With the barrel lifted clear of the engine unit, invert it and then place it on the work surface. It will be seen that the reed valve is mounted within the barrel to crankcase mating surface and resting the barrel on this valve will distort its curved stopper plate.

6   Prise one of the gudgeon pin circlips out of position, then press the gudgeon pin out of the small-end bearing through the piston boss. If the pin is a tight fit, it may be necessary to warm the piston so that the grip on the gudgeon pin is released. A rag soaked in warm water and wrapped around the piston should suffice. The piston may be detached from the connecting rod once the gudgeon pin is clear of the small-end bearing.

7   If the gudgeon pin is still a tight fit after warming the piston, it can be lightly tapped out of position with a hammer and soft

7.1 Examine the cylinder barrel inlet stub (185 and 250)

7.2 Detach each oil feed pipe from its retaining stub

7.5 Take care not to damage the reed valve assembly

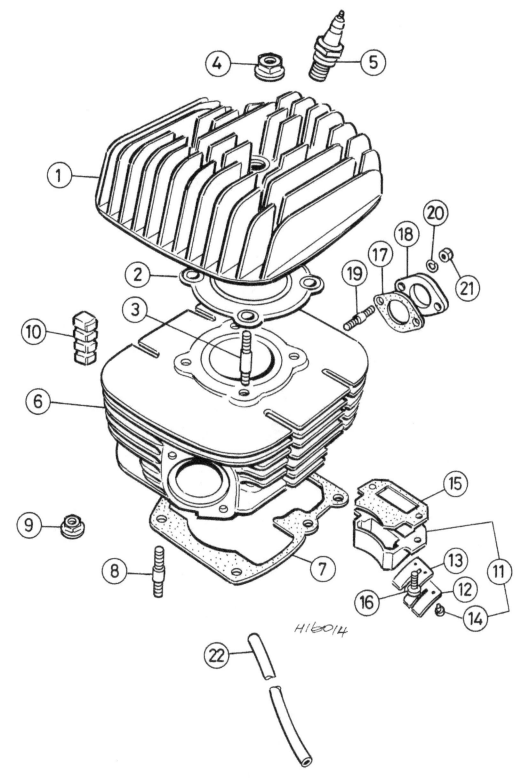

H16014

Fig. 1.1 Cylinder head – 100 and 125 models

1	Cylinder head	7	Cylinder base gasket	13	Valve reeds	18 Inlet stud
2	Cylinder head gasket	8	Stud – 6 off	14	Screw – 2 off	19 Stud – 2 off
3	Stud – 4 off	9	Flange nut – 6 off	15	Gasket	20 Spring washer – 2 off
4	Flange nut – 4 off	10	Damping block	16	Screw – 2 off	21 Nut – 2 off
5	Spark plug	11	Reed valve assembly	17	Gasket	22 Oil feed pipe
6	Cylinder barrel	12	Stopper plate			

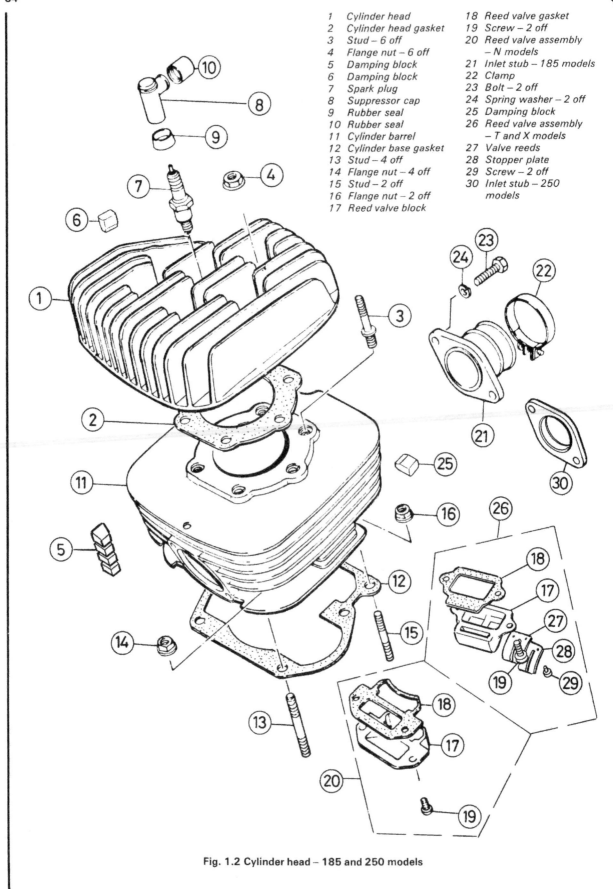

1 Cylinder head
2 Cylinder head gasket
3 Stud – 6 off
4 Flange nut – 6 off
5 Damping block
6 Damping block
7 Spark plug
8 Suppressor cap
9 Rubber seal
10 Rubber seal
11 Cylinder barrel
12 Cylinder base gasket
13 Stud – 4 off
14 Flange nut – 4 off
15 Stud – 2 off
16 Flange nut – 2 off
17 Reed valve block
18 Reed valve gasket
19 Screw – 2 off
20 Reed valve assembly
   – N models
21 Inlet stub – 185 models
22 Clamp
23 Bolt – 2 off
24 Spring washer – 2 off
25 Damping block
26 Reed valve assembly
   – T and X models
27 Valve reeds
28 Stopper plate
29 Screw – 2 off
30 Inlet stub – 250
   models

Fig. 1.2 Cylinder head – 185 and 250 models

metal drift. **Do not** use excess force and make sure the connecting rod is supported during this operation, or there is a risk of its bending.

8   With the piston free of the connecting rod, remove the second circlip and fully withdraw the gudgeon pin from the piston. Place the piston and gudgeon pin aside for further attention. On no account reuse the circlips, they should be discarded and new ones fitted during rebuilding.

9   Push the small-end needle roller bearing out of the connecting rod eye and place it to one side, ready for cleaning and examination. Finally, remove and discard the cylinder barrel base gasket.

8   Dismantling the engine/gearbox unit: removing the flywheel generator assembly

1   The flywheel generator rotor is mounted on the left-hand end of the crankshaft. The stator assembly, which includes the contact breaker points or pulser coil (depending on the type of ignition system fitted), is attached to the crankcase beneath the rotor. The complete flywheel generator assembly may be removed with the engine unit in or out of the frame.

2   Before removal of the rotor can be attempted, the crankshaft must be prevented from rotating. This is achieved by passing a close-fitting bar through the small-end eye and allowing the ends of the bar to rest on wooden blocks placed on each side of the crankcase mouth. Never allow the ends of the bar to come into direct contact with the jointing face. If the rotor is to be removed with the engine in the frame, crankshaft rotation may be prevented by selecting top gear and applying the rear brake.

3   Remove the rotor retaining nut, spring washer and plain washer. The rotor is a tapered fit in the crankshaft end and is located by a Woodruff key; it therefore requires pulling from position. Because the type of rotor fitted varies between the model types covered in this Manual, then the method of rotor removal must also vary.

TS100 and 125 models

4   The rotor fitted to these models has a boss which is threaded internally to take the special Suzuki service tool No 09930-30102 with an attachment, No 09930-30161. If this tool cannot be acquired, then it is possible to remove the rotor by careful use of a two-legged puller.

5   If it is decided to make use of a two-legged puller, great care must be taken to ensure that both the threaded end of the crankshaft and the rotor itself remain free from damage. To protect the crankshaft end, refit the rotor retaining nut and screw it on until its outer face lies flush with the end of the crankshaft. Assemble the puller, checking that its feet are fitted through the two larger slots in the rotor face and are resting securely on the strengthened hub. Remember that the closer the feet of the puller are to the centre of the rotor then the less chance there is of the rotor becoming distorted. Gradually tighten the centre bolt of the puller to apply pressure to the rotor. Do not overtighten this bolt but apply reasonable pressure and then strike the end of the bolt with a hammer to break the taper joint. If this fails at first, tighten the centre bolt to apply a little more pressure and then try shocking the rotor free again.

TS185 and 250 models

6   The rotor fitted to these model types has three threaded holes tapped in its strengthened hub section. These holes are positioned so as to accommodate the special Suzuki service tool No 09930-30713, which is the recommended tool with which to pull the rotor from position. In the absence of this tool, an extractor may be made by using the following procedure.

7   Obtain a piece of mild steel plate which is approximately 7 mm ($\frac{1}{4}$ inch) thick. Obtain three high tensile steel bolts which are of the correct thread type and diameter to fit into the holes tapped in the rotor. Finally, obtain a suitably sized bolt and nut

8.2 Remove the flywheel generator rotor retaining nut

8.10 Release the electrical lead from the neutral indicator switch

to fit through the centre of the extractor as shown in the figure accompanying this text.

8   Referring to the figure, accurately mark the centres of the four holes shown to correspond with the holes in the flywheel rotor. Drill these holes and fit the bolts through the plate to check that their ends align with their locations in the rotor. To finish the job, tack weld the nut to the centre of the plate and grind a taper on the end of the centre bolt so that it locates in the machined centre of the crankshaft.

9   Arrange the extractor on the rotor whilst ensuring that the three bolts are screwed fully home into the rotor hub section. Screw down the centre bolt to draw the rotor off its taper. If the rotor refuses to move, do not overtighten the centre bolt but apply reasonable pressure and then strike the head of the bolt with a hammer to break the taper joint. If this fails at first, tighten the bolt a little further and try again.

All models

10 With the rotor removed from the crankshaft end, release the single electrical lead from its location on the neutral indicator switch. Using a sharp scribing tool or a centre punch, carefully mark the position of the stator plate in relation to the crankcase; this will serve as a reference when refitting the plate. Remove the three screws that serve to retain the stator plate to the crankcase and lift the stator plate from position, threading the

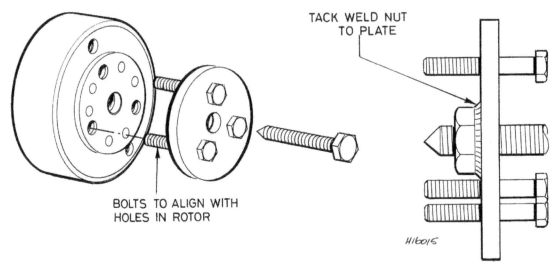

TACK WELD NUT
TO PLATE

BOLTS TO ALIGN WITH
HOLES IN ROTOR

Fig. 1.3 Fabricated tool for flywheel generator rotor removal – 185 and 250 models

electrical leads clear of their retaining grommet whilst doing so.

11 Finally, ease the Woodruff key from its location in the crankshaft taper and place it, together with the flywheel generator assembly, in safe storage until required for reassembly.

9 Dismantling the engine/gearbox unit: removing the neutral indicator switch assembly and the gearbox sprocket

1 On all but the TS250 models, the neutral indicator switch takes the form of a white plastic cover which is sited over the left-hand end of the gearchange selector drum. To remove this switch, unscrew its two retaining screws and lift it clear of the selector drum end. It will be seen that an O-ring is fitted to the inside face of the switch, this should be discarded and replaced with a new item before refitting the switch. Using a pair of long-nose pliers, carefully withdraw the contact pin from its location in the end of the gearchange drum and then hook the spring out of the pin location. Place the spring together with the other switch components, in a clean, dry storage space. If the spring is seen to be damaged or seems to have become fatigued, then it must be replaced with a new item.

2 The type of neutral switch fitted to the TS250 models differs in that the switch casing is of black plastic and of a protruding design. Switch location and removal is as described in the preceding paragraph of this Section. Removal of the casing will give access to the indicator switch arm which can be detached from the end of the gearchange drum by removing the countersunk screw. It will be seen that this arm is keyed to the drum end to avoid confusion when refitting.

3 The gearbox sprocket is secured to the splined end of the gearbox output shaft by a large nut, the removal of which will necessitate locking the sprocket. With the engine in the frame, this may be accomplished by applying the rear brake, thus immobilising the sprocket via the final drive chain. With the engine on the bench, it will be necessary to select top gear and lock the crankshaft in position by passing a close-fitting bar through the small-end eye and allowing the ends of the bar to rest on wooden blocks placed on each side of the crankcase mouth. Never allow the ends of the bar to come into direct contact with the jointing face.

4 Having locked the gearbox sprocket in position, knock back the tab of the lock washer from the nut and unscrew the nut. With the nut and its lock washer removed, pull the sprocket off the output shaft and follow this with the spacer which is inserted into the shaft seal. It should be noted that an O-ring is located between the inboard end of the spacer and the shaft bearing.

10 Dismantling the engine/gearbox unit: removing the oil pump assembly

1 The oil pump is mounted in a compartment to the rear of the left-hand crankcase, and is retained by two mounting screws. The pump can be removed with the engine unit installed, or removed from, the frame. In the case of the former, it will be necessary to detach the gearbox sprocket cover and oil pump cover, to gain access to the pump.

2 If the pump is to be removed with the engine in the frame, then provision must be made to catch the oil that will issue from both the feed and delivery pipes once they are disconnected from the pump. To prevent complete draining of the oil tank, the feed pipe should be plugged as soon as it is detached; a clean screw or bolt of the appropriate thread diameter is ideal for this purpose.

3 To remove the pump, slacken and remove both of its retaining screws. Lift the pump clear of its drive shaft and whilst threading the oil feed pipe(s) to the cylinder barrel clear of the crankcase. Discard the pump base gasket and replace it with a new item on reassembly.

4 On TS185 models, note the keyed drive attachment which connects the pump to the kickstart drive pinion assembly. On TS250 models, note the fitted position of the pump to crankcase spacer washer.

11 Dismantling the engine/gearbox unit: removing the right-hand crankcase cover, primary drive pinion and clutch assembly

1 Free the kickstart lever from its shaft end by unscrewing its retaining bolt and pulling it free of its locating splines. It will be found that the screws which retain the cover to the crankcase are of varying lengths; because of this, it is well worth making up a cardboard template of the cover into which each screw may be inserted as it is removed. This will serve to both keep the screws from becoming lost and to provide a reference as to the correct fitted position of each screw.

2 Proceed to loosen the cover retaining screws. This should be done evenly and in a diagonal sequence to prevent the cover from becoming distorted. Pull the cover clear of its locating dowels, remove and discard the cover gasket and pull the two dowels from their location in the crankcase. The dowels should be placed in safe storage until required for reassembly.

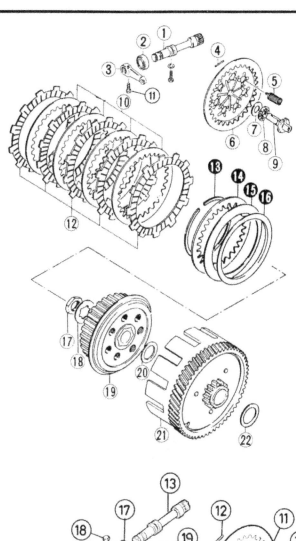

Fig. 1.4 Clutch – 100 and 125 models

1   Operating shaft
2   Oil seal
3   Operating lever
4   Anchor pin – 7 off
5   Spring – 7 off
6   Pressure plate
7   Thrust washer
8   Thrust bearing
9   Release rod
10   Plain plate – 4 off
11   Bolt
12   Friction plate – 5 off
13   Retaining ring
14   Plain plate
15   Wave plate
16   Wave plate seat
17   Nut
18   Tab washer
19   Clutch hub
20   Thrust washer
21   Clutch drum
22   Thrust washer

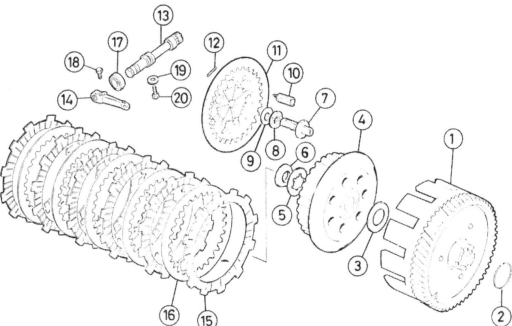

Fig. 1.5 Clutch – 185 models

1	Clutch drum	6	Nut	11	Pressure plate	16	Plain plate – 5 off
2	Retaining ring	7	Release rod	12	Anchor pin – 7 off	17	Oil seal
3	Thrust washer	8	Thrust bearing	13	Operating shaft	18	Bolt
4	Clutch hub	9	Thrust washer	14	Operating lever	19	Washer
5	Tab washer	10	Spring – 7 off	15	Friction plate – 6 off	20	Bolt

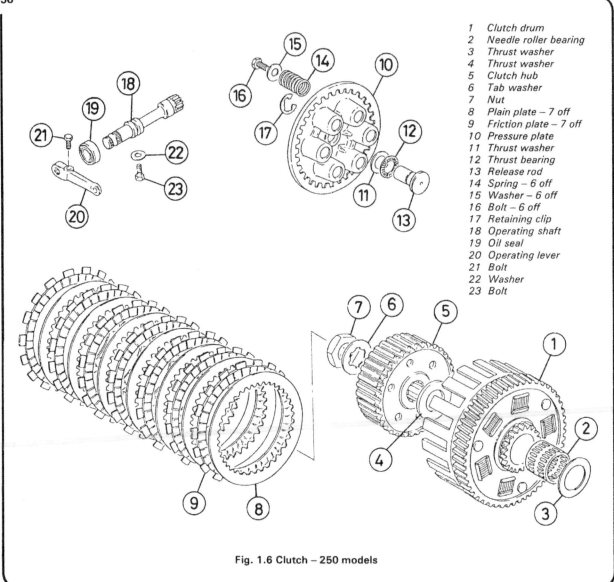

1   Clutch drum
2   Needle roller bearing
3   Thrust washer
4   Thrust washer
5   Clutch hub
6   Tab washer
7   Nut
8   Plain plate – 7 off
9   Friction plate – 7 off
10 Pressure plate
11 Thrust washer
12 Thrust bearing
13 Release rod
14 Spring – 6 off
15 Washer – 6 off
16 Bolt – 6 off
17 Retaining clip
18 Operating shaft
19 Oil seal
20 Operating lever
21 Bolt
22 Washer
23 Bolt

Fig. 1.6 Clutch – 250 models

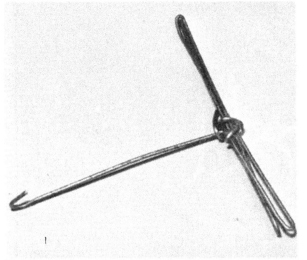

11.3 Manufacture a clutch spring releasing tool ...

11.4 ... and use it to release spring pressure from each anchor pin (100, 125 and 185)

TS100, 125 and 185 models

3   Before attempting to dismantle the type of clutch assembly fitted to these models, it will first be necessary to manufacture a tool with which to ease the pressure placed by the clutch springs on their anchor pins. Suzuki manufacture a special tool No 09920-20310 specifically for this purpose. It was found that a length of small-gauge welding rod bent to the shape shown in the accompanying photograph was perfectly adequate for the job.

4   Proceed to release the pressure plate from the clutch assembly by placing the hooked end of the tool through the end of one of the clutch springs; pull the spring end out from the pressure plate so that it releases its grip on the anchor pin and draw the pin out through the spring end by using a pair of long-nose pliers. Repeat this procedure until all the pins are removed and the pressure plate can be lifted clear. Lift the clutch friction and plain plates out of the clutch drum as an assembly and place them on the work surface ready for examination.

5   Remove the small central thrust bearing together with the clutch release rod and the thrust washer.

TS250 models

6   With the type of clutch assembly fitted to the TS250 models, commence the clutch removal sequence by unscrewing the six spring retaining bolts. Take care to avoid placing any undue strain upon the pressure plate by unscrewing these bolts in small increments and in a diagonal sequence until they can be removed along with their plate washers. The pressure plate can now be removed along with the clutch springs. Detach the clutch release rod from the pressure plate by removing its retaining clip. With the rod thus freed, remove the thrust bearing and thrust washer from its shank. Remove the plain and friction plates from their location within the clutch drum and place them, in the order of removal, on a clean area of work surface so that they can be inspected and if serviceable, refitted in the same order.

All models

7   It will now be necessary to devise some means of locking the gearbox input shaft in position whilst the clutch hub retaining nut is slackened. The method used to do this was to lock the gearbox output shaft by utilising the gearbox sprocket, an old length of final drive chain, a length of metal tubing and a metal drift of a small enough diameter to pass through the links of the chain, and then to place the unit in gear. Commence by refitting the gearbox sprocket over the output shaft end and securing it in position by refitting its retaining nut, finger-tight. Engage the centre section of the length of chain over the sprocket and pass its ends through the length of tube. The metal

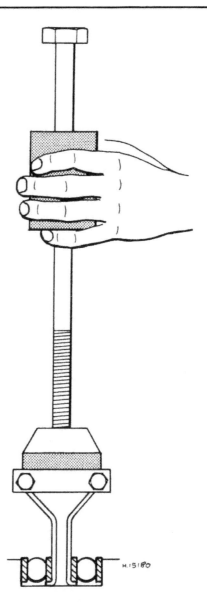

Fig. 1.7 Use of slide hammer when removing input shaft bearing – 125, 185 and 250 models

Note: TS185 model is fitted with needle roller bearing in place of the ball bearing shown

drift should now be passed through both ends of the chain at a point where it will also abut against the end of the tube (see accompanying photograph).

8   With an assistant holding the output shaft locking tool in positon, place the unit in gear and knock back the tab of the lock washer from the clutch hub retaining nut. Support the engine unit and slacken the retaining nut. In practice, this nut was found to be very tight and great care had to be taken to prevent the engine unit from rotating around its axis due to the opposing forces placed upon it. With the nut thus slackened, run it off the input shaft thread and pull the clutch hub from position, together with the lock washer. Remove the thrust washer which will have remained on the input shaft.

9   On all but the TS185 models, the clutch drum can now be lifted off the input shaft and the thrust washer located beneath it also removed. The clutch drum fitted to TS185 models is held, by means of a retaining ring, to the input shaft. If pulling of the

11.7 Lock the gearbox input shaft in position to release the clutch hub retaining nut

drum by finger pressure fails to remove it, then take the weight of the engine unit by lifting the drum a few inches and then strike the end of the input shaft with a soft-faced hammer. This will cause the retaining ring to free and the drum to slide off the shaft.

10 The locking assembly should now be removed from the gearbox sprocket and the sprocket removed from the output shaft end. With this done, lock the crankshaft in position by passing a close-fitting bar through the small-end eye and allowing the ends of the bar to rest on wooden blocks placed on each side of the crankcase mouth. Never allow the ends of the bar to come into contact with the jointing face.

11 Knock back the tab of the lock washer from the primary drive pinion retaining nut and slacken the nut. Run the nut off the crankshaft thread, remove the lock washer, pull off the pinion and remove the Woodruff key. On TS100 and 125 models it will be seen that there is a spacer ring fitted between the pinion and the main bearing; this ring should also be removed. For reference when refitting, note that the lipped side of the pinion fitted to all model types must face the crankcase side.

12 Dismantling the engine/gearbox unit: removing the kickstart drive and idler pinions and the oil pump drive – TS100, 125 and 250 models

1   Remove the kickstart drive pinion from the kickstart shaft simply by lifting it up and off the shaft. TS100 and 125 models have a thrust washer fitted over this pinion, whereas TS250 models are fitted with a spacer.
2   Remove the kickstart idler pinion by releasing its retaining circlip and then drawing the pinion off its shaft. Note, for reference when refitting, the fitted position of the thrust washer which accompanies this pinion.
3   The oil pump drive pinion together with the drive shaft can now be withdrawn from their location in the crankcase.

13 Dismantling the engine/gearbox unit: removing the kickstart return spring assembly – TS185 models

1   The kickstart return spring assembly fitted to the TS185 models differs from that fitted to the other models covered in this Manual in that it is located outside of the crankcase halves. As a result access may be made to the spring for renewal without first having to separate the crankcase halves. To remove this assembly, first pull the spring thrust plate off the kickstart shaft and follow this by withdrawing the plastic guide from within the spring. Finally, use a pair of flat-nosed pliers to unhook carefully the return spring from its location in the crankcase and restrain the spring as it unwinds. Do not attempt this by hand as the results can be very painful. Detach the spring from the kickstart shaft.

14 Dismantling the engine/gearbox unit: removing the gearchange shaft and pawl mechanism – TS100, 125 and 250 models

1   The gearchange shaft assembly can be removed from the crankcase by pulling it free of its spring locating pin and drawing it out of position.
2   The pawl mechanism can now be released from the crankcase by unscrewing the pawl lifter plate and cam guide plate securing screw. It was found in practice that all these screws were extremely tight and needed the use of an impact driver to free them. If it is found that only one of the two screws

securing each plate can be freed with the driver, then moving the freed end of the plate may well serve to release the remaining screw. Take care when removing the pawl lifter plate to retain the two pawls in position in their housings, otherwise their operating springs will cause both the pawls and their operating pins to be ejected.
3   Remove the pawl mechanism and place it in safe storage until required for examination and reassembly. It is a good idea to wrap a rubber band around the unit to stop the pawls and pins from becoming separated and lost.
4   On TS250 models only, remove the gearbox detent plunger housing from its location on the underside of the right-hand crankcase half. This housing takes the form of a hexagon-headed plug which is identical in appearance to the gearbox oil drain plug situated in front of it. Take this opportunity to examine and clean the detent plunger assembly. The spring must be renewed if it shows signs of fatigue or failure. The sealing washer should be renewed as a matter of course; always replace the used washer with a new one of an identical type; never omit to fit the washer or replace it with any form of sealant as this will cause the plunger assembly to malfunction because of the incorrect pressure placed upon the spring. Once cleaned, the assembly may be stored in a safe place ready for refitting. Note that if it is intended not to separate the crankcase halves then great care should be taken to avoid movement of the gearchange components whilst the detent plunger assembly is removed.

15 Dismantling the engine/gearbox unit: removing the gearchange mechanism – TS185 models

1   Commence removal of the gearchange mechanism by using a pair of pliers to unhook the cam stopper arm return spring from the eye in the gearbox input shaft bearing retaining plate. Unscrew the stopper arm pivot bolt and remove the arm together with the bolt, the washer and the spring.
2   The retaining plate for the gearchange pins is secured to the end of the gearchange drum by a single countersunk screw. In practice, this screw was found to be very tight and required the use of an impact driver to free it. Remove the screw together with the plate and then proceed to remove each gearchange pin. It will be seen that one of these pins is different from the other four in that it is shorter and has a lipped shank. Make a note of the hole into which this pin fits by marking the end of the gearchange drum.
3   With the pin retaining plate removed, pull the cam stopper pawl off its pivot. To remove the main gearchange arm and selector pawl, grasp the gearchange arm at its pivot point on the gearchange shaft and, making sure that the selector pawl is free from its channel in the change drum end, gently pull the shaft until it is free of the engine unit.
4   Remove the two screws which hold the cam guide plate in position and remove the plate. If renewal of the gearbox input shaft bearing is contemplated, then remove the bearing retaining plate by unscrewing its two securing screws.
5   Remove the gearbox detent plunger housing from its location on the upper surface of the right-hand crankcase half, adjacent to the tachometer drive cable attachment. This housing takes the form of a hexadon-headed plug. Take this opportunity to examine and clean the detent plunger assembly. The spring must be renewed if it shows signs of fatigue or failure. The sealing washer should be renewed as a matter of course; always replace the used washer with a new one of an identical type, never omit to fit the washer or replace it with any form of sealant as this will cause the plunger assembly to malfunction because of the incorrect pressure placed upon the spring. Once cleaned, the assembly may be stored in a safe place ready for refitting. Note that if it is intended not to separate the crankcase halves then great care should be taken to avoid movement of the gearchange components whilst the detent plunger assembly is removed.

15.1 Remove the cam stopper arm ...

15.2a ... followed by the retaining plate for the gearchange pins

15.2b ... each gearchange pin (whilst noting their positions) and the cam stopper pawl ...

15.3 ... the main gearchange arm and selector pawl assembly ...

15.4 ... and finally, the cam guide plate (185 only)

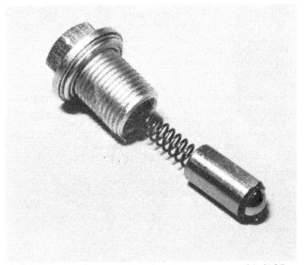
15.5 Remove and examine the detent plunger assembly (185 only)

16 Dismantling the engine/gearbox unit: separating the crankcase halves

1   Prepare the crankcase for separation by arranging it on the work surface so that it is well supported on wooden blocks with the heads of the crankcase securing screws facing uppermost. It will be found that these screws are of various lengths. To prevent them from becoming refitted in the wrong locations, it is a good idea to make up a cardboard template of the crankcase through which the screws can be fitted as they are removed. An impact driver will be required to free the screws before they can be removed.

2   When removing these screws, take care to work in a diagonal sequence, loosening the screws in small increments at first until they are all completely free. This simple precaution will obviate any risk of the crankcase halves becoming distorted. Note that on TS250 models, the two screws which hold the oil reservoir plate in position also serve to secure the crankcase halves.

3   Check that all the screws are removed and then attempt to separate the crankcase halves by lifting the upper half away. On TS100 and 125 models, the crankcase halves have been sealed along their mating surfaces with a sealing compound and it is likely that this compound will prevent easy separation of the two components. it was found that tapping around the joint with a soft-faced mallet served to help break this seal.

4   Resist the temptation to lever the casing halves apart with a screwdriver, as this almost invariably leads to damaged sealing faces. It is imponticularly important that the crankcase joint remains absolutely airtight on a two-stroke engine, as an air leak can cause loss of secondary compression, and can upset the fuel air mixture. If the crankcase halves refuse to separate easily, then proceed as follows.

5   Select the nut which fits over the upper facing crankshaft end and fit it so that its upper face is flush with that of the crankshaft end. This will protect the threaded end of the crankshaft from damage. With an assistant gripping the engine unit by its upper facing crankcase half and at the same time lifting it clear of the work surface, strike the protected end of the crankshaft sharply with a soft-faced hammer. It was found that one good blow was sufficient to cause initial separation of the crankcase halves and that subsequent blows caused further separation until finally, the crankshaft was driven clear of its bearing.

6   This task proved to be fairly easy and required no excessive force to complete it. It must be realised that striking the crankshaft and with excessive force will almost certainly cause the crankshaft assembly to become distorted. In practice, it was found that separation of the crankcase halves on the TS185 model was hindered by the rear section of casing tending to 'hang up' on both the gearbox shafts and then the surface locating dowels. Alternate tapping apart of each end of the casings eventually solved this problem.

7   It should be noted that Suzuki recommend the use of a special service tool (No 09910-80115) for separating the crankcase halves. If difficulty is experienced when carrying out the procedure described in the above paragraphs of this Section, then it is advisable either to attempt to hire this tool from a local Suzuki service agent or to return the complete crankcase unit to the agent so that it may be pulled apart and the crankshaft and associated bearings inspected and renovated.

8   With the crankcase halves thus separated, discard the sealing gasket (TS185 and 250 models only) and then remove the two large locating dowels from their locations in the mating surface and place them in safe storage until required for reassembly.

17 Dismantling the engine/gearbox unit: removing the kickstart shaft assembly, gearbox components, tachometer drive gear and crankshaft

1   On TS100 and 125 models, remove the kickstart shaft assembly from its location in the right-hand crankcase half by first withdrawing the plastic guide from within the kickstart return spring. Unlatch the return spring from its location in the shaft by using a pair of long-nose pliers. Remove the spring, rotate the shaft so that its ratchet stop clears its retaining plate and draw the shaft from position.

2   To remove the kickstart shaft assembly from the TS250 models, follow the procedure given for the TS100 and 125 models whilst noting that the plastic guide cannot be withdrawn from the return spring until its retaining circlip has been removed.

3   The kickstart shaft assembly fitted to the TS185 models cannot be removed until the tachometer drive gear has been withdrawn. This operation should now be carried out on all model types to which a tachometer is fitted by unscrewing the single retaining screw from the side of the drive gear housing and then withdrawing the complete assembly from position. Relocate the retaining screw in the housing and tighten it finger-tight so that it does not become separated from the crankcase until required for reassembly. Note that there should be a plain washer fitted beneath the head of this screw. The kickstart shaft assembly can now be removed by rotating the shaft to clear its ratchet stop retaining plate and then withdrawing the complete assembly from position.

4   It will be found that once the crankcase halves are separated, the gearbox components will almost always remain with the crankshaft assembly in the crankcase half which remains supported on the work surface. Proceed to remove these components as follows.

TS100, 125 and 250 models

5   Before disturbing any of the components fitted in the gearbox units of these models, it is well worth making a quick sketch of the fitted position of all three selector forks before they are disturbed; this simple precaution will save considerable anxiety at a later stage if the forks should become separated from their shafts during the cleaning and examination procedures.

6   Withdraw each selector fork shaft, detaching the fork(s) from the gear clusters and relocating each one in its correct fitted position on the shaft directly after doing so. Place the reassembled components on a clean work surface.

7   The detent plunger assembly will have already been removed from the right-hand crankcase half for the purpose of draining the gearbox oil. On TS100 and 125 models it is now possible to pull the gearchange drum free of its housing. On TS250 models however, it is first necessary to remove the cam stopper arm from within the gearbox housing by releasing its return spring and then unscrewing its pivot bolt.

8   Withdraw both the input and output shaft assemblies from the crankcase half as a complete unit, taking great care to keep both assemblies together. Place the complete unit next to the selector fork shafts and cover all the gearbox components with a piece of clean rag or paper to prevent the ingress of dirt into the bearing surfaces. If none of the assemblies are to be dismantled, it is advisable to secure them in their correct relative positions with elastic bands before proceeding further. It was found in practice that the output shaft required some gentle persuasion to free it from its bearing. This persuasion took the form of protecting the end of the shaft by refitting the gearbox sprocket retaining nut and then gently tapping the end of the shaft with a soft-faced hammer to drift it from position. Take note of the thin O-ring that will remain on the shaft.

TS185 models

9   Whilst the information given for removal of the gearbox components from the engine units of the TS100, 125 and 250

17.3 Withdraw the tachometer drive gear to allow removal of the kickstart shaft assembly (185 only)

17.11 Each kickstart plate retaining screw must be locked in position with the tab washer

models holds true for the gearbox components fitted to the TS185 models, there is, however, one major difference in the method of removal in that the complete component assembly must be withdrawn from the gearbox housing as one unit and then, if found to be necessary, dismantled on the work surface.
10 On these models, it is now necessary to remove the kickstart driven gear from the centre of the gearbox input shaft bearing located in the right-hand crankcase half. This component may prove to be stubborn, in which case the area around the bearing should be well supported before using a soft-metal drift and hammer to drive carefully the gear from position.

All models
11 Both crankcase halves should now be free of all removable components except bearing retaining plates and, where applicable, the kickstart ratchet stop retaining plate and the gearchange shaft return spring stop. None of these components need be removed unless they are found to be damaged or bearing renewal is required. Note that the tab washer located beneath the heads of the two kickstart plate retaining screws must be renewed if these screws are removed.

12 The crankcase half containing the crankshaft should now be supported on wooden blocks on the work surface so that the crankshaft can be pushed clear of its bearing without its end coming into contact with the work surface. Great care must be taken to ensure that the crankcase half is well supported at all points on its mating surface, as the crankshaft will have to be struck quite sharply to free it and this simple precaution will lessen the risk of the casing being distorted.
13 The method used to free the crankshaft was to place a heavy copper drift against the crankshaft and sharply strike the drift with a heavy hammer. It was found that the crankshaft moved on the first blow and continued to move under the force of subsequent lighter blows until completely free. In order to protect the end of the crankshaft during this procedure, the appropriate nut was refitted and run down its thread until the end of the shaft appeared flush with the upper side of the nut.
14 If it is thought that excessive force is being required to move the crankshaft and there is a serious risk of the crankcase becoming damaged, it is recommended that advice and aid be sought from an official Suzuki service agent. Alternatively, Suzuki provide a special tool No 09920-13111 for pressing the shaft from position, although this tool is shown only for use on the TS250 models.

18 Examination and renovation: general

1   Before examining the parts of the dismantled engine unit for wear it is essential that they should be cleaned thoroughly. Use a petrol/paraffin mix or a high flash-point solvent to remove all traces of old oil and sludge which may have accumulated within the engine. Where petrol is included in the cleaning agent normal fire precautions should be taken and cleaning should be carried out in a well ventilated place.
2   Examine the crankcase castings for cracks or other signs of damage. If a crack is discovered it will require a specialist repair.
3   Examine carefully each part to determine the extent of wear, checking with the tolerance figures listed in the Specifications Section of this Chapter or in the main text. If there is any doubt about the condition of a particular component, play safe and renew.
4   Use a clean lint free rag for cleaning and drying the various components. This will obviate the risk of small particles obstructing the internal oilways, and causing the lubrication system to fail.
5   Should any studs or internal threads require repair, now is the appropriate time to attend to them. Where internal threads are stripped or badly worn, it is preferable to use a thread insert, rather than tap oversize. Most dealers can provide a thread reclaiming service by the use of Helicoil thread inserts. They enable the original component to be re-used.
6   Sheared studs or screws can usually be removed with screw extractors, which consist of tapered, left-hand thread screws, of very hard steel. These are inserted by screwing anticlockwise, into a pre-drilled hole in the stud, and usually succeed in dislodging the most stubborn stud or screw. The only alternative to this is spark erosion, but as this is a very limited specialised facility, it will probably be unavailable to most owners. It is wise, however, to consult a professional engineering firm before condemning an otherwise sound casing. Many of these firms advertise regularly in the motorcycle papers.

19 Examination and renewal: crankcase and gearbox oil seals

1   Any failure in the crankcase oil seals on a two-stroke engine will have a profound effect on the performance and efficiency of the unit. When the seals begin to wear, air is admitted into the crankcase; this air serves to dilute the incoming fuel/air mixture, thereby causing uneven running and difficulty in starting.
2   Examine each seal carefully, paying particular attention to

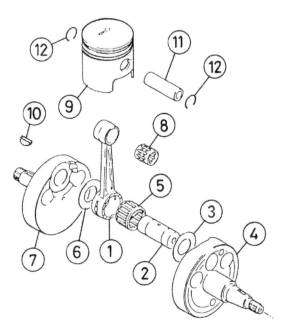

Fig. 1.8 Crankshaft

1	Connecting rod	7	Right-hand flywheel
2	Crank pin	8	Small-end bearing
3	Thrust washer	9	Piston
4	Left-hand flywheel	10	Woodruff key
5	Big-end bearing	11	Gudgeon pin
6	Thrust washer	12	Circlip – 2 off

19.3a Seals can be prised from position with the flat of a screwdriver

its sealing lip. This area performs the sealing function, and the seal should be renewed without question if it is scored or marked in any way. If the seal is found to have become hardened, due to the machine having been in use for a considerable amount of time, it should be renewed without question. In view of the important part that every seal plays, it is considered good practice to renew them as a set as a matter of course whilst the engine is dismantled.

3   Each seal can be removed by prising it out of position with the flat of a screwdriver. Great care should be taken, however, to ensure that the alloy in or around the seal housings is not damaged during this operation. Equal care should be taken when fitting a new seal. Ensure that the seal is facing in the

19.3b New seals should be carefully drifted into their locations

right direction and push it as far as possible into its housing with hand pressure whilst ensuring that it enters squarely. Select a socket or length of metal tube that has an external diameter slightly less than that of the seal and with the casing well supported around the seal housing, gently drift the seal into position. Note that the two crankcase oil seals fitted to the TS100 and 125 engine units must have a thread locking compound applied to their outer edges before insertion otherwise they will tend to rotate out of position during use.

20 Examination and renovation: crankshaft main bearings and gearbox bearings

1   Failure of the crankshaft main bearings is usually evident in the form of an audible rumble from the bottom end of the engine, accompanied by vibration. The vibration will be most noticeable through the footrests. Main bearing failure will immediately be obvious when the bearings are inspected, after the old oil has been washed out. If any play is evident or if the bearings do not run freely, renewal is essential.

2   Each bearing can be removed by first removing its oil seal or retaining plate, where appropriate, and then applying heat to the engine casing. The application of heat will cause the aluminium alloy of the casing to expand at a faster rate than the steel of the bearing thus allowing the bearing to become loose. The safest way of doing this is to place the casing in an oven, heating it to approximately 80-100°C, or to immerse the casing in boiling water. The casing may then be sharply tapped on a wooden surface, face down, to jar the bearing free. Alternatively, the bearing may be drifted out of position by using a hammer in conjunction with a length of metal tube of the appropriate diameter; take care when doing this to ensure that the casing is properly supported and that the bearing is made to leave the casing squarely. Note that care should always be exercised when heating alloy casings as excessive or localised heat can easily cause warpage.

3   Each new bearing can be drifted into position, after heating the casing as for removal, by using a socket or length of metal tube. The diameter of the tube should be equal to that of the outer race of the bearings. It is essential that the component to which the bearing is to be fitted is well supported during the fitting procedure and that the bearing enters the casing squarely. Check, before fitting each bearing, that any oil feed drilling in the bearing housing is clear of sludge or contamination.

20.2 Where necessary, remove the bearing retaining plate

20.3 Each new bearing can be drifted into position

20.4 Bushes should be dealt with in a manner similar to that described for ball bearings

4   The procedure for examination, removal and fitting of the gearbox bearings is as listed in the above paragraphs of this Section. The only bearing that will require different treatment is that fitted to the left-hand end of the input shaft. Because this bearing is fitted into a blind hole it cannot be driven out in the normal manner. On 125, 185 and 250 models a slide hammer arrangement similar to that shown in the accompanying figure should be used for removal. On 100 models, where a plain bush is fitted, extraction can only easily take place using a special threaded attachment. It is recommended that the casing be returned to a motorcycle engineer for bush removal to be carried out safely.

21 Examination and renovation: crankshaft assembly

1   The crankshaft assembly comprises two full flywheels, two mainshafts, a crankpin and caged needle roller big-end bearing, a connecting rod, and a caged needle roller small-end bearing. The general condition of the big-end bearing may be established with the assembly removed from the engine, or with just the cylinder head and barrel removed, as would be the case during a normal decoke. In this way it is possible to decide whether big-end renewal is necessary, without a great deal of exploratory dismantling.
2   Big-end failure is characterised by a pronounced knock which will be most noticeable when the engine is worked hard. The usual causes of failure are normal wear, or a failure of the lubrication supply. In the case of the latter, big-end wear will become apparent very suddenly, and will rapidly worsen. Check for wear with the crankshaft set in the TDC (top dead centre) position, by pushing and pulling the connecting rod. No discernible movement will be evident in an unworn bearing, but care must be taken not to confuse end float, which is normal, and bearing wear. If a dial gauge is readily available, a further test may be carried out by setting the gauge pointer so that is abuts against the upper edge of the periphery of the small-end eye. Measurement may then be taken of the amount of side-to-side deflection of the connecting rod. If this measurement exceeds the service limit of 3.0 mm (0.12 in) then the big-end bearing must be renewed.
3   Any measurement of the crankshaft deflection can only be made with the crankshaft assembly removed from the crankcase and set up on V-blocks which themselves have been positioned on a completely flat surface. The amount of deflection should be measured with a dial gauge at a point just inboard of the threaded portion of each mainshaft end. If, in either case, the amount of deflection shown by the gauge needle exceeds the service limit of 0.05 mm (0.002 in) then the assembly must be replaced with a serviceable item.
4   Like the big-end bearing, the small-end bearing is of the caged needle roller type and will seldom give trouble unless a lubrication failure has occurred. Push the bearing into the small-end eye of the connecting rod and push the gudgeon pin through the bearing. Hold the connecting rod steady and feel for any discernible movement between it and the gudgeon pin. If movement is felt, do not automatically assume that the bearing is worn but check that the bore of the small-end eye and the outer diameter of the gudgeon pin are not worn beyond their service limits. If both the connecting rod and gudgeon pin are found to be within limits, discard the bearing and replace it with a new item. Close inspection of the bearing will show if the roller cage is beginning to crack or wear, in which case the bearing must be renewed.
5   If any fault is found or suspected in any of the components comprising the crankshaft assembly, it is recommended that the complete crankshaft assembly is taken to a Suzuki service agent, who will be able to confirm the worst, and supply a new or service-exchange assembly. The task of dismantling and reconditioning the big-end assembly is a specialist task, and it is considered to be beyond the scope and facilities of the average owner.

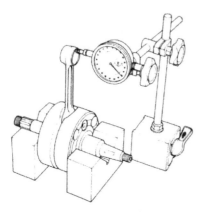

Fig. 1.9 Measuring big-end bearing wear

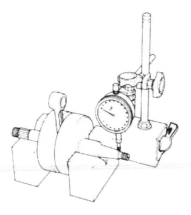

Fig. 1.10 Measuring the amount of crankshaft deflection

21.6 The oil guide plate can be detached from the crankshaft flywheel (185 and 250)

TS 185 and 250 models

6   On these models, it will be seen that an oil guide plate is fitted to the outboard side of the left-hand flywheel. This plate should not be disturbed unless renewal of either the crankshaft assembly or the plate itself is necessary. Great caution should be exercised if removing this plate as it is very easily distorted. Identify the point at which the stub on the face of the plate is inserted into the flywheel. Carefully push the flat of a medium sized screwdriver between the plate and flywheel at this point and ease the plate retaining stub out of its recess. Moving the

screwdriver to any other point on the edge of the plate will only result in the plate becoming distorted. Note the spacer which accompanies the plate.

22 Examination and renovation: decarbonising

1   Decarbonising must take place as part of any major overhaul, in addition to being a normal routine maintenance function. In the case of the latter, the operation can be undertaken with minimal dismantling, namely removal of the cylinder head, cylinder barrel and exhaust silencr baffle. Carbon build up in a two-stroke engine is more rapid than that of its four-stroke counterpart, due to the oily nature of the combustion mixture. It is however, rather softer and is therefore more easily removed.
2   The object of the exercise is to remove all traces of carbon whilst avoiding the removal of the metal surface on which it is deposited. It follows that care must be taken when dealing with the relatively soft alloy cylinder head and piston. Never use a steel scraper or screwdriver for carbon removal. A hardwood, brass or aluminium scraper is the ideal tool as these are harder than the carbon, but no harder that the underlying metal. Once the bulk of the carbon has been removed, a brass wire brush of the type used to clean suede shoes can be used to good effect.
3   The whole of the combustion chamber should be cleaned, as should the piston crown. It is recommended that as smooth a finish as possible is obtained, as this will show the subsequent build up of carbon in this area will restrict the flow of exhaust gases from the cylinder. Take care to remove traces of debris from the cylinder and ports, prior to reassembly, by washing the components thoroughly in petrol or paraffin whilst taking care to observe the necessary fire precautions.
4   Full details of decarbonising the silencer assembly are contained in Section 18 of the following Chapter.

22.3 Each part in the cylinder barrel should be cleaned of carbon

23 Examination and renovation: cylinder head

1   Check that the cylinder head fins are not clogged with oil or road dirt, otherwise the engine will overheat. If necessary, use a degreasing agent and brush to clean between the fins. Check that no cracks are evident, especially in the vicinity of the holes through which the holding down studs and bolts pass, and near the spark plug threads.
2   Check the condition of the thread in the spark plug hole. If it is damaged an effective repair can be made using a Helicoil thread insert. This service is available from most Suzuki service

23.3 Check the cylinder head for distortion

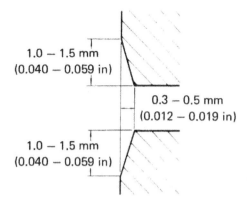

1.0 – 1.5 mm
(0.040 – 0.059 in)

0.3 – 0.5 mm
(0.012 – 0.019 in)

1.0 – 1.5 mm
(0.040 – 0.059 in)

Fig. 1.11 Cylinder bore port chamfer measurements

agents. The cause of a damaged thread can usually be traced to overtightening of the plug or using a plug of too long a reach. Always use the correct plug and do not overtighten.

3   If leakage problems have been experienced between the cylinder head and cylinder barrel mating surfaces, the cylinder head should be checked for distortion by placing a straight-edge across several places on the mating surface and attempting to slide a 0.05 mm (0.002 in) feeler gauge between the straight-edge and the mating surface.

4   If the cylinder head is found to be warped beyond this limit, grind it flat by placing a sheet of emery paper on a surface plate or sheet of plate glass and rubbing the cylinder head mating surface against it, in a slow, circular motion. Commence with 200 grade paper and finish with 400 grade paper and oil. If it is found necessary to remove a substantial amount of metal before the mating surface becomes completely flat, obtain advice from a Suzuki service agent as to whether it is necessary to obtain a new head.

5   Note that most cases of cylinder head distortion can be traced to unequal tensioning of the cylinder head securing nuts and by tightening them in the incorrect sequence.

24 Examination and renovation: cylinder barrel

1   The usual indication of a badly worn cylinder barrel and piston is known as piston slap, a metallic rattle that occurs when there is little or no load on the engine.

2   Commence by cleaning the outside of the cylinder barrel, taking care to remove any accumulation of dirt from between the cooling fins. Carefully remove the ring of carbon from the mouth of the bore, so that an accurate assessment of bore wear can be made.

3   A close visual examination of the bore surface must be made, to check for scoring or any other damage, particularly if broken piston rings were encountered during the stripdown. Any damage of this nature will necessitate reboring and a new piston, as it is impossible to obtain a satisfactory seal if the bore is not perfectly finished.

4   There will probably be a lip at the uppermost end of the cylinder bore which marks the limit of travel of the top of the piston ring. The depth of the lip will give some indication of the amount of bore wear that has taken place even though the amount of wear is not evenly distributed.

5   The most accurate method of measuring bore wear is by the use of a cylinder bore DTI (Dial Test Indicator) or a bore micrometer. Suzuki supply a special tool, No 09900-20508, specifically for this purpose. Measurement should be taken at a point 20 mm (0.8 in) from the top of the cylinder bore and the reading obtained compared with the service limit for bore wear given in the Specifications Section of this Chapter. Note that it is necessary to rotate the measuring instrument so that the point of greatest wear in the bore is found.

6   It is however, most unlikely that the average owner will have the above listed items of equipment readily available. A slightly less accurate but more practical method is as follows. It is possible to determine the amount of bore wear by inserting the piston, without rings, so that it is just below the ridge at the top of the bore. Measure the distance between the bore wall and the side of the piston using feeler gauges. Move the piston down to the bottom of the bore and repeat the measurement. Doing this, and subtracting the lesser measurement from the greater, will give the difference between the bore diameter in an area where the greatest amount of wear is likely to occur and an area in which there should be little or no wear. If the difference found exceeds 0.10 mm (0.004 in), then remedial action will have to be taken. Note that the curvature of the gap being measured will tend to preclude measurement by this method but it should be possible to gain a good indication of whether a rebore is necessary.

7   If it is found that the amount of wear in the bore exceeds those limits given, then it will be necessary to have the cylinder barrel rebored to the next oversize and the appropriate oversize piston fitted. Suzuki supply pistons in two oversizes: +0.5 mm (0.196 in) and +1.0 mm (0.0394 in).

8   On receiving the cylinder barrel back from the service agent who has carried out the rebore, check the edges of the ports at the bore end to ensure that they have been correctly chamfered to the measurements given in the figure accompanying this text. This work must be done before the cylinder barrel is refitted to the machine, otherwise there is a distinct possibility that the piston rings will catch on the unchamfered edge of each port and break, thus necessitating a further strip down and rebore. Chamfering of the port edges can be carried out by very careful use of a scraper but it is essential to ensure that the wall of the bore does not become damaged in the process. Finish off the process by polishing the cut edges with fine emery paper.

9   Establishment of the clearance between the cylinder bore and the piston can be made either by direct measurement of the cylinder bore and piston diameters and then by subtracting the latter figure from the former, or by actual measurement of the gap using a feeler gauge. In either case, if the clearance exceeds the maximum wear limit there is evidence that a new piston or a rebore and new piston is required. Note that if the method of direct measurement of the piston and bore is decided upon, then the measurement for piston diameter should be made at about 25 mm (1.0 in) up from the base of the piston skirt. Note that any measurement for piston diameter must be made at right-angles to the gudgeon pin hole, whereas the measurement for cylinder bore diameter should be made at various points around a line 20 mm (0.80 in) from the top of the bore.

10 Finally, check the cylinder barrel to cylinder head mating surface for distortion by placing a straight-edge across several places on it and attempting to slide a 0.05 mm (0.002 in) feeler gauge between the straight-edge and mating surface. If the cylinder barrel proves to be warped beyond this limit, grind it flat by placing a sheet of emery paper on a surface plate or sheet of plate glass and rubbing the mating surface against it, in a slow, circular motion. Commence this operation with 200 grade paper and finish with 400 grade paper and oil. If it is thought necessary to remove a substantial amount of metal in order to bring the mating surface back to within limits, obtain advice from a Suzuki service agent as to whether it is necessary to obtain a replacement barrel.

25 Examination and renovation: piston and piston rings

1   If a rebore is necessary, the existing piston and rings can be disregarded because they will be replaced with their oversize equivalents as a matter of course.
2   Remove the piston rings by pushing the ends apart with the thumbs whilst gently easing each ring from its groove. Great care is necessary throughout this operation because the rings are brittle and will break easily if overstressed. If the rings are gummed in their grooves, three strips of tin can be used, to ease them free, as shown in the accompanying illustration. Take great care to keep the rings separate and the right way up so that they can be refitted in their correct positions.
3   Piston wear usually occurs at the skirt or lower end of the piston and takes the form of vertical streaks or score marks on the thrust side. There may also be some variation in the thickness of the skirt. Measurement for piston diameter should be taken at about 25 mm (1.0 in) up from the base of the piston skirt. This measurement must be made at right-angles to the gudgeon pin hole. If the measurement obtained is found to be less than the service limit given in the Specifications Section of this Chapter, then the piston must be renewed.
4   Check that the piston and bore are not scored, particularly if the engine has tightened up or seized. If the bore is badly scored, it will require a rebore and oversize piston. If the scoring is not too severe or the piston has just picked up, it is possible to remove the piston high spots by careful use of a fine swiss file. Application of chalk to the file teeth will help prevent clogging of the teeth, and the subsequent risk of scoring.
5   Check for any build up of carbon in the piston ring grooves. Any carbon should be carefully removed by using a section of broken piston ring or similar. The piston ring grooves may have become enlarged in use, thus allowing each ring to have a greater side float than is permissible. To measure this side float, insert each piston ring in its cleaned groove and measure the clearance between the side of the ring and the groove with a feeler gauge. If the measurement obtained exceeds that clearance given in the Specifications Section of this Chapter, then the piston is due for renewal.
6   When cleaning the piston ring grooves, check that the ring locating peg located in each groove is not loose or worn. If in doubt as to the condition of these pegs, seek professional advice from a Suzuki service agent and renew the piston if necessary.
7   Examine the gudgeon pin for any scoring on its bearing surface. If damage is apparent on its contact surface with the holes in the piston, examine the surface of these holes and reject the piston if similar damage is found. The gudgeon pin should be a firm press fit in the piston. If any excessive looseness is felt between the two components, check the pin to piston clearance and compare the reading obtained with the service limit given in the Specifications Section of this Chapter. If this reading is found to be excessive, measure the gudgeon pin outer diameter and the piston hole bore diameter before rejecting one or both of the components, as necessary.
8   Check that the gudgeon pin circlip retaining grooves in the piston are free from damage. If there is the slightest doubt as to

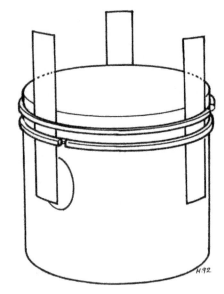

Fig. 1.12 Method of removing gummed piston rings

25.5 Check the piston ring to groove clearance

the condition of these grooves, seek further advice from a Suzuki service agent who will determine whether or not the piston should be renewed. Remember that if one of these rings should become detached in use, certain damage will be caused to both the piston and the surface of the bore.
9   Examine the working surface of each piston ring. If discoloured areas are evident, the ring should be renewed because these areas indicate the blow-by of gas.
10 Piston ring wear is measured by inserting each ring in a part of the cylinder bore which is not normally subject to wear. Ideally, the ring should be inserted into the cylinder bore so that it is positioned approximately 10 mm (0.4 in) from the bore spigot cutouts. Use the crown of the piston as a means of locating the ring squarely in the bore and measure the gap between the ring ends with a feeler gauge. If the gap measured is found to be greater than the service limit given in the Specifications Section of this Chapter, then the ring must be renewed. Note that the ring end gap service limits are given for when each ring is in its free state. These are given as a means of determining the spring tension of each ring. Measure this end gap with the ring placed flat on a clean work surface and compare the reading obtained with the appropriate reading given in the Specifications Section of this Chapter. Note that

each piston ring has a letter stamped on its upper surface adjacent to the ring end gap. These letters vary according to the model type and are as follows:

Model type	Stamp mark
TS100	N or T
TS125	N or RN
TS185 and 250	R

11 It must be appreciated that, if a set of new rings is to be fitted to a used piston which is to run in a part worn bore, then the upper of the two rings should be stepped to clear the ridge left in the bore by the previous top ring. If a new ring is fitted, it will hit the ridge and break. This is because the new ring will not have worn in the same way as the old, which will have worn in unison with the ridge.

12 Suzuki do not supply a stepped upper ring to fit any of the machines covered in this Manual. It is therefore necessary to ensure that the wear ridge at the top of the bore is completely removed before fitting a set of new rings. This may be incorporated in the 'glaze busting' process, which should be carried out in order to break down the surface glaze on the used bore so that the new rings are able to bed in to the bore surface. This combined process of ridge removal and 'glaze busting' can be carried out by many competent engineering firms who advertise regularly in the national motorcycle press; your local Suzuki service agent could well offer a similar service.

13 Bear in mind, when refitting used rings to the piston, that they must be fitted in exactly the same positions as noted during removal. When fitting either new or old rings, take note of the following points. Each ring should be fitted to the piston by pulling its ends apart just enough to allow it to pass over the piston crown and into its groove. Always fit each top and second ring with its marked (N, T, RN or R) surface uppermost; the two rings are identical when new. Check that each ring presses easily into its groove and that its ends locate correctly over the locating pin.

14 Finally, do not automatically assume when fitting new rings, that their end gaps will be correct. As with part worn rings, the end gap must be measured. It may be necessary to enlarge the gap, in which case this should be done by careful use of a needle file. With the rings fitted to the piston, place the assembly to one side ready for engine reassembly.

25.13 The ends of each piston ring must locate correctly over the locating pin

26 Examination and renovation: gearbox components

1 Examine each of the gear pinions to ensure that there are no chipped or broken teeth and that the dogs on the end of the pinions are not rounded. Gear pinions with these defects must be renewed; there is no satisfactory method of reclaiming them. If damage or wear warrants renewal of any gear pinions the assemblies may be stripped down, displacing the various shims and circlips as necessary.

2 If, on TS100, 125 and 185 models, dismantling of the input shaft assembly proves to be necessary, the 2nd gear pinion will have to be pressed from position by using a hydraulic press; no other method of removal is possible. As it is unlikely that this type of tool will be readily available, it is recommended that the complete shaft assembly be returned to an official Suzuki service agent who will be able to remove the pinion, renew any worn or damaged components and return the shaft assembly complete.

3 If a hydraulic press is available and it is recommended to attempt removal of the 2nd gear pinion from the input shaft, it is very important to fully realise the dangers involved when using such a tool. Both the tool and the shaft assembly must be set up so that there is no danger of either item slipping. The tool must be correctly assembled in accordance with the maker's instructions as the force exerted by the tool is considerable and perfectly capable of stripping any threads from holding studs or inflicting other damage upon itself and the mainshaft. Always wear proper eye protection in case a component should fail and shatter before it becomes free, as will happen if the component is flawed. The type of press used in practice is shown in the photograph accompanying this text, some considerable force was needed to break the seal between the pinion and the shaft due to the application of a locking compound between the two components during assembly.

4 The accompanying illustrations show how both clusters of the gearbox are assembled on their respective shafts. It is imperative that the gear clusters, including the thrust washers, are assembled in **exactly** the correct sequence, otherwise constant gear selection problems will occur. In order to eliminate the risk of misplacement, make rough sketches as the clusters are dismantled. Also strip and rebuild as soon as possible to reduce any confusion which might occur at a later date.

5 Examine the selector forks carefully, ensuring that there is no sign of scoring on the bearing surface of either their fork ends, their bores or their gearchange drum locating pins. Check for any signs of cracking around the edges of the bores or at the base of the fork arms. Refer to the Specifications Section at the beginning of this Chapter and measure the thickness of the fork claw ends; renew the fork if the measurement obtained is less than the limit given.

6 Place each selector fork in its respective pinion groove and, using a feeler gauge, measure the claw end to pinion groove clearance. If the measurement obtained exceeds the given service limit, then it must be decided whether it is necessary to renew one or both the components. The acceptable limits for pinion groove width are given in the Specifications Section of this Chapter.

7 Check each selector fork shaft for straightness by rolling it on a sheet of plate glass and checking for any clearance between the shaft and the glass with feeler gauges. A bent shaft will cause difficulty in selecting gears. There should be no sign of any scoring on the bearing surface of the shaft or any discernible play between each shaft and its selector fork(s). On TS185 models, the surface area of the gearchange drum which acts as a bearing surface for each selector fork should also be examined for scoring and wear.

8 The tracks in the gearchange drum, which co-ordinate the movement of the selector forks, should not show signs of undue wear or damage. Check also that the contact surfaces of the detent plunger and drum are not worn or damaged.

9 Note that certain pinions have a bush fitted within their centres. If any one of these bushes appears to be overworn or in any way damaged, then the pinion should be returned to an official Suzuki service agent, who will be able to advise on which course of action to take as to its renewal.

10 Finally, carefully inspect the splines on both shafts and

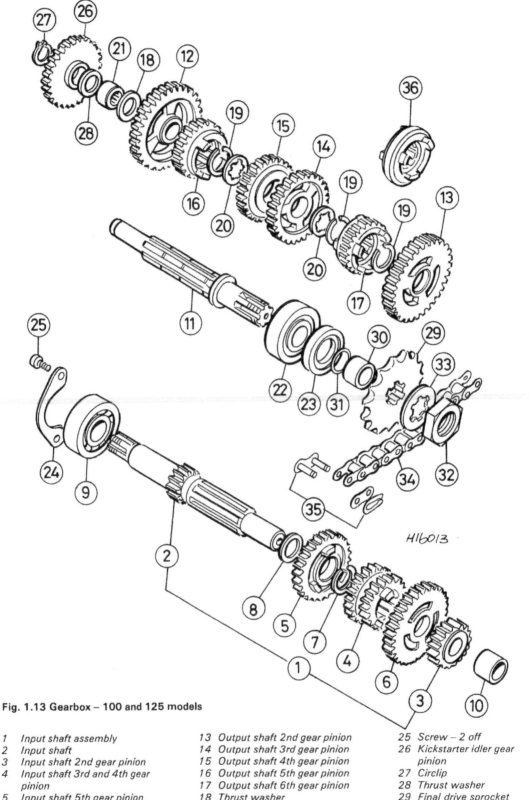

Fig. 1.13 Gearbox – 100 and 125 models

1 Input shaft assembly
2 Input shaft
3 Input shaft 2nd gear pinion
4 Input shaft 3rd and 4th gear pinion
5 Input shaft 5th gear pinion
6 Input shaft 6th gear pinion – 125 model only
7 Circlip
8 Thrust washer
9 Input shaft right-hand bearing
10 Bush
11 Output shaft
12 Output shaft 1st gear pinion

13 Output shaft 2nd gear pinion
14 Output shaft 3rd gear pinion
15 Output shaft 4th gear pinion
16 Output shaft 5th gear pinion
17 Output shaft 6th gear pinion
18 Thrust washer
19 Circlip – 3 off
20 Splined washer – 2 off
21 Output shaft right-hand bearing
22 Output shaft left-hand bearing
23 Oil seal
24 Bearing retaining plate

25 Screw – 2 off
26 Kickstarter idler gear pinion
27 Circlip
28 Thrust washer
29 Final drive sprocket
30 Spacer
31 O-ring
32 Nut
33 Tab washer
34 Final drive chain
35 Master link
36 Output shaft gear wheel – 100 model only

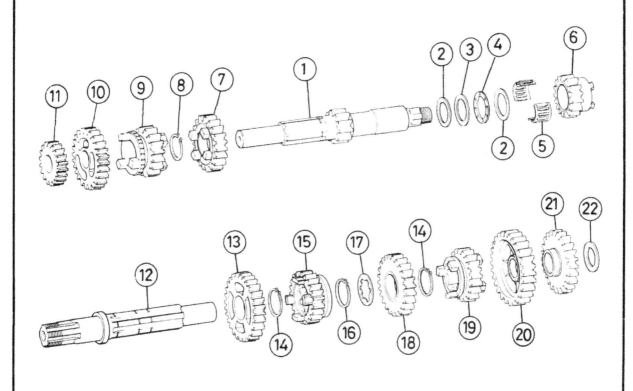

Fig. 1.14 Gearbox – 185 models

1	Input shaft	8	Circlip	16	Circlip
2	Thrust washer – 2 off	9	Input shaft 3rd gear pinion	17	Splined washer
3	Thrust bearing half	10	Input shaft 5th gear pinion	18	Output shaft 3rd gear pinion
4	Thrust bearing half	11	Input shaft 2nd gear pinion	19	Output shaft 4th gear pinion
5	Input shaft right-hand bearing	12	Output shaft	20	Output shaft 1st gear pinion
6	Kickstarter driven gear	13	Output shaft 2nd gear pinion	21	Kickstarter idler gear
7	Input shaft 4th gear pinion	14	Circlip – 2 off	22	Thrust washer
		15	Output shaft 5th gear pinion		

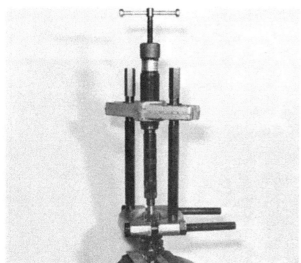

26.3 Use a hydraulic press to remove the 2nd gear pinion from the gearbox input shaft (100, 125 and 185)

26.5 Carefully examine each selector fork

26.6 Check the claw end to pinion groove clearance of each selector fork

26.9 Certain pinions have a bush fitted within their centres

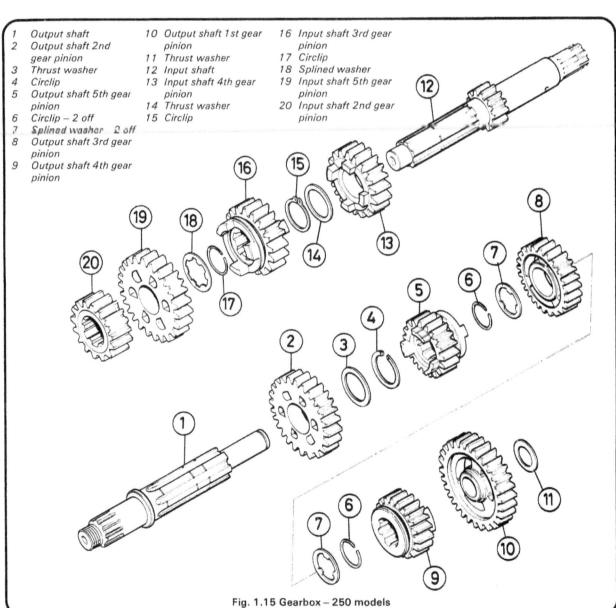

1   Output shaft
2   Output shaft 2nd
    gear pinion
3   Thrust washer
4   Circlip
5   Output shaft 5th gear
    pinion
6   Circlip – 2 off
7   Splined washer – 2 off
8   Output shaft 3rd gear
    pinion
9   Output shaft 4th gear
    pinion
10 Output shaft 1st gear
    pinion
11 Thrust washer
12 Input shaft
13 Input shaft 4th gear
    pinion
14 Thrust washer
15 Circlip
16 Input shaft 3rd gear
    pinion
17 Circlip
18 Splined washer
19 Input shaft 5th gear
    pinion
20 Input shaft 2nd gear
    pinion

Fig. 1.15 Gearbox – 250 models

pinions for any signs of wear, hairline cracks or breaking down of the hardened surface finish. If any one of these defects is apparent, then the offending component must be renewed. It should be noted that damage and wear rarely occur in a gearbox which has been properly used and correctly lubricated, unless very high mileages have been covered.

11 It should be assumed that the gearbox sprocket constitutes part of the output shaft assembly and should therefore be examined along with the rest of that assembly. Clean the sprocket thoroughly and examine it closely, paying particular attention to the condition of the teeth. The sprocket should be renewed if the teeth are hooked, chipped, broken or badly worn. It is considered bad practice to renew one sprocket on its own; both drive sprockets should be renewed as a pair, preferably with a new final drive chain. If this recommendation is not observed, rapid wear resulting from the running of old and new parts together will necessitate even earlier replacement on the next occasion. Examine the splined centre of the sprocket for signs of wear. If any wear is found, renew the sprocket as slight wear between the sprocket and shaft will rapidly increase due to the torsional forces involved. Remember that the output shaft will probably wear in unison with the sprocket, it is therefore necessary to carry out a close inspection of the shaft splines.

27.4 Dismantle the gearchange pawl mechanism (100, 125 and 250)

27 Examination and renovation: gearchange shaft and pawl mechanisms – TS 100, 125 and 250 models

1 Unless the gearchange shaft and pawl mechanism have been abused in any way or the assemblies have suffered from lack of lubrication, the chances of either component malfunctioning are relatively slight until the machine has been in use for a considerable amount of time.

2 Commence by checking the gearchange shaft for straightness. This may be done by laying the shaft on a sheet of plate glass and attempting to slide a feeler gauge between it and the glass. A bent shaft will almost certainly result in some difficulty in changing gear. Check the condition of the shaft return spring; if it is broken or is showing obvious signs of fatigue, then it must be renewed. Note that the legs of the spring must be placed one each side of the locating pin cast in the shaft endplate. If the detachable spacer placed between the spring and shaft (TS 100 and 125 models only) is beginning to show signs of wear or damage, then it too should be renewed. On TS 250 models this spacer forms a permanent part of the shaft assembly. It will therefore be necessary to renew the shaft as a whole if the surface of the spacer is seriously damaged or worn.

3 Inspect the teeth on the gearchange shaft endplate for damage or wear and look closely for hairline cracks around the base of the teeth. Carry out a similar inspection on the teeth of the pawl mechanism housing. These components will have worn in unison and should therefore be renewed as a matched pair.

4 Prepare a clean piece of rag or paper on a work surface. Dismantle the pawl mechanism, placing each component part on the clean surface as it is removed. Take care when doing this to ensure that each part is positioned as as to obviate any chance of its being incorrectly refitted or transposed with its opposite number.

5 Closely inspect each pawl operating spring for signs of fatigue or failure. These springs must be renewed as a pair, as must both the pawl operating pins and the pawls themselves if either set of component is found to be worn or damaged.

6 Inspect the groove in the pawl housing where it comes into contact with both the pawl lifter and cam guide plates. If the contact surface is seen to be scored or badly worn, seek further advice from an official Suzuki service agent as to whether the component should be renewed. A similar inspection should be carried out on the corresponding contact surfaces of the two plates.

7 Reassemble the renovated pawl mechanism assembly,

noting that the pawls are handed and must be positioned with the spring recesses offset towards the inside face of the pawl housing as shown in the accompanying photographs; re-wrap it in a rubber band and place it in a safe storage unit required for engine unit reassembly.

8 On TS 250 models inspect for wear the cam stopper arm end roller and the cam stopper plate with which the roller locates. If necessary the stopper plate can be removed from the change drum end where it is secured by three screws. Removal of the screws, which are staked in position, will require the use of an impact screwdriver. When fitting the plate note that the punch mark on the outside face must be placed in alignment with the notch in the change drum end. This is vital to maintain correct gear selection indexing. Refit the screws having applied locking compound to the screw threads.

28 Examination and renovation: gearchange mechanism – TS 185 models

1 Unless the components which make up the gearchange mechanism fitted to these models have suffered from misuse or lack of lubrication, then they should not require renewal until the machine has covered a considerable mileage.

2 Commence the examination sequence by refitting each gearchange pin into its location in the end of the gearchange drum. If excessive movement between the two components is felt, then both the drum and the pin must be closely examined for wear and each component renewed as necessary. Note that if one pin is renewed, then the other four should also be renewed. Examine the pin retaining holes in the pin retaining plate for any signs of ovality and renew the plate if necessary.

3 Examine the bearing surfaces of the cam stopper pawl and the cam guide plate where they contact the gearchange drum assembly. Where excessive wear is found, renew the component concerned. The cam stopper pawl must be a good fit on its pivot. Feel for excessive movement between the two components; if movement is detected, then one or both items will need to be renewed.

4 The cam stopper arm return spring should show no signs of failure or fatigue. Examine closely the contact points between the spring and hooks and the stopper arm and bearing retaining plate. In extreme cases, fretting caused by movement of the spring will have caused the holes in both the arm and plate to have become slotted, thus reducing the effectiveness of the spring. If it appears that the spring has taken a set to a longer length, compare it with a new item and renew it as necessary.

Examine the pivot hole in the cam stopper arm for ovality and the shank of the pivot bolt for wear. Examine also the contact surface of the arm roller.

5   Finally, examine the gearchange shaft and its associated components as follows. Note the fitted position of the shaft return spring before removing it from the shaft. Check the condition of the spring; if it is broken or showing obvious signs of fatigue, then it must be renewed. If the detachable spacer placed between the spring and shaft is beginning to show signs of wear or damage, then it too should be renewed.

6   Check the gearchange shaft for straightness by laying it on a sheet of plate glass and attempting to slide a feeler gauge between it and the glass. A bent shaft will almost certainly result in some difficulty when changing gear.

7   The spring between the gearchange arm and shaft should show no signs of fatigue or failure. Do not omit to examine its locating holes for wear. To complete examination of the shaft, inspect its contact points with the stop which is screwed into the crankcase and check that the pivot between the gearchange arm and the shaft shows no sign of excessive wear.

8   If in the slightest doubt as to the condition of any of the components which make up the gearchange mechanism, then return the component concerned to an official Suzuki service agent who will be able to give an expert opinion and supply a replacement item if necessary.

29.2 Check for wear of the clutch friction plates

29 Examination and renovation: clutch assembly and primary drive

1   Lay the various component parts of the clutch assembly out on a clean work surface, carefully cleaning each item before doing so. The plain and friction plates should all be given a thorough wash in a petrol/paraffin mix to remove all traces of friction material debris and oil sludge; follow this by cleaning the inside of the clutch drum in a similar manner. If this action is not taken, a gradual build-up of contamination will occur and eventually affect the clutch action. On TS100 and 125 models, take care when removing the retaining ring from the base of the clutch hub not to distort the ring or to cause damage to the surface of the hub itself. A pair of small flat-ended screwdrivers with any sharp edges removed from their flats was found to be ideal tools for this job. With the ring removed, the last of the plain plates, together with the wave plate and plate seat, can be detached for cleaning.

2   Provided the clutch has been reasonably treated and run in the correct type of oil, the bonded linings of the friction plates should last for a considerable time. The obvious sign of the linings having worn beyond their service limit is the advent of clutch slip. To check the degree of wear present on the friction plates, measure the thickness of each plate across the faces of the bonded linings and compare the measurement obtained with the information given in the Specifications Section of this Chapter. If the linings have worn beyond the service limit, then the plate must be renewed.

3   Check the condition of the tangs around the periphery of each friction plate, at the same time checking the slots in the clutch drum wall. In an extreme case, clutch chatter may have caused the plate tangs to make indentations in the slots; these indentations will trap the plates as they are freed, thereby impairing clutch action. If the damage found is only slight, the indentations can be removed by careful work with a fine file and any burrs removed from the plate tangs in a similar fashion. More extensive damage will necessitate renewal of the parts concerned. Note that there is a definite limit to the amount of material that can be removed from each plate tang, the minimum tang width allowable being given in the Specifications Section of this Chapter.

4   Carry out a check on both the friction and plain plates for any signs of warpage. This can be achieved by laying each plate on a completely flat surface, such as a sheet of plate glass, and attempting to pass a feeler gauge between the plate and the

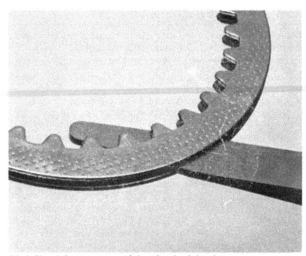

29.4 Check for warpage of the clutch plain plates

surface. Both the plate and the surface must be cleaned of all contamination. The maximum allowance warpage for each friction plate is 0.4 mm (0.016 in), whereas the maximum allowable warpage for each plain plate is 0.1 mm (0.004 in).

5   The plain plates should be free from scoring and any signs of overheating which will be apparent in the form of blueing. Check the condition of both the tangs in the inner periphery of each plain plate and the slots of the clutch centre. Any slight damage found on either of these components should be removed by using a method similar to that described for the friction plates and clutch drum. The final check for the plain plates is to measure each plate for thickness. The standard thickness for each plain plate is given in the Specifications Section of this Chapter.

6   Inspect the clutch pressure plate for both cracking and signs of overheating. Hairline cracks are most likely to occur around the central bearing hole in the plate and each one of the seven holes that accommodate the spring anchor pins (TS 100, 125 and 185 models only). On TS250 models, check closely around the base of each spring retaining hole. Check the pressure plate for warpage as if it were a plain or friction plate. Suzuki give no warpage limit for this component, so it is advisable to seek expert advice from an official Suzuki service agent if any warpage found seems excessive.

7   Check the condition of the small central thrust bearing, the

29.7 Examine the clutch central thrust bearing

29.8a Measure the free length of each clutch spring ...

29.8b ... and refit each serviceable spring into the clutch centre (shown for 100, 125 and 185)

clutch release rod over which it fits and the thrust washer. On TS250 models only, examine the rod retaining clip for signs of fatigue or failure and renew it if necessary. Any wear in the bearing will be indicated by roughness felt in the rollers as they are rotated and by cracking of the bearing cage. The release rod should be renewed if its bearing surfaces show signs of scoring or excessive wear, as should the thrust washer.

8   On TS100, 125 and 185 models, measure the free length of each clutch spring. If any one spring has taken a permanent set to more than the service limit given in the Specifications Section of this Chapter, then the complete set of springs must be renewed. Check each spring for excessive wear at its contact point with its anchor pin, bearing in mind that the two components will have worn in unison. Refit the serviceable springs into the clutch centre, noting that they must be screwed through the casting from its rear face as attempting to insert them from the front will only result in the springs locking in their threads due to their opening action. Note that it is most important that no one spring is allowed to project from the face of the clutch hub. A projecting spring will cause serious damage to the face of the clutch drum once the clutch is reassembled and the machine ridden.

9   On TS250 models, measure the free length of each spring and determine whether it has taken a permanent set to less than the service limit given in the Specifications Setion of this Chapter. All springs must be comparable in length and, if necessary, renewed as a complete set.

10   Closely examine both of the thrust washers in the clutch assembly for any signs of scoring or overheating and renew each one as necessary. On TS185 models, where one of these washers is replaced by a clutch drum to gearbox input shaft retaining ring, check this ring for signs of fatigue or failure and renew it as necessary.

11   The clutch operating mechanism is located in the right-hand crankcase cover and can be removed for inspection by unscrewing its single retaining bolt and then pulling it from position; although it should be borne in mind that this mechanism should not normally require any attention other than periodic examination for wear. Wear in the mechanism can normally be felt by grasping the end of the shaft between the fingers of one hand, steadying the crankcase cover with the other hand and then attempting to move the shaft from side to side.

12   Leakage oil from the seal through which the clutch operating shaft passes will necessitate removal of the shaft and renewal of the seal. The seal will be displaced from its housing as the shaft is pulled from position. After having refitted the shaft, the new seal should be carefully drifted into position by using a socket or length of metal tube of the appropriate diameter in conjunction with a hammer. Take care to ensure that the seal remains square to its housing at all times during its fitting. The shaft should be inspected for damage to its end teeth and splines; any serious damage or excessive wear will cause erratic operation of the clutch which in turn will constitute a serious danger to the rider. The shaft must be renewed if so damaged and the clutch release arm and rod also examined.

13   TS250 models have a needle roller bearing contained within the bore of the clutch drum. Any wear in this bearing will be indicated by roughness felt in the rollers as they are rotated and by failure of the bearing cage. The bearing must be in excellent condition, as its collapse during use may cause damage to both the clutch drum and gearbox input shaft.

14   Primary drive is by means of a crankshaft-mounted pinion driving the clutch by way of a toothed damper unit which is riveted to the rear face of the clutch drum. Because of its method of attachment to the clutch drum, this damper unit is effectively a permanent part of the drum and cannot be dismantled for either examination or servicing. Examine both sets of pinion teeth for signs of excessive wear or damage. Note that both sets of teeth will have worn in unison and should therefore be renewed as a matched pair.

15   Note the condition of the Woodruff key which serves to

29.12a Fit the serviceable clutch operating shaft into the clutch cover ...

retain the drive pinion to the crankshaft end. If this key shows any signs of wear or damage, then it must be renewed. Wear or damage on the key will invariably mean that the keyways in the pinion and crankshaft will be similarly affected. Apart from renewing these components, the only satisfactory answer to the problem of worn or damaged keyways is either to have them recut to an acceptable larger size and have a new key fitted or to have each keyway refilled by either a welding or metal spraying process and then recut to accommodate a key of the original size. Either one of these alternatives requires a high degree of skill in the use of specialist equipment and the work should, therefore, be placed in the hands of a light engineering company which specialises in such a task. It must be realised that fitting a new key into worn keyways will only, at the best, effect a temporary cure and the turning forces imposed on the crankshaft and pinion will soon reduce any new key to the state of the one removed whilst at the same time wearing the keyways to such a degree that any reclamation or repair may well be impossible.

16 Finally, note the details given in the following Section of this Chapter on the examination of the kickstart driven pinion, which in the case of TS100, 125 and 250 models, is like the primary drive pinion, effectively part of the clutch drum.

30 Examination and renovation: kickstart assembly

1 Assemble the various component parts that comprise the kickstart assembly and lay them out on a clean work surface ready for inspection. Thoroughly clean and examine each individual component whilst paying particular attention to the following points.
2 If any slipping has been encountered when attempting to turn the engine by means of the kickstart lever, a worn ratchet or drive pinion will invariably be traced as the cause.

TS100, 125 and 250 models
3 If, on these models, the ratchet is found to be damaged or worn, then both halves of the assembly must be renewed as a pair; that is, the drive pinion and the ratchet stop.

TS185 models
4 The ratchet assembly on these models comprises a spring-loaded pawl which is located in the kickstart shaft boss and which comes into contact with the bore of the kickstart drive

29.12b ... and retain the shaft with the single bolt

29.12c With the shaft in position, fit the new seal

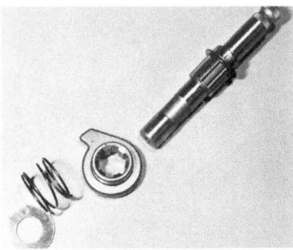

30.3 Examine the kickstart shaft assembly (100, 125 and 250)

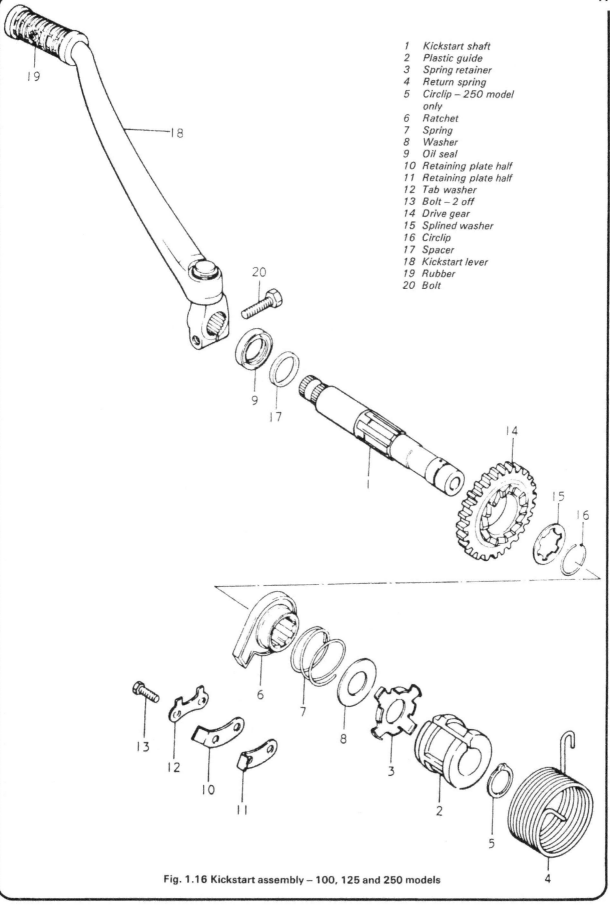

1 Kickstart shaft
2 Plastic guide
3 Spring retainer
4 Return spring
5 Circlip – 250 model only
6 Ratchet
7 Spring
8 Washer
9 Oil seal
10 Retaining plate half
11 Retaining plate half
12 Tab washer
13 Bolt – 2 off
14 Drive gear
15 Splined washer
16 Circlip
17 Spacer
18 Kickstart lever
19 Rubber
20 Bolt

Fig. 1.16 Kickstart assembly – 100, 125 and 250 models

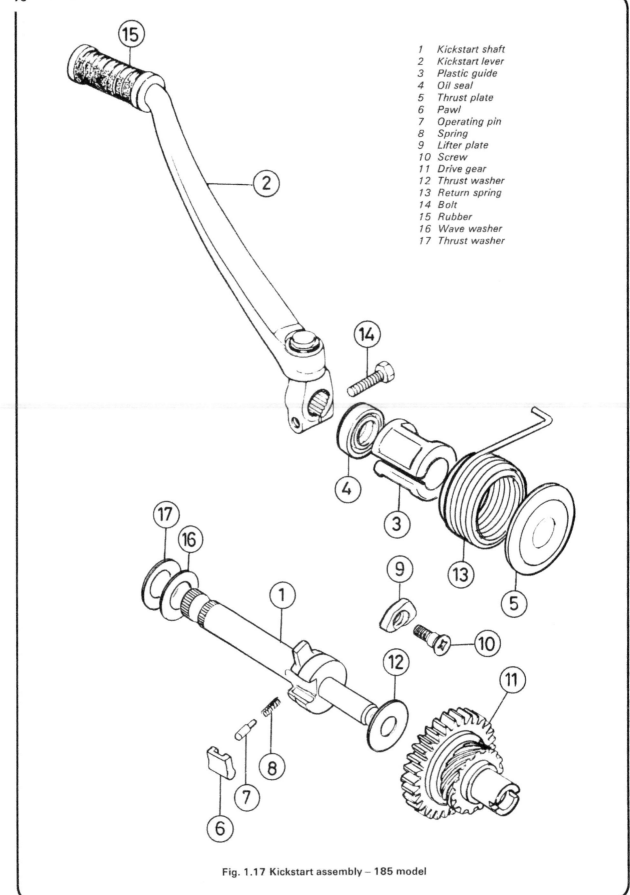

1   Kickstart shaft
2   Kickstart lever
3   Plastic guide
4   Oil seal
5   Thrust plate
6   Pawl
7   Operating pin
8   Spring
9   Lifter plate
10 Screw
11 Drive gear
12 Thrust washer
13 Return spring
14 Bolt
15 Rubber
16 Wave washer
17 Thrust washer

Fig. 1.17 Kickstart assembly – 185 model

30.4a Examine the kickstart shaft assembly ...

30.4b ... including the component parts of the ratchet (185 only)

30.6 Check for wear in the cutouts of the kickstart driven pinion (185 only)

pinion. Clean and examine each component part of the pawl mechanism. The spring must be renewed if it is seen to be showing signs of fatigue or failure. Any damage to the operating pin and the pawl will necessitate renewal of each component as necessary. Excessive wear or damage to the contact surfaces of both the shaft boss cutout and the bore of the drive pinion will again necessitate renewal of one or both components. If the pawl lifter plate is worn, it may be removed from its crankcase location by removing its single retaining screw.

All models

5   If the teeth of the kickstart drive pinion are seen to be worn, then they must be compared with the condition of the teeth on both the idle pinion and the driven pinion. This is because all three pinions will wear in unison and should therefore be renewed as a set. It will be appreciated that any great degree of wear is unlikely to occur between these pinions until the machine has been in use for a considerable amount of time, assuming that the mechanism has always been correctly lubricated and has not been subject to misuse. Any damaged pinion teeth will of course necessitate renewal of the pinion concerned.

6   On TS185 models, inspect the clutch drum to driven pinion contact surfaces. Fit the pinion cutouts into those of the clutch drum and feed for movement between the two components. Wear on the edges of these cutouts will necessitate the renewal of one or both components. Check also that the driven pinion is a good firm fit in the centre of the gearbox input shaft bearing.

7   Inspect all pinion to shaft bearing surfaces for obvious wear or damage. If any one shaft is seen to be stepped or is, for example, deeply scored, then it will have to be renewed. The same applies to any pinion which has a worn or damaged bore. This does, however, only apply to those pinions which do not have a bushed centre as it may be possible to renovate a bushed pinion by employing a competent motor engineer or acquiring the services of a light engineering company who are willing to do such a small job. Suzuki do not supply separate bushes to fit any of the pinions fitted to the machines covered in this Manual. It may, however, be worth confirming that this, at the time, is the case by consulting an official Suzuki service agent.

8   Examine the kickstart shaft for wear or damage, paying particular attention to the condition of its splined sections. On TS100, 125 and 250 models, examine the central splined section in conjunction with the ratchet stop and splined washer that fit over it, checking for obvious wear of the splines which will be indicated by a breaking down of the hardened surfaces. Carry out a similar examination on the shaft end and kickstart lever of all models.

9   It will appear obvious if the spring(s) within the assembly are either broken or fatigued. Other obviously worn or damaged components should be renewed as a matter of course.

31 Examination and renovation: oil pump assembly

1   Because no replacement parts are obtainable for the oil pump fitted to the machines covered in this Manual, the unit is therefore effectively sealed. The only maintenance that can be done is to ensure that the unit body is thoroughly cleaned and given a good, close inspection for any hairline cracks that may be apparent in the parts of the pump body which are subject to stress, that is, around the securing screw holes, pipe union, etc.

2   In normal circumstances, the pump unit can be expected to give long service whilst requiring no maintenance. Do not omit to fit a new gasket between the pump body and crankcase mating surfaces when refitting the pump and always fit new sealing washers when reconnecting the pipe union.

3   On TS100, 125 and 250 models, wear or damage in either

the oil pump drive pinion or its shaft will be obvious on inspection, necessitating the immediate renewal of the component concerned. The shaft pin should also be carefully examined.

4   On TS185 models, examine the keyed drive attachment which connects the pump to the end of the kickstart drive pinion unit. Check for wear in the slots of the pinion unit and for corresponding wear in the projections of the drive attachment. Renew each component as necessary.

5   On TS250 models, check the condition of the spacer washer and renew it if necessary.

32 Examination and renovation: tachometer drive assembly

1   With the complete tachometer drive assembly removed from the crankcase, separate the component parts and place them on a clean work surface ready for inspection. As the assembly operates under ideal conditions, aided by the fact that

31.1 Examine the body of the oil pump and its lever return spring

31.4 Examine the keyed drive attachment of the oil pump (185 only)

the shaft rotates within a self-lubricating nylon sleeve, it is unlikely that any degree of wear will be observed on any of the component parts until the engine has been running for a considerable amount of time. Once wear is apparent, the part concerned must be renewed.

2   Check the condition of the O-ring fitted to the nylon sleeve. If the ring is damaged or has in any way deteriorated, then it must be renewed; this should be done as a matter of course when rebuilding the complete engine/gearbox unit.

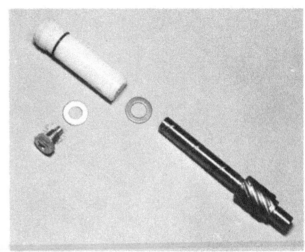

32.1 Inspect the tachometer drive assembly (125 shown)

33 Examination and renovation: reed valve assembly

Reference should be made to the appropriate Section in Chapter 2 for full details of the examination and renovation procedures listed for the reed valve assembly.

34 Engine reassembly: general

1   Before reassembly of the engine/gearbox unit is commenced, the various component parts should be cleaned thoroughly and placed on a sheet of clean paper, close to the working area.

2   Make sure all traces of old gaskets have been removed and that the mating surfaces are clean and undamaged. Great care should be taken when removing old gasket compound not to damage the mating surface. Most gasket compounds can be softened using a suitable solvent such as methylated spirit, acetone or cellulose thinner. The type of solvent required will depend on the type of compound used. Gasket compound of the non-hardening type can be removed using a soft brass-wire brush of the type used for cleaning suede shoes. A considerable amount of scrubbing can take place without fear of harming the mating surfaces. Some difficulty may be encountered when attempting to remove gaskets of the self-vulcanising type, the use of which is becoming widespread, particularly as cylinder head and base gaskets. The gasket should be pared from the mating surface using a scalpel or a small chisel with a finely honed edge. Do not, however, resort to scraping with a sharp instrument unless necessary.

3   Gather together all the necessary tools and have available an oil can filled with clean engine oil. Make sure that all new

gaskets and oil seals are to hand, also all replacement parts required. Nothing is more frustrating than having to stop in the middle of a reassembly sequence because a vital gasket or replacement has been overlooked. As a general rule each moving engine component should be lubricated thoroughly as it is fitted into position.

4 Make sure that the reassembly area is clean and that there is adequate working space. Refer to the torque and clearance setting wherever they are given. Many of the smaller bolts are easily sheared if overtightened. Always use the correct size screwdriver bit for the cross-head screws and never an ordinary screwdriver or punch. If the existing screws show evidence of maltreatment in the past, it is advisable to renew them as a complete set.

5 If the purchase of a replacement set of screws is being contemplated, it is worthwhile considering a set of socket or Allen screws. These are invariably much more robust than the originals, and can be obtained in sets for most machines, in either black or nickel plated finishes. The manufacturers of these screw sets advertise regularly in the motorcycle press.

35 Reassembling the engine/gearbox unit: reassembling the gearbox components

1 Having examined and renewed the gearbox components as necessary, the gear clusters can be built up and assembled as a complete unit ready for fitting into the appropriate half.

2 Refer to the line drawings accompanying this text and proceed to assemble both the input and output shaft components in the exact order shown whilst noting the following points.

3 Each component must be carefully inspected for any signs of contamination by dirt or grit during assembly and, if necessary, cleaned before being liberally coated with clean oil on its mating surfaces. When seating each circlip in its groove, take care to ensure that its ends are correctly located in relation to the splines of the shaft; that is, the gap between the circlip ends should fall directly in line with the base of any one of the channels in the splined shaft, as indicated in the various photographs accompanying this text.

4 Upon refitting the 2nd gear pinion to the input shaft of TS100, 125 and 185 models, thoroughly degrease both the surfaces of the shaft and the pinion where they contact each other. Coat the surface of the pinion with Suzuki Lock Super or a good quality bearing locking compound and, using a suitable socket or length of thick-walled tube in conjunction with a hammer, drift the pinion onto the shaft. Ensure, whilst doing

1   Spacer
2   Stop bolt
3   Return spring
4   Gearchange arm
5   Oil seal
6   Cam guide plate
7   Pawl
8   Pawl
9   Pin – 2 off
10   Spring – 2 off
11   Pawl mechanism
12   Pawl lifter plate
13   Selector fork – 2 off
14   Selector fork shaft
15   Gearchange lever rubber
16   Gearchange lever
17   Gearchange drum
18   Spring
19   Contact pin
20   O-ring
21   Neutral switch cover
22   Selector fork
23   Selector fork shaft
24   Detent plunger
25   Spring
26   Detent plunger housing

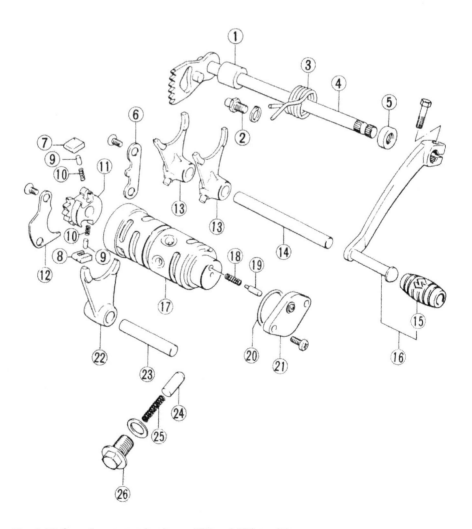

Fig. 1.18 Gearchange mechanism – 100 and 125 models

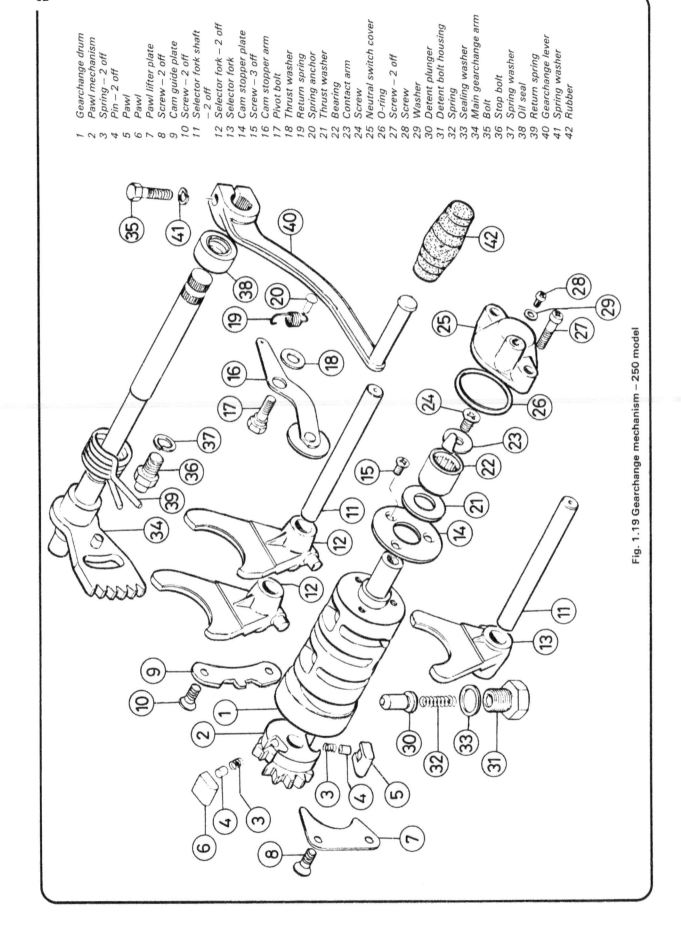

1 Gearchange drum
2 Pawl mechanism
3 Spring – 2 off
4 Pin – 2 off
5 Pawl
6 Pawl
7 Pawl lifter plate
8 Screw – 2 off
9 Cam guide plate
10 Screw – 2 off
11 Selector fork shaft – 2 off
12 Selector fork – 2 off
13 Selector fork
14 Cam stopper plate
15 Screw – 3 off
16 Cam stopper arm
17 Pivot bolt
18 Thrust washer
19 Return spring
20 Spring anchor
21 Thrust washer
22 Bearing
23 Contact arm
24 Screw
25 Neutral switch cover
26 O-ring
27 Screw – 2 off
28 Screw
29 Washer
30 Detent plunger
31 Detent bolt housing
32 Spring
33 Sealing washer
34 Main gearchange arm
35 Bolt
36 Stop bolt
37 Spring washer
38 Oil seal
39 Return spring
40 Gearchange lever
41 Spring washer
42 Rubber

Fig. 1.19 Gearchange mechanism – 250 model

83

1   Selector fork
2   Selector fork
3   Selector fork
4   Pin – 2 off
5   Roller – 3 off
6   Split pin – 2 off
7   Spring
8   Gearchange drum
9   Gearchange pin – 4 off
10 gearchange pin
11 Pin retaining plate
12 Screw
13 Contact pin
14 Cam stopper pawl pivot
15 Screw – 2 off
16 Cam stopper pawl
17 Gearchange lever
18 Cam stopper arm
19 Return spring

20 Washer
21 Pivot bolt
22 Detent plunger
23 Spring
24 Washer
25 Detent plunger housing
26 Selector fork shaft
27 Main gearchange arm
28 Spring
29 Return spring
30 Spacer
31 Oil seal
32 Stop bolt

33 Spring washer
34 Gearchange lever rubber
35 Bolt
36 Neutral cover
37 O-ring
38 Screw – 2 off
39 Spring washer

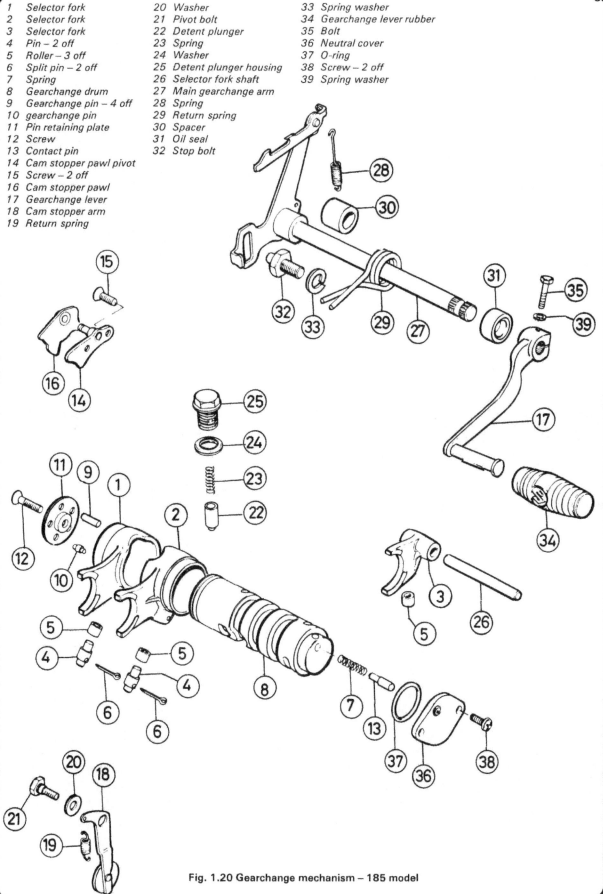

Fig. 1.20 Gearchange mechanism – 185 model

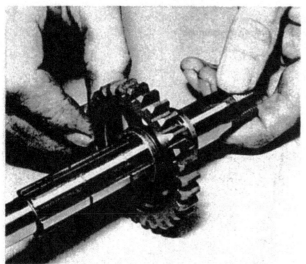

35.2a On 125 models, commence reassembly of the gear clusters by sliding the 2nd gear pinion onto the output shaft ...

35.2b ... and retaining the pinion in position with the circlip

35.2c Follow this by lifting the 6th gear pinion ...

35.2d ... followed by the circlip and the thrust washer

35.2e Fit the 3rd gear pinion to the output shaft ...

35.2f ... followed by the 4th gear pinion and thrust washer

35.2g Correctly position the circlip in its retaining groove ...

35.2h ... before fitting the 5th gear pinion ...

35.2i ... followed by the 1st gear pinion and thrust washer

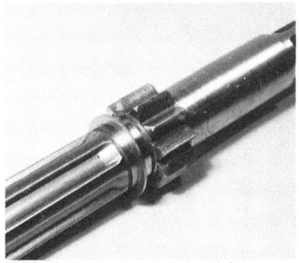

35.2j Position the thrust washer on the input shaft ...

35.2k ... followed by the 5th gear pinion and circlip

35.2l Fit the 3rd and 4th gear pinion ...

35.2m ... followed by the 6th gear pinion

35.2n After the application of locking compound, push the 2nd gear pinion onto the pinion shaft ...

35.2o ... and drift it down the shaft ...

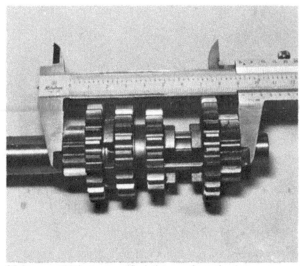

35.2p ... until it is the specified distance from the 1st gear pinion

35.2q On 185 models, commence reassembly of the gear clusters by sliding the 2nd gear pinion onto the output shaft

35.2r Ensure the circlip retaining the 2nd gear pinion is correctly positioned ...

35.2s ... before fitting the 5th gear pinion ...

35.2t ... followed by its retaining circlip ...

35.2u ... and its thrust washer

35.2v Fit the 3rd gear pinion with its circlip ...

35.2w ... followed by the 4th gear pinion ...

35.2x ... the 1st gear pinion ...

35.2y ... and the kickstart idler pinion with its thrust washer

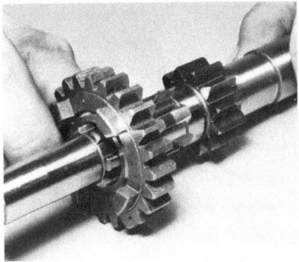

35.2z Position the 4th gear pinion on the input shaft ...

35.2aa ... and retain it in position with the circlip

35.2bb Fit the 3rd gear pinion ...

35.2cc ... followed by the 5th gear pinion

35.2dd Apply a locking compound ...

35.2ee ... before drifting the 2nd gear pinion onto the input shaft ...

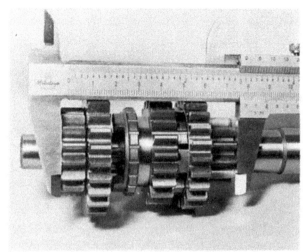

35.2ff ... until it is the specified distance from the 1st gear pinion

35.2gg Position the thrust washer against the 2nd gear pinion ...

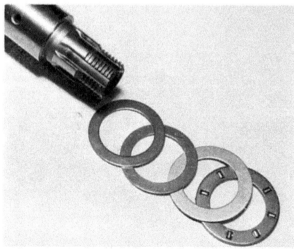

35.2hh ... and the combination of washers against the 1st gear pinion

this, that the threaded end of the shaft is protected from damage by placing it on a wooden surface and that the pinion is kept square to the shaft. Using a micrometer or vernier gauge, keep a close and accurate observation on the distance between the outer faces of the pinion being fitted and the 1st gear pinion which forms a permanent part of the mainshaft (see accompanying photograph). The pinion is correctly fitted when this distance is seen to be between the limits given in the following table:

Model type	1st to 2nd pinion distance
TS100 and 125	91.8 – 91.9 mm (3.614 – 3.618 in)
TS185	77.8 – 78.3 mm (3.062 – 3.087 in)

5   Place the assembled gearbox components on a clean piece of card or rag and cover them to protect any ingress of dirt or grit into the assembly during the time taken to prepare the crankcase for their installation.

36 Reassembling the engine/gearbox unit: fitting the crankcase components and joining the crankcase halves

1   Commence fitting of the crankcase components by refitting any bearing retaining plate removed during the examination and renovation procedure. It is recommended that a thread locking compound be applied to the threads of each one of the plate securing screws. Check tighten the screws of any plate that has been left in situ in the crankcase.
2   Coat the lip of each oil seal with a good quality high melting point grease. Doing this will help to prevent damage to the seal lip as the appropriate shaft is inserted through it.
3   Select the crankcase half from which the crankshaft was finally removed and set it up on wooden blocks on a sturdy work surface so that once the crankshaft is fitted, its end will remain clear of the surface. Ensure that the crankcase half is well supported around the area of the main bearing and then insert the crankshaft into the bearing as far as it will go with hand pressure. To obviate any risk of the crankshaft becoming distorted during fitting, select an open-ended spanner or similar that will be a good firm fit between the two flywheels and push it into position at a point directly opposite the crankpin. Remember to lubricate each main and gearbox bearing.
4   The method used to drift the crankshaft into position was to place a long socket or length of thick-walled tube over its end and to tap sharply the end of the socket or tube with a soft-faced hammer whilst taking great care to keep the crankshaft

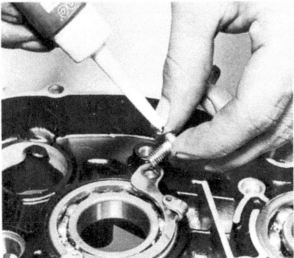

36.1 Apply a locking compound to the screws of each bearing retaining plate

36.3a Lubricate each main and gearbox bearing

36.3b Insert the crankshaft into the main bearing ...

36.3c ... and use a spanner to support the crankshaft whilst drifting it into position

36.6a Drift the kickstart driven pinion into the gearbox input shaft bearing (185 only)

36.6b Note the fitted position of the split needle roller bearing over the gearbox input shaft (185 only)

36.7a On 125 models, position both gearbox shaft assemblies into their housing

36.7b Fit the single selector fork ...

36.7c ... followed by the gearchange drum

36.7d Position the pair of selector forks ...

36.7e ... and retain them in position by fitting their shaft

36.7f Finally, fit the shaft through the single fork

square to the crankcase. Using a tube as a drift will serve to spread the shock imposed by the hammer evenly over a larger area of flywheel than would an ordinary drift placed on the crankshaft end. Indeed the use of an ordinary drift is highly inadvisable as it will not only damage the threaded end of the shaft but prove almost impossible to keep square to the flywheel. Only a moderate amount of force was needed to tap the crankshaft fully home; bear in mind that heating the crankcase half may make fitting easier but will also dry out the bearings and affect the oil seal. If it is thought that excessive force is being required to fit the crankshaft and that there is a serious risk of the crankshaft or crankcase casting becoming damaged, then it is recommended that aid and advice be sought from an official Suzuki service agent who may also be able to supply the correct Suzuki service tool for fitting the crankshaft.

5   The assembled gear clusters, together with the gearchange drum, forks and shaft(s) should now be fitted in the crankcase half. Use a method of fitting which is a direct reversal of that used for removal of the gearbox components whilst noting the following points.

6   On TS185 models only, carefully tap the kickstart driven pinion into the centre of the input shaft bearing whilst taking care to keep the pinion square to the bearing at all times. Do this by using a socket or length of metal tube of the appropriate diameter in conjunction with a soft-faced hammer and make sure that the area of crankcase around the bearing is well supported. Note that the split needle roller bearing which is fitted between the input shaft and the kickstart driven pinion must be fitted to the shaft with its stepped (the longer) side facing the clutch end of the shaft. Use a good quality high melting point grease to retain this bearing to the shaft during the insertion of the shaft into the crankcase.

7   On all models, ensure that any thrust washers remain in position on the gearbox shaft ends as the shafts are lowered into position. Retain them with grease if necessary. Ensure also that the gears are correctly meshed before lowering the shaft assemblies into their crankcase location.

8   With the gearchange drum fitted in position, rotate the drum until its neutral indent is aligned with the hole for the detent plunger assembly in the crankcase. Refit and tighten the detent plunger assembly, ensuring that the sealing washer remains beneath the housing head. Check that each selector fork end is located correctly into its respective pinion groove and that each fork pin is located correctly in its channel in the gearchange drum. On TS185 models, do not omit to check that the split-pin retaining each fork pin in position is correctly fitted. On TS250 models fit the cam stopper arm into the casing, securing it with its shouldered bolt. Ensure the cam stopper arm is free to pivot.

9   Remember to refer closely to both the figures and photograph sequences which accompany this text whilst reassembling the gearbox components. A mistake made now will cause much frustration during the later stages of engine reassembly.

10   It is now necessary to fit the renovated kickstart assembly into the right-hand crankcase half. Proceed as follows.

TS100, 125 and 250 models

11   Before fitting the kickstart shaft assembly to these models, ensure that the ratchet stop retaining plate securing bolts are properly tightened and locked in position with their tab washer. Slide the splined thrust washer onto the kickstart shaft and secure it in position with the circlip. Note the punch mark at the end of one of the shaft spline channels; this mark must align with the corresponding mark on the edge of the ratchet stop boss once the stop is slid into position. Slide the spring over the boss of the ratchet stop and follow this with the thick plate washer. Note that although these various component parts will have been cleaned during the examination procedure, care must be taken to prevent any further contamination during assembly; do not omit to apply lubricant between any bearing surfaces as each part is fitted.

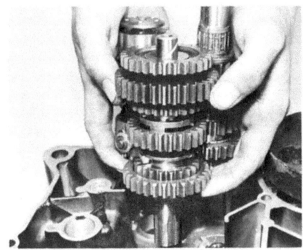

36.7g On 185 models, position the complete gearbox assembly into its housing ...

36.7h ... and check that all component parts are correctly aligned

36.8 Ensure the selector fork pin is correctly retained (185 only)

36.11a On 100, 125 and 250 models, retain the thrust washer in position on the kickstart shaft with the circlip

36.11b Check alignment of the shaft and ratchet stop ...

36.12a ... before inserting the shaft into the crankcase ...

36.12b ... fitting the kickstart return spring ...

36.12c ... and inserting the plastic guide between the spring and shaft

36.14 On TS 185 models only, insert the kickstart shaft assembly into the crankcase ...

12 Insert the assembled kickstart shaft into the crankcase half and locate the ratchet stop between its retaining plate and the crankcase casing. On TS250 models only, position the spring guide plate over the shaft. On all models, position the return spring end over the locating boss in the crankcase and, using a pair of long-nose pliers, grip the inner end of the spring and rotate it through approximately 90° in a clockwise direction until it can be inserted into the shaft hole. In practice, it was found that this job was made far easier if an assistant was employed to steady the crankcase half so that the two hands were free to grip the pliers. With the return spring properly located, push the plastic guide into position between the shaft and spring. On TS250 models, retain the guide in position with the circlip.

TS185 models

13 Before fitting the kickstart shaft assembly to these models, check that the pawl lifter plate is securely fitted in its crankcase location. If this plate has been removed for renewal then the threads of its securing screw must be degreased and coated with a thread locking compound before being used to secure the plate in position.

14 Refer to the figure accompanying this text and fit both the plain and wave washers to the kickstart shaft in the positions shown. Slide the thick thrust washer over the shaft end with the smaller diameter and follow this with the kickstart drive pinion. Check that the ratchet assembly operates correctly before inserting the kickstart shaft through its crankcase location. Do not allow either the plain or wave washer to fall from position as the shaft is inserted.

15 Move to the outer side of the crankcase half and fit the spring thrust plate over the shaft end. Note that this plate should be fitted with its concave side facing the lower end of the shaft. Place the inner facing end of the return spring into its location in the kickstart shaft. Grip the other end of the spring with a pair of pliers and rotate it in a clockwise direction through approximately 90° until the hook on the spring end can be located in its crankcase slot. With the spring thus located, push the plastic guide into position between the spring and shaft.

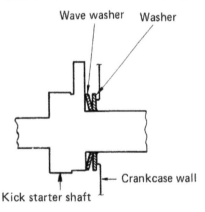

Fig. 1.21 Kickstarter shaft washer positions – 185 model

All models

16 Lightly grease the length and teeth of the tachometer drive shaft before inserting it into its nylon sleeve. Check that the new O-ring is correctly located in its retaining groove and slide the complete assembly into the crankcase, pushing it fully home. Align the hole in the sleeve with that for the retaining screw and fit and tighten the single retaining screw and plain washer. Note that it may be necessary to rotate the shaft during insertion so that its teeth mesh correctly with those of the drive pinion.

17 Both crankcase half assemblies are now complete and ready for joining. Carry out a final check of each assembly to confirm that each component part is in its correct location and then carry out a final lubrication of any components that may have been left dry.

36.15a ... invert the crankcase half and fit the return spring thrust plate over the shaft ..

36.15b ... followed by the return spring ...

36.15c ... and the plastic guide

36.16a Fit the tachometer drive assembly (100, 125 and 250)

36.16b Fit the tachometer drive assembly (185 only)

36.18a Apply a bead of sealing compound to one crankcase mating surface (100 and 125)

36.18b Fit a new crankcase sealing gasket (185 and 250)

36.19 Carefully join the two crankcase halves

18 Thoroughly degrease the mating surface of each crankcase half and then firmly push both of the two large locating dowels into their locations in one of the mating surfaces. On TS100 and 125 models, apply a thin coat of sealing compound (Suzuki Bond No 4 or equivalent) to one of the mating surfaces whilst taking care not to omit any area of the surface. On TS185 and 250 models, fit the new sealing gasket over the two locating dowels.

19 With the crankcase half containing the gearbox components properly supported on the work surface, lower the other crankcase half onto it whilst taking care to guide the crankshaft and gearbox shaft ends into their respective locations. Push the crankcase halves together with hand pressure as far as they will go whilst noting that the two locating dowels are correctly aligned with their locations in the mating surface. It will now be necessary to tap the left-hand crankcase half with a soft-faced hammer in order to bring the mating surfaces together. On no account should excessive force be used when joining the crankcase halves. On TS185 models, it will be found necessary to align the teeth of the kickstart drive pinion with those of the idler pinion by slowly turning the kickstart shaft as the crankcase halves are pushed together.

20 If difficulty is experienced in fitting the crankshaft into the second of the two crankcase halves, then it is advisable to

consult an official Suzuki service agent as to the availability of Suzuki service tool No 09910-32812, which can be used to draw the crankshaft through the main bearing (listed for TS100, 125 and 250 models only).

21 With the crankcase halves pressed properly together, insert all the crankcase securing screws into their previously noted locations. Use an impact driver to tighten these screws whilst working in a diagonal sequence so as to obviate any risk of the crankcase halves becoming distorted.

22 With all the crankcase retaining screws fully tightened, wipe away any excess sealing compound from around the mating surfaces (TS100 and 125 models) and check the free running and operation of the crankshaft and the gearbox components. Any tightness or malfunction will necessitate separation of the crankcases so that the problem may be located and rectified. When satisfied with the operation of the crankcase components, support the crankcase on the work surface so that its right-hand side is facing uppermost ready to receive the gearchange components.

37.1a Fit the gearchange pawl mechanism ...

37 Reassembling the engine/gearbox unit: fitting the gearchange shaft and pawl mechanism – TS100, 125 and 250 models

1   Remove the rubber band from around the assembled pawl mechanism, and insert it into the gearchange drum end. Place both the lifter plate and guide plate in position and fit and tighten their securing screws. Note that all four of these screws must have their threads degreased and coated with a thread locking compound before insertion.

2   Lightly grease the length of the gearchange shaft and check that the legs of its spring are placed one each side of the anchor pin. Insert the shaft into its crankcase location, taking care to align its teeth with those of the pawl (as shown in the accompanying photograph) before pushing it fully home. Failure to do this will mean, at the very most, an inefficient gearchange operation.

37.1b ... followed by the lifter plate and guide plate (100, 125 and 250)

38 Reassembling the engine/gearbox unit: fitting the gearchange mechanism – TS185 models

1   Commence fitting of the gearchange mechanism to these models by positioning the cam guide plate so that its end locates correctly in the end groove of the gearchange drum. Align the two screw holes in the plate with those in the crankcase and fit and tighten the two retaining screws after having degreased their threads and coated them with locking compound.

2   Lightly grease the length of the gearchange shaft and with the shaft return spring and its spacer correctly fitted, carefully slide the shaft into its crankcase location. Remember that the legs of the spring must be placed one each side of the anchor pin. Engage the gearchange arm in the end groove of the gearchange drum and check that the arm return spring functions correctly.

3   Fit the gearchange pins into their previously noted locations in the end of the gearchange drum. The odd pin out, ie the shouldered pin, can be fitted only in one position. Lightly grease the pivot of the cam guide plate and push the cam stopper pawl over it. Reference to the series of photographs accompanying the text of Section 15 will show the position of the pawl end in relation to the pins in the gearchange drum. Position the retaining plate for these pins over the drum end so that the hole in the periphery of the plate aligns over the lipped pin. Degrease the threads of the plate retaining screw and coat them with a thread locking compound before fitting and tightening the screw.

37.2 Note the fitted position of the gearchange shaft spring and endplate (100, 125 and 250)

38.2 Correctly position the legs of the gearchange shaft return spring (185 only)

4   Place the cam stopper arm in position and fit and tighten its pivot bolt. Take note of the figure accompanying this text before positioning the washer between the arm and the crankcase and make sure that the bolt is seated correctly through the arm. Degrease the threads of the bolt and coat them with a thread locking compound before placing the bolt in position. Finally, use a pair of pliers to hook the cam stopper arm return spring into position between the arm and the gearbox input shaft bearing retaining plate and then check that the arm pivots freely and is returned by the spring.

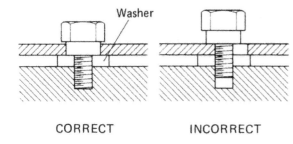

Washer

CORRECT   INCORRECT

Fig. 1.22 Correct fitting of cam stopper arm – 185 model

39 Reassembling the engine/gearbox unit: fitting the kickstart drive and idler pinions and the oil pump drive – TS100, 125 and 250 models

1   Lubricate the centre of the kickstart drive pinion and position it over the kickstart shaft. Place the thrust washer (TS100 and 125) or spacer (TS250) in position on top of the pinion.
2   On TS100 and 125 models, place the thrust washer over the end of the gearbox output shaft. On all models, lubricate the centre of the kickstart idler pinion and position it over the gearbox output shaft so that its teeth mesh with those of the drive pinion. Note that the boss of this pinion must face the crankcase. On TS250 models, place the thrust washer over the pinion and, on all models, fit the pinion retaining circlip and check that it is correctly located in its retaining groove.
3   The oil pump drive assembly may now be fitted by lightly lubricating its shaft with clean engine oil before carefully inserting it into its crankcase location. Check that the teeth of the pump drive pinion mesh correctly with those of the kickstart drive pinion and that the shaft pin is correctly located.

39.2a Place the thrust washer over the end of the gearbox output shaft (100 and 125)

39.2b Fit the kickstart idler pinion over the gearbox output shaft (100, 125 and 250)

39.3 Retain the idler pinion with the circlip and fit the oil pump drive assembly (100, 125 and 250)

40 Reassembling the engine/gearbox unit: fitting the primary drive pinion, clutch assembly and right-hand crankcase cover

1   Lock the crankshaft in position by using the method described for removal of the primary drive pinion. On TS100 and 125 models, slide the spacer ring over the crankshaft so that it abuts against the main bearing. On all models, fit the Woodruff key into the slot in the crankshaft and then slide the primary drive pinion into position over the crankshaft end so that the Woodruff key enters the slot in the pinion. Remember to fit the pinion with its lipped side facing the crankcase. Using a soft-faced hammer in conjunction with a socket or length of metal tube of the appropriate diameter, tap the outer surface of the pinion to seat it on the crankshaft. Position a new lock washer over the pinion whilst making sure that the tab of the washer is located in the end of the pinion slot. Fit the retaining nut and tighten it to the specified torque loading. Lock this nut in position by bending the lock washer against one of its flats.

TS185 models
2   On these models, check that the retaining ring is fitted correctly to the groove in the clutch drum boss; it is best if the end gap of this ring is positioned away from any gap between the projections of the boss. Position the drum over the gearbox input shaft so that the teeth of the primary drive pinion align with those of the drum and the projections of the drum boss locate the projections in the end of the kickstart driven pinion. Push sharply down on the drum to lock it in position. With this done, place the thrust washer over the gearbox input shaft.

TS100, 125 and 250 models
3   With the clutch assembly fitted to these models, position the thrust washer over the end of the gearbox input shaft and slide it down onto the shaft bearing. Lubricate both the washer and bearing with clean engine oil and place the clutch drum in position over the shaft. Ensure that the teeth of the primary drive pinion mesh with those of the drum by rotating both pinion and drum as the drum is placed into position. Fit the second of the two thrust washers onto the input shaft.

TS100, 125 and 185 models
4   Before fitting the clutch hub to these models, check that none of the clutch spring ends are protruding above the base of the hub. Any protruding springs will come into contact with the face of the clutch drum thereby causing severe clutch drag and damage to the drum casting.

40.1b Fit the primary drive pinion with its lock washer

40.1c Tighten the pinion retaining nut to the specified torque loading ...

40.1a Fit the spacer ring (arrowed) over the crankshaft before fitting the Woodruff key (100 and 125)

40.1d ... and lock the nut in position by bending the lock washer against one of its flats

40.2a Check the retaining ring is correctly fitted to the boss of the clutch drum ...

40.2b ... before pushing the clutch drum into position (185 only)

40.3 Position the thrust washer over the gearbox input shaft (100, 125 and 250)

40.5a Fit the thrust washer over the gearbox input shaft ...

40.5b ... before fitting the clutch hub, lock washer and retaining nut

40.5c Once tightened, the clutch retaining nut must be locked in position

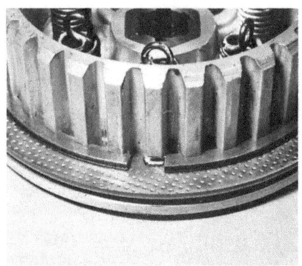

40.6a The clutch plate retaining ring must be correctly located in its groove (100 and 125)

40.6b Fit the clutch plates in the correct indicated sequence

40.7 Insert the release rod assembly through the centre of the clutch pressure plate

40.8a Align the pressure plate with the clutch hub ...

40.8b ... before fitting the plate and tensioning the clutch springs (100, 125 and 185)

40.11 Refit the right-hand crankcase cover with its new gasket and the two locating dowels

All models

5   Place the clutch hub in position over the gearbox input shaft, fit a new lock washer and fit and tighten the hub retaining nut to the specified torque loading whilst holding the input shaft in position by employing the method used for removal of the nut. Bend the lock washer against one of the flats of the nut to lock the nut in position.

6   Refer to Figures 1.4, 1.5 and 1.6 and fit the clutch plates into the clutch drum. On TS100 and 125 models, ensure that the retaining ring fitted to the base of the clutch hub is located correctly in its groove.

7   Using a good quality high melting point grease, lubricate the clutch thrust bearing and fit it to the clutch release rod. Fit the thrust washer over the bearing and insert the complete rod assembly through the centre of the pressure plate.

TS100, 125 and 185 models

8   Place the pressure plate in position so that the alignment mark on the periphery of the plate aligns with the boss cast in the wall of the clutch hub. Using the special spring tensioning tool mentioned in paragraph 3 of Section 11, pull each spring end out through the pressure plate and insert its anchor pin into position. Take great care to ensure that these pins locate properly into the recesses provided in the plate whilst gripping each one securely with a pair of long-nose pliers so as to obviate any chance of its being dropped into the clutch assembly.

TS250 model

9   On these models, retain the clutch release rod assembly in position, before fitting the pressure plate, by fitting its retaining clip. With the pressure plate in position and with its spring retaining bosses aligned with the threaded holes in the clutch hub, fit, finger tight, each spring with its plate washer and retaining bolt. Tighten the six spring retaining bolts, in small increments and in a diagonal sequence. This method of tightening will avoid any risk of the pressure plate becoming cracked or distorted.

All models

10 Carry out a final check to ensure that all the components contained within the right-hand crankcase cover have been correctly assembled and, where necessary, locked in position. Rotate the clutch release rod so that its toothed edge faces the bottom of the crankcase. This will ensure correct engagement with the clutch operating mechanism contained in the crankcase cover.

11 Degrease the mating surfaces of both the crankcase and the cover and insert the two locating dowels into position in the crankcase. Smear the splined end of the kickstart shaft and the lip of the cover oil seal with grease; this will lessen the risk of the sealing being damaged as the shaft passes through it. Place a new gasket over the mating surface and push the cover into position over the dowels, tapping it lightly with a soft-faced hammer to seat it properly. Insert each retaining screw into its previously noted position in the cover, tightening it finger-tight. With all the screws in position, proceed to tighten them fully, working in a diagonal sequence to prevent the cover from becoming distorted. Finally, refit the kickstart lever to its shaft end, ensuring that it is correctly positioned and that its retaining bolt is fully tightened.

41 Reassembling the engine/gearbox unit: fitting the oil pump assembly

1   Clean both the oil pump and crankcase mating surfaces and position a new gasket on the pump surface. On TS185 models, ensure that the keyed drive attachment is correctly fitted into the end of the drive pinion unit. On TS250 models, note the position of the spacer washer which should be fitted over the drive shaft.

41.2 Fit the oil pump into its crankcase location (125 shown)

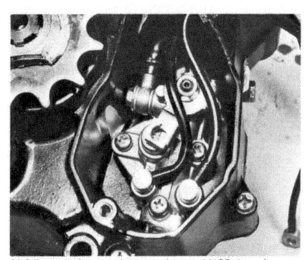

41.3 Each oil pipe must be correctly routed (185 shown)

42.2a Insert the spring of the neutral indicator switch assembly into the end of the gearchange drum ...

42.2b ... before inserting the contact pin (100, 125 and 185)

42.4a The O-ring must be correctly located in its retaining groove ...

42.4b ... before the switch cover is fitted

42.6a Carefully fit the O-ring over the gearbox output shaft ...

42.6b ... before pushing it through the oil seal with the spacer ...

42.6c ... and then fitting the gearbox sprocket, lockwasher and retaining nut

2   On all models, align the central driven spigot of the pump with its drive slot and then fit the pump into its crankcase housing. With the pump properly seated, fit and tighten its two retaining screws. Note that each of these screws must have a serviceable spring washer located beneath its head.

3   If the pump was removed with the engine in the frame and the oil feed and delivery pipes were detached from the pump during removal of the assembly, then they should now be unplugged and reconnected. Ensure that each pipe is a good push fit on its respective stub and, where applicable, is correctly retained by its spring clip. Route each pipe through its original location whilst taking care to ensure that it is neither twisted nor crimped between any engine components.

42 Reassembling the engine/gearbox unit: fitting the neutral indicator switch assembly and gearbox sprocket

1   Remove the component parts of the neutral indicator switch from their storage and lay them out on a clean area of work surface.

TS100, 125 and 185 models
2   With the type of switch fitted to these models, insert the spring into its location in the end of the gearchange drum and follow this with the contact pin. The stepped end of this pin must locate correctly into the end of the spring.

TS250 models
3   With the type of switch fitted to these models, locate the switch contact arm on the end of the gearchange drum so that the protrusion of the arm fits into the notch of the drum. With the arm thus aligned, fit and tighten its single securing screw.

All models
4   Check that the new O-ring is located correctly in its retaining groove and then fit the switch cover into position over the drum end so that its contact screw is nearest to the upper surface of the crankcase. Secure the cover in position by fitting and tightening its two retaining screws. Finally, connect the single electrical lead to the switch contact, making sure that the lead is correctly routed before finally tightening the contact screw.

5   Grease the splined end of the output shaft over which the gearbox sprocket fits. This will lessen the risk of the O-ring being damaged as it passes over the threaded end and splines of the shaft. Lightly grease the lip of the shaft oil seal and the surface of the sprocket spacer.

6   Carefully ease the O-ring over the shaft end and push it as far up the shaft splines as possible. Fit the spacer over the shaft and use its end to push the O-ring through the oil seal. With the spacer pushed fully home, wipe any excess grease from the threads of the shaft and fit the gearbox sprocket followed by a new lock washer and the retaining nut. Lock the layshaft in position by using the method described for sprocket removal and tighten the nut to the specified torque loading. Lock the nut in position by bending the lock washer against one of its flats.

43 Reassembling the engine/gearbox unit: fitting the fly-wheel generator assembly

1   Position the flywheel generator stator plate on the crankcase and fits its three retaining screws finger tight. Route the electrical leads from the stator through their retaining grommet and position the end of the loom on the top of the crankcase. Take care to ensure that the single lead to the neutral indicator switch is correctly routed and that none of the leads are allowed to pass over the crankcase to cover mating surfaces.

2   Rotate the stator plate until the alignment marks made on removal align exactly and then lock the plate in position by fully

43.1 Route the electrical leads from the generator stator through their retaining grommet in the crankcase

43.2 Align the stator plate before tightening its securing screws

43.4a Insert the Woodruff key into the crankshaft keyway

43.4b With the rotor fitted, apply a thread locking compound ...

43.4c ... before fitting the plain washer, lock washer and fitting
and tightening the retaining nut

tightening the three retaining screws. If these marks were not
made or the stator being fitted is a replacement item, note the
index line cast in the plate adjacent to one of the retaining
screw slots and align this line with the centre of the appropriate
screw before locking the plate in position.

3   The ignition timing and, where applicable, the contact
breaker gap, should be checked as a matter of course before the
engine is started. The relevant details will be found in Chapter
3. Make these checks before fitting the rotor for the final time,
using locking fluid and torque wrench as described in the
following paragraph.

4   Clean and degrease both the taper of the crankshaft and the
bore of the flywheel generator rotor where the two components
come into contact. Insert the Woodruff key into the crankshaft
keyway and push the rotor onto the crankshaft. Gently tap the
centre of the rotor with a soft-faced hammer to seat it on the
crankshaft taper and then fit the plain washer followed by the
spring washer. Clean and degrease the threads of the rotor
retaining nut and of the crankshaft end. Apply a thread locking
compound to these threads and fit and tighten the nut, finger-
tight. Lock the crankshaft in position by employing the method
used for rotor removal and tighten the rotor retaining nut to the
specified torque loading.

44 Reassembling the engine/gearbox unit: fitting the small-end bearing, piston, cylinder barrel and cylinder head

1   Position the engine/gearbox unit so that it is upright on the
work surface. Rotate the crankshaft to raise the connecting rod
to its highest point and thoroughly lubricate the big-end bearing
with clean engine oil. On TS185 and 250 models, trim the
excess gasket material from across the mouth of the crankcase.
On all models, wipe clean the crankcase to cylinder barrel
mating surface before easing the new base gasket into position
over the barrel retaining studs and packing the crankcase mouth
with clean rag in order to prevent any component parts from
falling   into   the   crankcase   during   the   following   fitting
procedures. On TS185 and 250 models, check that the gasket
is not blocking the oil feed hole in each mating surface.

2   Lubricate the small-end eye of the connecting rod and the
small-end bearing itself with clean engine oil before pushing the
bearing into position. Place the piston over the connecting rod
so that the arrow cast in the piston crown faces forward and
slide the gudgeon pin into position. The pin should be a light
sliding fit but if it proves to be tight, warm the piston in hot
water to expand the metal around the gudgeon pin bosses. Use
new circlips to retain the gudgeon pin, and double check to
ensure that each is correctly located in the piston boss groove.
If a circlip works loose, it will cause serious engine damage. The
circlips should be fitted so that the gap between the circlip ends
is well away from the cutout to the side of the gudgeon pin hole.
Finally, check that the piston rings have not been disturbed from
the positions quoted in Section 25 of this Chapter.

3   Lubricate both the cylinder bore and piston rings with clean
engine oil. Position two blocks of wood across the crankcase
mouth, one each side of the connecting rod, and carefully lower
the piston onto the blocks. This will provide positive support to
the piston whilst easing the rings into the bore.

4   Carry out a close inspection of the reed valve assembly to
check that there is no contamination between the valve reeds
and their stopper plate. Any contamination found must be
removed if the valve is to operate efficiently. Place the cylinder
barrel in position over its retaining studs and proceed to lower
it carefully down over the piston. Guide the piston crown into
the bore and push in on each side of the piston rings so that
they slide into the bore. There is a generous lead in on the base
of the bore which will aid this operation. Take care that the ring
ends stay each side of the ring pegs; if the rings ride up over the
pegs breakage is certain. With the rings safely inserted into the
cylinder bore, remove the blocks from underneath the piston
and the rag from the crankcase mouth. Push the cylinder barrel
firmly down onto the crankcase and tap it lightly around its
upper surface with a soft-faced hammer to ensure that it is
properly seated. Fit the six cylinder barrel retaining nuts. Tighten
these nuts evenly and in a diagonal sequence to their specified
torque loading. Route the oil feed pipe(s) from the oil pump to
the union(s) at the base of the cylinder barrel. Ensure that each
pipe is neither twisted nor routed in such a way that it is likely
to become trapped between two mating surfaces. Leave the
pipe(s) disconnected until the oil pump has been bled of air.

5   Clean the mating surfaces of both the cylinder barrel and
cylinder head. Carefully ease a new cylinder head gasket over
the retaining studs and press it into position on the barrel. Note
that the gasket fitted to TS185 and 200 models is marked
'Top'. This marked side of the gasket must face uppermost.
TS185 gaskets also have a cutout in one edge. This cutout must
face the front of the engine.

6   Push the cylinder head into position over its gasket and fit
and tighten its retaining nut, finger tight. These nuts should now
be tightened to their specified torque loading whilst tightening
in even increments and working in a diagonal sequence. This
method of tightening will prevent the cylinder head from
becoming distorted.

7   It should be noted that on no account should any form of
jointing compound be applied to the surfaces of either the

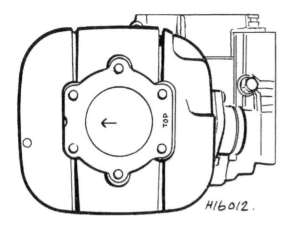

Fig. 1.23 Correct fitting of cylinder head gasket – 185 and 250 models

Note: Only 185 model has cut-out at front of gasket

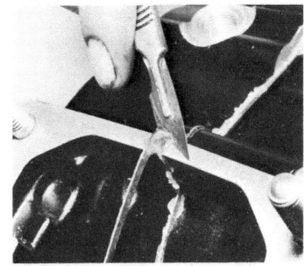

44.1a Trim the excess gasket material from across the crankcase mouth (185 and 250)

44.1b Position the cylinder barrel base gasket (185 shown). Note the oil feed hole (arrowed)

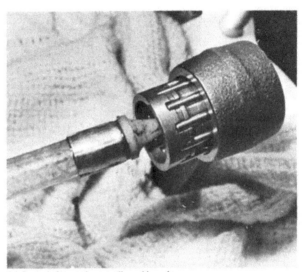

44.2a Lubricate the small-end bearing

44.2b Fit the piston with the arrow facing forward and insert the gudgeon pin

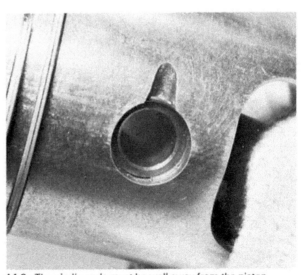

44.2c The circlip ends must be well away from the piston cutout

44.4 Support the piston whilst fitting the cylinder barrel

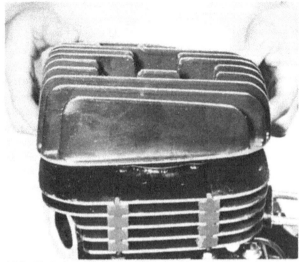
44.6a Push the cylinder head into position over its gasket ...

44.6b ... and tighten its retaining nuts to the specified torque loading

44.9a Fit the carburettor to barrel gasket ...

44.9b ... followed by the spacer plate ...

44.9c ... and a serviceable O-ring (200 and 125)

cylinder head gasket or cylinder barrel base gasket.

8   Refit the carburettor to the cylinder barrel by reversing the procedure used for removal and note the following points.

TS100 and 125 models

9   With these models, check the O-ring and gasket for any signs of damage or deterioration and renew each component as necessary. Any air allowed to be drawn in through the carburettor to cylinder barrel joint will produce a marked effect on engine performance. If either of the spring washers placed over the carburettor retaining studs is seen to be flattened, then both washers must be renewed as a pair. With the carburettor, its O-ring and spacer plate with gasket all positioned correctly, fit the two retaining nuts with their spring washers and tighten the nuts evenly to avoid distortion of the carburettor mating face.

TS185 and 250 models

10   If the inlet stub fitted to the cylinder barrel of these models has been removed, then it should now be refitted. Ensure the stub to barrel mating surfaces are both clean and degreased before placing the stub in position on the barrel. Note that if these mating surfaces are undamaged then there should be no problem with air being drawn between them. If, however, there is some doubt as to whether the joint will be air tight, then lightly coat one of the surfaces with a jointing compound such as Blue Hylomar.

11   Check that the spring washers placed beneath the heads of the inlet stub retaining bolts are not flattened. If necessary, renew these washers as a pair before fitting and tightening the retaining bolts. Tighten these bolts evenly to avoid distortion of the stub mating face.

12   The carburettor can now be fitted to the inlet stub by pushing it into position in the stub, making sure it is correctly positioned and then tightening its retaining clamp. Note that if the stub has been renewed, then the rubber sleeve will be fairly inflexible and some difficulty may be experienced in pushing the carburettor into it. If this situation occurs, then smear a small amount of washing up liquid around the lip of the sleeve. This will greatly lessen the resistance of the rubber to the carburettor and therefore make fitting of the component much easier.

All models

13   Finally, if the engine unit is to be left unattended for any period of time, prevent the ingress of dirt and moisture into either the spark plug hole or the carburettor mouth by plugging each orifice with a wad of clean rag. Make the presence of this rag noticeable so that it is not inadvertently left in position at a later stage during refitting of the engine unit into the frame.

45 Fitting the engine/gearbox unit into the frame

1   It is well worth checking at this stage that no component part has been omitted during the various rebuilding sequences. It is better to discover any left-over items now rather than just before the engine is to be started.

2   Fitting of the engine/gearbox unit into the frame is, generally speaking, a direct reversal of the removal procedure. Ideally, the help of one person will be needed in the initial stages of fitting. If this help is not available, then make arrangements to protect the frame tubes that surround the engine so that the engine unit can be rested on them whilst its mounting bolts are being fitted.

3   Check around the frame to ensure that nothing will impede the progress of the engine unit whilst it is being lifted into position. Ease the unit into the frame from the left-hand side of the machine and, once it is aligned with its mounting points, insert from the left-hand side the upper of the two rear mounting bolts. The engine unit may now be pivoted around the bolt so that the remaining mounting bolts and plate(s) can be fitted. Do not tighten any one bolt or nut until they have all been fitted and the position of their washers checked. With the engine unit thus positioned in the frame, remove any protective

material from the frame tubes and then proceed to tighten each mounting bolts to the specified torque loading.

4   On TS100 and 125 models, remember to relocate the engine crankcase guard before fitting and tightening the lower of the front engine mounting bolts. With the guard in position, the four bolts with washers that retain both the guard and the footrest mounting bracket can now be tightened. If any one of the spring washers fitted to these bolts is seen to be flattened, then it must be replaced with a new item.

5   Check the condition and gap of the spark plug before fitting it to the cylinder head. Do not omit to check that the aluminium crush washer is still attached to the plug; this washer ensures an effective seal between the plug and head casting and serves to keep the plug electrodes the required height from the piston crown. Ideally, a new spark plug should be fitted after a full engine rebuild. Applying a smear of graphite grease to the threads of the plug will greatly lessen the chances of the plug becoming seized in the cylinder head during engine use.

6   Relocate the air inlet hose over the mouth of the carburettor and retain it in position by fitting and tightening its clamp. If this hose has been renewed, some difficulty will be experienced in easing it over the carburettor mouth due to its not being very pliable. Do not use any sharp instrument, such as a screwdriver, with which to lever it into position, as this will only serve to damage the alloy of the carburettor and, if it slips, to tear the hose. Instead, lubricate the lip of the hose with washing up liquid and form a tool out of a piece of hard wood with which to ease the hose into position (half a clothes-peg with its edges rounded off is ideal).

7   Carefully slide the throttle valve and jet needle into the carburettor body and tighten the top of the mixing chamber. Whilst working on the TS185 model used as a project bike for this Manual, some difficulty was experienced in getting the jet needle to locate properly in the needle jet. If this problem occurs, do not attempt to force the needle into position but loosen the carburettor retaining clamps so that it may be canted to the left. With the carburettor so positioned, ease the throttle valve return spring against the carburettor top and place the tip of one finger on top of the needle to keep it centralised. The valve may now be slid into position and the tip of the needle moved to align with the hole of the jet before finally pushed home. With the top of the mixing chamber tightened, reposition the carburettor so that it is vertical and retighten its retaining clamps.

45.4 Relocate the engine crankcase guard before fitting the front mounting bolts (100 and 125)

45.7a Carefully fit the throttle valve assembly (125 shown)

45.7b Carburettor hose retaining clamps must be fully tightened

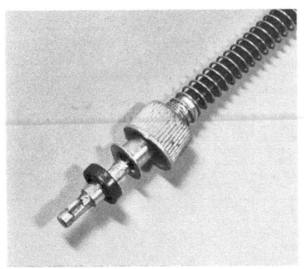

45.9a Check the tachometer cable and components before reconnecting the cable ...

45.9b ... and making sure the cable is correctly secured with its retaining ring (125 shown)

45.10 Reconnect the oil pump control cable

45.11 View the throttle valve indicator mark (arrowed) through the wall of the carburettor mixing chamber

8 It is now necessary to ensure that the throttle cable is correctly adjusted and functions smoothly over its full operating range. Adjustment of this cable is correct when there is 0.5 – 1.0 mm (0.02 – 0.04 in) of free movement in the cable outer when it is pulled out of its adjuster at the carburettor top. If cable adjustment is found to be incorrect, then loosen the adjuster locknut and rotate the adjuster the required amount before retightening the locknut. Initial adjustment should always be carried out at the carburettor end of the cable. Where the machine has an adjuster at the throttle twistgrip, then use this adjuster for any fine adjustment that may be necessary.

9 Reconnect the tachometer drive cable to its location at the top of the gearbox housing. Ensure that the end of the cable inner is correctly located in the drive gear assembly before fitting and tightening the knurled retaining ring.

10 Move to the left-hand side of the engine and insert the oil pump control cable through its adjuster at the crankcase end and push up on the end of the pump lever to allow insertion of the cable nipple into its nylon holder. The oil pump must now be checked for correct adjustment in accordance with the following procedure.

11 Remove the single screw with sealing washer from the wall of the carburettor mixing chamber. Check the condition of the sealing washer and renew it if necessary. On TS185 and 250 models, this screw is located on the left-hand side of the carburettor which makes a sighting comparison between the throttle valve and the oil pump relatively easy. Unfortunately, this is not the case with TS100 and 125 models, because the screw is located on the right-hand side of the carburettor and therefore faces away from the oil pump. In this instance, it is best to recruit the help of an assistant to sight the position of the throttle valve whilst the oil pump control cable is adjusted.

12 Rotate the throttle twistgrip until the circular indicator mark on the side of the throttle valve comes into alignment with the upper edge of the screw hole in the wall of the mixing chamber. With the throttle set in this position, check that the mark scribed on the pump lever boss is in exact alignment with the mark cast in the pump body. If this is not the case, then the marks should be made to align by rotating the control cable adjuster, after having released its locknut. On completion of the adjustment procedure, retighten the locknut whilst holding the cable adjuster in position and slide the rubber sealing cap down the cable to cover the adjuster. It should be noted that any adjustment of the oil pump control cable may well affect the adjustment of the throttle cable. It is therefore, necessary to check the throttle cable for correct alignment before proceeding further. Refit and tighten the screw, with its sealing washer, to the carburettor.

13 Reconnect the oil feed pipe to its retaining stub on the oil pump. Ensure that the pipe is a good push fit on the stub and, where applicable, is retained by its spring clip. Check also that the pipe is routed correctly through its original location and is not twisted or likely to be trapped between any engine components. Refill the oil tank with oil of the specified type and proceed to check that both the feed pipe and pump are primed. This can be accomplished by loosening the cross-headed screw on the side of the pump body and waiting for a steady stream of oil to emerge before retightening the screw. Failure to carry out this priming procedure will mean that the engine will run dry when first started, with the subsequent risk of seizure.

14 It is now necessary to prime and reconnect the oil feed pipe(s) running from the pump to the base of the cylinder barrel. To do this, remove the protective tape from each pipe and its retaining stub and, using a syringe or a similar container full of clean engine oil, inject oil into the pipe until it is full. Prevent any loss of oil from the pipe by pinching its end between two fingers and then push it over its retaining stub. Ensure that each pipe is retained in position by its spring clip. The priming operation is now complete.

15 Place the oil pump cover plate in position and fit and tighten its two retaining screws.

16 Fully tighten the spark plug and reconnect its suppressor cap whilst routing the HT lead so that it cannot chafe on any

45.12 Align the oil pump lever with the mark cast in the pump body (arrowed)

45.14a Each oil feed pipe must be primed with oil ...

45.14b ... before it is reconnected and secured with its spring clip

45.19a Fit a new exhaust pipe sealing ring

45.19b The exhaust pipe retaining spring must be in good condition (185 and 250)

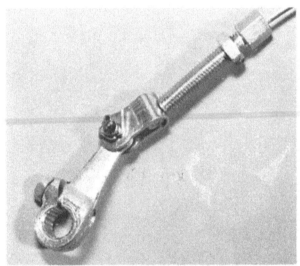

45.20 The clutch cable adjuster and clutch operating lever assembly should now be fitted

45.22a Route the clutch cable through its exhaust-mounted guide (100 and 125)

45.22b ... or its cylinder barrel guide (185 and 250)

45.22c Refit the cover over the clutch operating assembly

frame or engine component parts.

17 Reconnect the electrical leads of the flywheel generator stator assembly, ensuring that they are correctly routed and clipped to their frame attachment points.

18 The battery should now be refitted to the machine and its leads reconnected. Observe the polarity markings of the battery and ensure that the earth lead is connected to the negative (-) terminal. Both terminal connections must be clean and free from corrosion. To prevent corrosion from occurring, smear both connections with a liberal coating of petroleum jelly; do not use ordinary grease. Take care to route the battery vent pipe correctly so that it is not trapped between any frame components and so that its end is placed well clear of the lower frame tubes.

19 Refitting of the exhaust system is a straightforward reversal of the removal procedure. If in doubt as to the fitted position of any one component part of the system, refer to the figure of the system type included in Chapter 2. Do not tighten any one fixing bolt until the complete system is fitted and torque load the exhaust pipe to cylinder barrel securing bolts to the recommended setting. Remember to renew any spring washers that have become flattened during use and also to fit a new sealing ring into the recess provided in the cylinder barrel. Before fitting the pipe retaining spring to TS185 and 250 models, check that it has not become fatigued or weakened through corrosion and renew the spring if thought necessary.

20 Relocate the clutch cable adjuster into the crankcase cover and refit the operating lever in its original position on the splined shaft. Secure the lever in position by tightening its pinch bolt. The clutch should now be adjusted so that there is 2 – 3 mm (0.08 – 0.12 in) of free play in the cable inner. The amount of free play is measured between the pivot end of the handlebar lever and its retaining clamp.

21 Carry out adjustment of the clutch cable by rotating the adjuster through its crankcase location until the amount of free play measured is correct. Hold the adjuster in position and tighten its locknut. Any subsequent fine adjustment to the amount of free play in the cable inner may now be achieved by rotating the knurled adjuster at the handlebar lever end. This adjuster is locked in position by its knurled lock ring.

22 On completion of adjustment, relocate the rubber sealing caps over the cable ends and check that the cable is held clear of the exhaust system and cylinder barrel by its guide. Wipe clean the mating surfaces of the clutch operating lever cover and the crankcase cover. Place the cover in position and secure it by fitting and tightening its two retaining screws.

23 With the gearbox sprocket fitted and locked in position with its lock washer, loop the final drive chain around the gearbox and rear wheel sprockets and reconnect the two ends with the split link. It is most important that the spring clip of this link is correctly fitted with the closed end facing the normal direction of chain travel. It is quite possible that the rear wheel will have to be moved forward in order to place enough slack in the chain to allow insertion of the link. In either case, the chain must be checked for correct tension and adjusted accordingly by referring to the instructions given in Chapter 5 of this Manual.

24 Remount the fuel tank on the machine. Carry out a check to ensure that no part of the tank is in direct contact with the frame as any metal contact will lead to eventual failure of the tank structure. Reconnect the fuel pipe to the fuel tap and retain it in position with the spring clip. Turn the tap lever to the 'On' position and carefully check both ends of the pipe for any signs of fuel leakage. On no account should fuel be allowed to come into contact with hot engine castings; if this is allowed to happen, fire may result causing serious personal injury.

25 Carry out a final check around the engine unit to ensure that all component parts have been correctly fitted and, where applicable, are functioning correctly. Check all electrical connections for security and check that all bolts, nuts, screws and fasteners have been tightened. Refit the seat and sidepanels.

26 On TS185 and 250 models, fit and tighten the gearbox oil drain plug with a serviceable sealing washer. On all models,

45.23 The closed end of the final drive chain spring clip must face the normal direction of chain travel

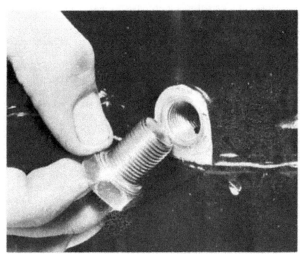

45.26a Fit the gearbox oil drain plug with its serviceable sealing washer (185 and 250)

45.26b Replenish the gearbox with oil ...

45.26c ... and refit the filler plug

46.4a Refit the flywheel generator rotor cover (125 shown)

check that the detent plunger housing has been properly tightened. Remove the plastic filler plug from the right-hand crankcase and replenish the gearbox with the specified quantity of SAE 20W/40 oil (see Specifications, Chapter 2). Refit the filler plug.

27 Note at this point that full details of setting the contact breaker gap (where applicable) and carrying out a static check of the ignition timing are contained within the relevant Sections of Chapter 3. Both of these procedures should have been carried out during generator assembly, the details of which were given in Section 43.

28 On TS250 models, check the condition of the spring washers which are fitted beneath the heads of the gearbox sprocket cover retaining screws. If any one of these washers is flattened, then it should be renewed. Wipe clean the mating surfaces of the cover and crankcase and place the cover in position, securing it with the three screws.

29 Finally, on all models, temporarily refit the gearchange lever and ensure that the machine is in neutral.

46 Starting and running the rebuilt engine

1 Attempt to start the engine using the usual procedure adopted for a cold engine. Do not be disillusioned if there is no sign of life initially. A certain amount of perseverance may prove necessary to coax the engine into activity even if new parts have not been fitted. Should the engine persist in not starting, check that the spark plug has not become fouled by the oil used during re-assembly. Failing this, go through the fault finding charts and work out what the problem is methodically.

2 When the engine does start, keep it running as slowly as possible to allow the oil to circulate. Open the choke as soon as the engine will run without it. During the initial running, a certain amount of smoke may be in evidence due to the oil used in the reassembly sequence being burnt away. The resulting smoke should gradually subside.

3 With the engine fully warmed up, switch the ignition off to stop it and set the equipment up for dynamic tuning as described in Chapter 3 of this Manual. Restart the engine and check that the ignition timing is correct.

4 With the ignition timing accurately set, clean the mating surfaces of the flywheel generator rotor cover and the crankcase and refit the cover. On TS100 and 125 models, fit the new gasket with this cover. Tighten the cover retaining screws evenly and in a diagonal sequence in order to lessen the risk of the cover becoming distorted. The gearchange lever can now be permanently refitted to its shaft by sliding it into position and

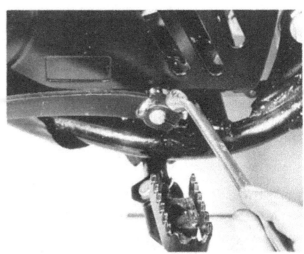

46.4b Position the gearchange lever correctly in relation to the footrest

fitting and tightening its retaining bolt. Ensure that this bolt passes through the channel in the gearchange shaft spline and also that the lever is positioned correctly in relation to the footrest.

5 Check the engine for blowing gaskets and oil leaks. Before using the machine on the road, check that all the gears select properly, and that the controls function correctly.

47 Taking the rebuilt machine on the road

1 Any rebuilt machine will need time to settle down, even if parts have been replaced in their original order. For this reason it is highly advisable to treat the machine gently for the first few miles to ensure oil has circulated throughout the lubrication system and that new parts fitted have begun to bed down.

2 Even greater care is necessary if the engine has been rebored or if a new crankshaft has been fitted. In the case of a rebore, the engine will have to be run in again, as if the machine were new. This means greater use of the gearbox and a restraining hand on the throttle until at least 500 miles have been covered. There is no point in keeping to any set speed limit; the main requirement is to keep a light loading on the

engine and to gradually work up performance until the 500 mile mark is reached. These recommendations can be lessened to an extent when only a new crankshaft is fitted. Experience is the best guide since it is easy to tell when an engine is running freely.

3   Remember that a good seal between the piston and the cylinder barrel is essential for the correct functioning of the engine. A rebored two-stroke engine will require more careful running-in, over a long period, than its four-stroke counterpart. There is a far greater risk of engine seizure during the first hundred miles if the engine is permitted to work hard.

4   If at any time a lubrication failure is suspected, stop the engine immediately and investigate the cause. If an engine is run without oil, even for a short period, irreparable engine damage is inevitable.

5   Do not on any account add oil to the petrol under the mistaken belief that a little extra oil will improve the engine lubrication. Apart from creating excess smoke, the addition of oil will make the mixture much weaker, with the consequent risk of overheating and engine seizure. The oil pump alone should provide full engine lubrication.

6   Do not tamper with the exhaust system or run the engine without the baffle fitted to the silencer. Unwarranted changes in the exhaust system will have a marked effect on engine performance, invariably for the worse. The same advice applies to dispensing with the air cleaner or the air cleaner element.

7   When the initial run has been completed allow the engine unit to cool and then check all the fittings and fasteners for security. Re-adjust any controls which may have settled down during initial use.

Chapter 2 Fuel system and lubrication

For modifications and information relating to later models, see Chapter 7

Contents

General description 1
Fuel tank: removal and fitting 2
Fuel tap: cleaning the collector bowl assembly 3
Fuel tap: removal, fitting and curing of leaks 4
Fuel feed pipe: examination 5
Carburettor: removal and fitting 6
Carburettor: dismantling, examination, renovation and reassembly 7
Carburettor adjustment and exhaust emissions: general note 8
Carburettor: adjustment 9
Carburettor: checking the settings 10
Reed valve induction system: mode of operation 11
Reed valve: removal, examination and fitting 12
Air filter element: removal, examination, cleaning and fitting 13
Engine lubrication system: general maintenance 14
Oil pump: removal and fitting 15
Oil pump: bleeding of air 16
Gearbox lubrication: general maintenance 17
Exhaust system: decarbonisation 18
Exhaust system: removal and fitting 19

Specifications

Fuel tank	**TS100**	**TS125**
Total capacity:		
UK N model	7.0 litre (1.5 Imp gal)	7.5 litre (1.6 Imp gal)
All other models	7.0 litre (1.8/1.5 US/Imp gal)	7.0 litre (1.7/1.5 US/Imp gal)
Reserve capacity:		
UK N model	1.5 litre (2.64 Imp pint)	0.9 litre (1.6 Imp pint)
All other models	1.5 litre (3.17/2.64 US/Imp pint)	1.5 litre (3.17/2.64 US/Imp pint)

Carburettor	**UK TS100 models**	**UK TS125 models**
Make	Mikuni	Mikuni
Type	VM 24 SH	VM 24 SH
ID no	48111	48111
Bore size	24 mm (0.945 in)	24 mm (0.945 in)
Main jet	85	85
Needle jet:		
ERN model	N-8	N-8
ERT and ERX model	N-6	N-6
Jet needle	4DH7-2	4DH7-2
Pilot jet	30	30
Pilot air screw	1½ turns out	1 turn out
Throttle valve cutaway:		
ERN model	1.5	1.5
All other models	2.0	2.0
Float height:		
ERN model	25.8 ± 0.5 mm (1.016 ± 0.020 in)	25.8 ± 0.5 mm (1.016 ± 0.020 in)
All other models	28 ± 1.0 mm (1.1023 ± 0.04 in)	28 ± 1.0 mm (1.1023 ± 0.04 in)
Engine idling speed	1300 ± 100 rpm	1300 ± 100 rpm

Carburettor	**US TS100 models**	**US TS125 models**
Make	Mikuni	Mikuni
Type	VM 24 SH	VM 24 SH
ID no	48160	48030
Bore size	24 mm (0.945 in)	24 mm (0.945 in)
Main jet	115	120
Needle jet	O-6	O-4
Jet needle	5E20	5F41
Pilot jet	17.5	17.5
Pilot air screw setting	Preset	Preset
Throttle valve cutaway	2.0	2.0
Float height	25.8 ± 1.0 mm (1.02 ± 0.04 in)	25.8 ± 1.0 mm (1.02 ± 0.04 in)
Engine idling speed	1300 ± 100 rpm	1300 ± 100 rpm

Air cleaner		
Element type	Oiled polyurethane foam	

Lubrication

Engine	**TS100** **TS125** Suzuki CCI – Pump fed total-loss system
Oil capacity	1.2 litre (2.6/2.3 US/Imp pint)
Oil type	Suzuki CCI, Suzuki CCI Super or any good quality non-diluent 2-stroke oil
Oil pump discharge rate (fully open) over 2 mins at 2000 rpm	1.30 – 1.58 cc (0.044 – 0.053/0.045 – 0.055 US Imp/oz)
Gearbox capacity:	
At oil change	800 cc 700 cc (1.69/1.41 US Imp pint) (1.48/1.24 US/Imp pint)
At engine overhaul: ERN model	900 cc 800 cc (1.90/1.58 US Imp pint) (1.69/1.41 US/Imp pint)
All other models	850 cc 750 cc (1.80/1.50 US/Imp pint) (1.58/1.32 US/Imp pint)
Oil type	SAE 20W/40 SAE 20W/40

Fuel tank

Total capacity	**TS185** **TS250** 7.0 litre 10.0 litre (1.8/1.5 US/Imp gal) (2.6/2.2 US/Imp gal)
Reserve capacity	1.5 litre 1.2 litre (3.17/2.64 US/Imp pint) (2.6/2.2 US/Imp pint)

Carburettor

	UK TS185 models	UK TS250 models
Make	Mikuni	Mikuni
Type	VM 29 SS	VM 29 SS
ID no	29910	30910
Bore size	29 mm (1.14 in)	29 mm (1.14 in)
Main jet	195	200
Needle jet	P-1	P-1
Jet needle	5DH48-3	5DH48-3
Pilot jet	25	27.5
Pilot air screw	1¼ turns out	1½ turns out
Air jet	0.7	1.0
Starter jet	80	80
Throttle valve cutaway	2.5	2.5
Float height	24 ± 1.0 mm (0.94 ± 0.04 in)	24 ± 1.0 mm (0.94 ± 0.04 in)
Engine idling speed	1300 ± 150rpm	1300 ± 150 rpm

	US TS185 models	US TS250 models
Make	Mikuni	Mikuni
Type	VM 29 SS	VM 29 SS
ID no	29900	30900
Bore size	29 mm (1.14 in)	29 mm (1.14 in)
Main jet	195	200
Needle jet	O-9	O-7
Jet needle	5DH55	5DH55
Pilot jet	27.5	25
Pilot air screw setting	Preset	Preset
Air jet	0.7	1.0
Starter jet	75	80
Throttle valve cutaway	2.5	2.5
Float height	24 ± 1.0 mm (0.94 ± 0.04 in)	24 ± 1.0 mm (0.94 ± 0.04 in)
Engine idling speed	1300 ± 100 rpm	1300 ± 100 rpm

Air cleaner

Element type	Oiled polyurethane foam

Lubrication

	TS185	TS250
Engine	Suzuki CCI – Pump fed total-loss system	
Oil capacity	1.2 litre (2.6/2.3 US/Imp pint)	
Oil type	Suzuki CCI, Suzuki CCI Super or any good quality non-diluent 2-stroke oil	
Oil pump discharge rate (fully open) over 2 mins at 2000 rpm	1.27 – 1.55 cc (0.043 – 0.045/ 0.052 – 0.055 US/Imp oz)	1.58 – 1.87 cc (0.053 – 0.056/ 0.063 – 0.066 US/Imp oz)

Gearbox oil capacity:

At oil change	700 cc (1.48/1.24 US/Imp pint) 750 cc (1.58/1.32 US/Imp pint) SAE 20W/40	850 cc (1.80/1.50 US/Imp pint) 900 cc (1.90/1.58 US/Imp pint) SAE 20W/40
At engine overhaul		
Oil type		

1   General description

The design and function of the fuel system fitted to the machines covered in this Manual is as follows. Fuel is gravity-fed from the frame-mounted fuel tank to the float chamber of the Mikuni carburettor via a three-position fuel tap. Air is drawn into the carburettor via an oil-impregnated polyurethane foam filter element which is contained in a housing mounted to the rear of the carburettor.

At low engine speeds, the proportions of air and atomised fuel which form the combustion mixture, are controlled by a pilot circuit, these being regulated by a combination of throttle stop and pilot mixture screw settings. As the twistgrip control is turned, the cylindrical throttle valve is lifted, allowing a greater volume of air to be drawn through the carburettor choke. The passage of air across the top of the needle jet causes fuel to be drawn up through the main jet and needle jet by venturi action.

The amount of fuel entering the engine is at this stage metered by the needle jet assembly, in which a tapered needle is drawn upwards with the throttle valve, allowing increasing amounts of fuel to enter the combustion mixture, as the throttle is opened. Eventually, the rate of flow of the fuel is restricted by the main jet, which has been selected to give the correct mixture at maximum throttle opening.

The point of induction on a two-stroke engine is normally controlled by the piston skirt, which covers and uncovers ports machined in the cylinder bore. On the Suzuki models covered in this Manual, a supplementary timing system is employed to enable more efficient induction timing. This device is known as a reed valve. This valve is mounted at the base of the cylinder barrel, within the barrel to crankcase mating surface, and serves to maintain an even flow of the combustion mixture as well as reducing the possibility of any blow-back of the combustible gases. This type of system contributes greatly towards an economical and powerful engine unit.

Engine lubrication is by a pump fed system known as Suzuki CCI. Two-stroke oil is gravity fed from a frame-mounted tank to the oil pump. The pump is driven by the engine via the primary and kickstart drive assemblies, and is also interconnected by a Bowden cable to the throttle twistgrip. Thus the amount of oil passed by the pump is varied according to the engine speed and throttle setting.

Oil from the pump is fed directly to the main and big-end bearings. The remaining engine components are lubricated by splash. Any residual oil is drawn into the combustion chamber along with the incoming fuel mixture, and is therefore burnt.

Lubrication for the transmission components is catered for by oil contained within the gearbox casing, isolated from the working parts of the engine proper.

2   Fuel tank: removal and fitting

1   Commence removal of the fuel tank by turning the fuel tap lever to its 'Off' position and releasing the fuel pipe retaining clip. This will allow the pipe to be pulled off the stub at the rear of the tap. Careful use of a small screwdriver may be necessary to help ease the pipe off the stub. Once the pipe is detached, allow any fuel in the pipe to drain into a small clean container. Unclip the tank cap breather pipe end attachment (where fitted) from the handlebar crossmember.
2   The tank may now be detached from the frame by lifting the seat and unscrewing the single retaining bolt at the rear of the tank. Grasp the rear of the tank and pull it up and rearwards off

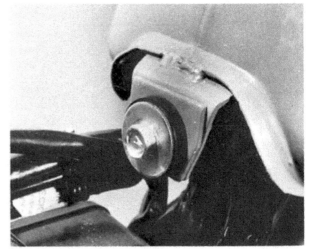

2.2 Remove the fuel tank retaining bolt

its front mounting rubbers. In practice, it was found that some degree of force was required to free the tank from these rubbers and it helped to have an assistant pushing the tank rearwards as it was moved from side to side.
3   Once removed, place the tank retaining bolt, together with any associated mounting components in a safe place ready for refitting. Inspect the mounting rubbers for signs of damage or deterioration and if necessary renew them before refitting of the tank is due to take place.
4   Store the tank in a safe place whilst it is removed from the machine, well away from any naked lights or flames. It will otherwise represent a considerable fire or explosion hazard. Check that the tap is not leaking and that it cannot be accidentally knocked into the 'On' position. It is well worth taking simple precautions to protect the paint finish of the tank, whilst in storage. Placing the tank on a soft protected surface and covering it with a protective cloth or mat may well avoid damage being caused to the finish by dirt, grit, dropped tools, etc.
5   To refit the tank, reverse the procedure adopted for its removal. Move it from side to side before it is fully home, so that the rubber buffers engage with the guide channels correctly. If difficulty is encountered in engaging the front of the tank with the rubber buffers, apply a small amount of petrol to the buffers to ease location. Secure the tank with the single retaining bolt whilst ensuring that the mounting components are correctly located and that there is no metal to metal contact between the tank and frame.
6   Finally, always carry out a leak check on the fuel pipe connections after fitting the tank and turning the tap lever to the 'On' position. Any leaks found must be cured; as well as wasting fuel, any petrol dropping onto hot engine castings may well result in a fire or explosion occurring.

3   Fuel tap: cleaning the collector bowl assembly

1   The tap incorporates a detachable plastic base which acts as a collecting bowl for any heavy sediment or any mixture that may find its way past the fuel tap filter stack. This bowl must be removed at intervals for cleaning and examination. Its contents

will provide a good guide to the cleanliness of the inside of the fuel tank.

2   Some machines will be found to incorporate a filter element within the collector bowl. The purpose of this element is to provide a back-up filter to the fuel tap filter stack in preventing any contamination in the fuel from being passed directly from the tank to the carburettor, thus precluding the likelihood of any one of the jets within the carburettor becoming blocked. Contamination caught by the filter is deposited in the bowl.

3   It should be realised that any loss in engine performance or a refusal of the engine to run for any more than a short period of time, might be attributable to fuel starvation caused by a blocked or partially blocked filter element. Suspected contamination of the fuel will therefore lead to the need to clean the element and bowl at more frequent intervals than that recommended, it may well also be necessary to remove and clean out the fuel tank.

4   To remove the collector bowl (and filter, where fitted) turn the tap lever to the 'Off' position, place an open-ended spanner over the squared end of the bowl and turn it anti-clockwise to unscrew the bowl. It was found in practice that the O-ring between the bowl and the tap casing had formed a semi-permanent seal between the two components and some effort was required to effect an initial release of the bowl.

5   Where a filter element is fitted, the element may be gently eased out of its location within the tap whilst taking care to note the positioning of the hole within the filter element in relation to the corresponding fuel line within the tap. Clean the filter element by rinsing it in clean fuel. Any stubborn traces of contamination may be removed from the element by gently brushing it with a small soft-bristled brush soaked in fuel; a used toothbrush is ideal. Remember to take the necessary fire precautions when cleaning the element and always wear eye protection against any fuel that may spray back from the brush.

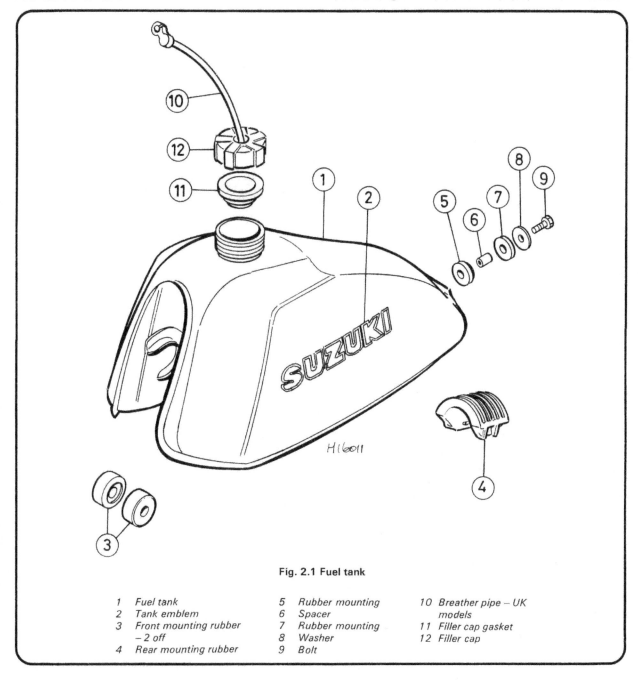

Fig. 2.1 Fuel tank

1	Fuel tank	5	Rubber mounting	10	Breather pipe – UK models
2	Tank emblem	6	Spacer		
3	Front mounting rubber – 2 off	7	Rubber mounting	11	Filler cap gasket
		8	Washer	12	Filler cap
4	Rear mounting rubber	9	Bolt		

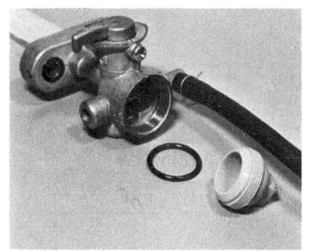

3.1 Unscrew the collector bowl from the body of the fuel tap

4.3a Carefully withdraw the fuel filter stack from the fuel tank ...

Once it is cleaned, closely inspect the gauze of the element for any splits or holes that will allow the passage of sediment through it onto the carburettor. Renew the element if it is in any way defective.

6   The collector bowl and the inside of the base of the fuel tap should be cleaning by using a procedure similar to that described in the preceding paragraph. Note and adhere to the safety precautions that accompany this procedure.

7   Inspect the condition of the O-ring; if it is flattened, perished, or in any way damaged, then it must be renewed.

8   Position the filter element (where fitted) in the base of the fuel tap. Remember to align the hole in the filter with the corresponding fuel line within the tap. Fit a serviceable O-ring and fit and tighten the collector bowl. Take care not to overtighten the bowl, it need only be nipped tight. Finally, turn the tap lever to the 'On' position and carry out a check for any leakage of fuel around the bowl to joint. Cure any leak found by nipping the bowl a little tighter. If this fails, remove the bowl and check that the O-ring has seated correctly.

---

4   Fuel tap: removal, fitting and curing of leaks

1   It will be found, should the fuel tap spring a leak, that the fuel will emit from any one of three points on the tap; the tap to tank joint, the tap lever joint or the collector bowl joint. The latter of these points has been fully dealt with in the preceding Section of this Chapter; however, the initial attempt at curing any one of these leaks remains the same, that of simply check-tightening the screws or bolts securing the component concerned. If this simple check fails to effect a cure, then the leak must be stopped by following the relevant procedure laid down in the following paragraphs.

2   Any leak from the tap to tank joint will be caused by the O-ring between the two components being defective. Renewal of the O-ring will call for removal of the tap from the tank. This may be done with the tank in situ but will necessitate draining the tank of as much fuel as possible. This is easily accomplished by detaching the fuel feed pipe from the carburettor and placing its end into a clean container placed next to the machine. Note that this container should be constructed of metal and have a removable top which incorporates an adequate seal; it should also be of an adequate capacity. Before turning the tap lever to the 'On' position, which will allow the fuel to drain, ensure that all necessary fire precautions have been taken; it is most important that the machine is positioned in an area where there is good ventilation.

3   With the tank thus drained of fuel, remove the tap by

4.3b ... and inspect the filter stack for signs of contamination

unscrewing the two retaining bolts with their sealing washers. Take care when drawing the filter stack out of the tank, not to damage the filter stack by letting it come into contact with the edge of the tank orifice. Place the tap unit on a clean work surface and ease the O-ring out of its retaining groove in the tap body. If the filter stack shows signs of being contaminated, then it should now be detached for cleaning. Clean the stack by rinsing it in clean fuel. Any stubborn traces of contamination may be removed by gently brushing the stack with a soft-bristled brush which has been soaked in fuel; a used toothbrush is ideal. Remember to take the necessary fire precautions when carrying out this cleaning procedure and always wear eye protection against any fuel that may spray back from the brush. On completion of cleaning, closely inspect the gauze area of the stack for any splits or holes that will allow the passage of sediment through it and into the fuel tap. Note that the filter stack is not listed as a separate item from the fuel tap body. It follows, therefore, that if the stack is in any way defective, then the complete body of the tap should be renewed.

4   Clean the O-ring retaining groove and the inner area of the tap before replacing the stack and pressing the new O-ring into position. Wipe clean the area around the orifice of the tank where the O-ring makes contact and carefully reinsert the tap into the tank. Fit a new sealing washer to each of the tap

retaining bolts and fit and tighten each bolt to secure the tap in position. The fuel feed pipe may now be reconnected to the carburettor and the tank replenished with fuel. Check any disturbed connections for fuel leaks before starting and riding the machine.

5   Any leak from the tap lever joint will be caused by the seal contained between the lever and tap body being defective. Renewal of this seal will call for draining of the tank prior to removal of the tap lever. The method best used for draining the tank is described in paragraph 2 of this Section.

6   To remove the tap lever, unscrew the small retaining screw which is located on the underside of the lever housing. With this screw removed, the lever may be pulled clear of the tap body, along with the tap valve spring. Remove the defective seal from its groove within the stub of the lever; clean its retaining groove of any sediment and then fit the new seal into position. The valve spring should be inspected for any signs of fatigue, failure and corrosion and renewed if thought necessary. Fit the spring and ease the lever back into its housing in the tap body; the seal may be lubricated with fuel to aid its insertion. With the slot in the lever stub aligned with the hole for the retaining bolt, refit and tighten the bolt.

7   With the tap lever thus fitted, check its operation and then reconnect the fuel feed pipe to the carburettor. Replenish the tank with fuel and carry out a check for fuel leaks on all the disturbed connections before starting and riding the machine.

5   Fuel feed pipe: examination

1   The fuel feed pipe is made from thin-walled synthetic rubber and is of the push-on type. It is only necessary to replace the pipe if it becomes hard or splits. It is unlikely that the retaining clip should need replacing due to fatigue as the main seal between the pipe and union is effected by an interference fit. Suzuki recommend that the fuel feed pipe be renewed every two years as a matter of course.

6   Carburettor: removal and fitting

1   Commence removal of the carburettor by unscrewing the clamp which retains the air inlet hose to the mouth of the carburettor. Ease the hose rearwards to clear the carburettor and leave the clamp pushed up over it to avoid loss. This hose should be closely examined during this procedure and renewed if split or perished.

2   If detachment of the inlet hose proves difficult and, as a result, greater access to the carburettor mouth is required, then it is advisable to remove both sidepanels from the machine and to disconnect the tachometer drive cable (where fitted) from the top of the gearbox housing by unscrewing its knurled retaining ring and pulling the cable from position.

3   Move to the left-hand side of the machine. Check that the lever of the fuel tap is turned to its 'Off' position and release the retaining clip from the fuel feed pipe at its carburettor end. This will allow the pipe to be pulled off its retaining stub; but before doing this, obtain a small clean container with which to catch any fuel contained in the pipe. Remember to observe the necessary fire precautions when draining fuel from the pipe. If the pipe proves stubborn, then carefully use the flat of a small screwdriver to ease it off its retaining stub.

4   Unscrew the top of the mixing chamber from the carburettor body and carefully lift the chamber top away from the carburettor whilst easing the throttle valve and jet needle out of their respective locations. Manoeuvre the assembly clear of the carburettor body and rest it on the cylinder head. If the carburettor is to be removed for some time, then it is a good idea to prevent contamination of the throttle valve assembly by wrapping it in a polythene bag. The carburettor body is now free to be detached from the cylinder barrel and manoeuvred clear of the machine.

TS100 and 125 models
5   On these models, the carburettor has a mounting flange which is part of the carburettor body. Remove the two nuts with spring washers that hold this flange against the cylinder and detach the carburettor from its two mounting studs. Ensure that the O-ring fitted in the flange of the carburettor is not misplaced.

TS185 and 250 models
6   The carburettor fitted to these models is retained to a separate inlet stub by means of a clamp. Release this clamp and ease the carburettor away from the stub. There is no need to remove the inlet stub unless it is seen to be damaged or its rubber section is split or perished, in which case it must be renewed. If necessary, detach the stub by removing its two retaining bolts with spring washers.

All models
7   When removing the carburettor, be careful to ensure that it does not snag on the oil pump to cylinder barrel oil feed pipe(s). The carburettor is best manoeuvred out to the left of the machine. Fitting the carburettor is a straightforward reversal of the removal procedure, whilst noting the following points.

6.5 Remove the carburettor flange securing nuts (100 and 125)

6.6 Examine the inlet stub for damage (185 and 250)

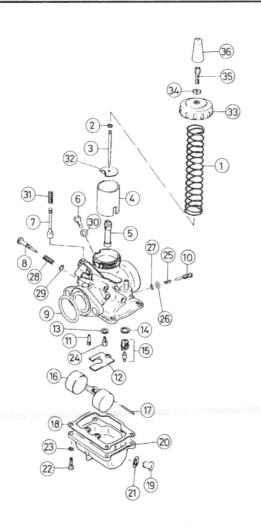

6.8a Place a serviceable O-ring in the groove of the carburettor flange ...

Fig. 2.2 Carburettor – 100 and 125 models

1	Return spring	19	Drain plug
2	Needle clip	20	Float chamber
3	Jet needle	21	Sealing washer
4	Throttle valve	22	Screw – 4 off
5	Needle jet	23	Spring washer
6	Throttle valve		– 4 off
	sighting plug	24	Main jet
7	Choke valve	25	Spring
8	Throttle stop screw	26	Washer
9	O-ring	27	O-ring
10	Pilot air screw	28	Spring
11	Pilot jet	29	O-ring
12	Baffle plate	30	Sealing washer
13	Sealing washer	31	Return spring
14	Sealing washer	32	Retaining plate
15	Float needle and seat	33	Carburettor top
16	Float	34	Locknut
17	Pivot pin	35	Cable adjuster
18	Float chamber gasket	36	Rubber cover

6.8b ... fit a serviceable gasket over the carburettor mounting studs ...

TS100 and 125 models

8   With these models, check that the serviceable O-ring is correctly located in the flange of the carburettor. Any air allowed to be drawn in through the carburettor to cylinder barrel joint will produce a marked effect on engine performance. If either of the spring washers placed over the carburettor retaining studs is seen to be flattened, then both washers must

6.8c ... and fit the spacer plate (100 and 125)

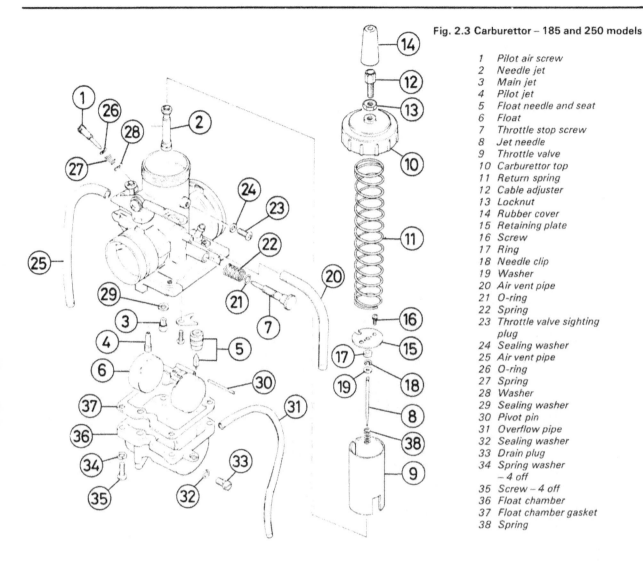

Fig. 2.3 Carburettor – 185 and 250 models

1 Pilot air screw
2 Needle jet
3 Main jet
4 Pilot jet
5 Float needle and seat
6 Float
7 Throttle stop screw
8 Jet needle
9 Throttle valve
10 Carburettor top
11 Return spring
12 Cable adjuster
13 Locknut
14 Rubber cover
15 Retaining plate
16 Screw
17 Ring
18 Needle clip
19 Washer
20 Air vent pipe
21 O-ring
22 Spring
23 Throttle valve sighting plug
24 Sealing washer
25 Air vent pipe
26 O-ring
27 Spring
28 Washer
29 Sealing washer
30 Pivot pin
31 Overflow pipe
32 Sealing washer
33 Drain plug
34 Spring washer – 4 off
35 Screw – 4 off
36 Float chamber
37 Float chamber gasket
38 Spring

be renewed as a pair. With the carburettor, its O-ring and spacer plate with gasket all positioned correctly, fit the two retaining nuts with their spring washers and tighten the nuts evenly to avoid distortion of the carburettor mating face.

TS185 and 250 models

9 If the inlet stub fitted to the cylinder barrel of these models has been removed, then it should now be refitted. Ensure the stub to barrel mating surfaces are both clean and degreased before placing the stub in position on the barrel. Note that if these mating surfaces are undamaged, then there should be no problem with air being drawn between them. If, however, there is some doubt as to whether the joint will be airtight, then lightly coat one of the surfaces with a jointing compound such as Blue Hylomar.

10 Check that the spring washers placed beneath the heads of the inlet stub retaining bolts are not flattened. If necessary, renew these washers as a pair before fitting and tightening the retaining bolts. Tighten these bolts evenly to avoid distortion of the stub mating face.

11 The carburettor can now be ftted to the inlet stub by pushing it into position in the stub, making sure it is correctly positioned and then tightening its retaining clamp. Note that if the stub has been renewed, then the rubber sleeve will not be very pliable and some difficulty may be experienced in pushing the carburettor into it. If this situation occurs, then smear a small amount of washing up liquid around the lip of the sleeve.

This will greatly lessen the resistance of the rubber to the carburettor and therefore make fitting of the component much easier.

All models

12 Relocate the air inlet hose over the mouth of the carburettor and retain it iñ position by fitting and tightening its clamp. If this hose has been renewed, some difficulty will be experienced in easing it over the carburettor mouth due to its being fairly inflexible. Do not use any sharp instrument, such as a screwdriver, with which to lever it into position, as this will only serve to damage the alloy of the carburettor and, if it slips, to tear the hose. Instead, lubricate the lip of the hose with washing up liquid and form a tool out of a piece of hard wood with which to ease the hose into position (half a clothes-peg with its edges rounded off is ideal).

13 Carefully slide the throttle valve and jet needle into the carburettor body and tighten the top of the mixing chamber. Whilst working on the TS185 model used as a project bike for this Manual, some difficulty was experienced in getting the jet needle to locate properly in the needle jet. If this problem occurs, do not attempt to force the needle into position but loosen the carburettor retaining clamps so that it may be canted to the left. With the carburettor so positioned, ease the throttle valve return spring against the carburettor top and place the tip of one finger on top of the needle to keep it centralised. The valve may now be slid into position and the tip of the needle

moved to align with the hole of the jet before being finally pushed home. With the top of the mixing chamber tightened, reposition the carburettor so that it is vertical and retighten its retaining clamps.

14 It is now necessary to ensure that the throttle cable is correctly adjusted and functions smoothly over its full operating range. Adjustment of this cable is correct when there is 0.5 – 1.0 mm (0.02 – 0.04 in) of free movement in the cable outer when it is pulled out of its adjuster at the carburettor top. If cable adjustment is found to be incorrect, then loosen the adjuster locknut and rotate the adjuster the required amount before retightening the locknut. Initial adjustment should always be carried out at the carburettor end of the cable. Where the machine has an adjuster at the throttle twistgrip, then use this adjuster for any fine adjustment that may be necessary.

15 Where necessary, reconnect the tachometer drive cable to its location at the top of the gearbox housing. Ensure that the end of the cable inner is correctly located in the drive gear assembly before fitting and tightening the knurled retaining ring. Refit both sidepanels.

16 Finally, reconnect the fuel pipe to the carburettor and retain it in position with the spring clip. Turn the tap lever to the 'On' position and carefully check both ends of the pipe for any signs of fuel leakage. On no account should fuel be allowed to come

into contact with hot engine castings; if this is allowed to happen, fire may result causing serious personal injury.

7 Carburettor: dismantling, examination, renovation and reassembly

1 Before dismantling the carburettor, cover an area of the work surface with clean paper or rag. This will not only prevent any components that are placed upon it from becoming contaminated with dirt, moisture or grit, but, by making them more visible, will also prevent the many small components removed from the carburettor body from becoming lost.

2 Proceed to dismantle the carburettor by removing the four screws and spring washers that retain the float chamber to the main body of the carburettor. In practice, it was found that the float chamber had become quite firmly adhered to the carburettor body and some gentle persuasion was needed to remove it. Tapping around the joint with a soft-faced hammer may serve to break this seal, otherwise it will be necessary to place the flat of a small screwdriver between the side of the chamber and the lip of the body in order to lever the components apart. Note that there may well be a slot cut in the mating surface of the

6.12 Clamp the air inlet hose to the carburettor mouth

6.13 Carefully slide the throttle valve into the carburettor body (125 shown)

6.14 Locate the throttle cable adjuster (100 and 125)

6.16 Retain the fuel feed pipe in position with its spring clip

carburettor body. If this is so, then the flat of the screwdriver should be inserted at this point. Take great care when using a screwdriver, not to place any great strain on the component castings: the two components should part fairly easily.

3 With the float chamber thus removed and placed to one side, pull the pivot pin from the twin float assembly and lift the floats from position. The float needle can now be displaced from its seating and should be put aside in a safe place for examination at a later stage. It is very small and easily lost if care is not taken to store in a safe place.

4 On TS100 and 125 models, unscrew and remove the float needle seat together with its sealing washer. Detach the baffle plate which the seat holds in position.

5 On TS185 and 250 models, unscrew and remove the single crosshead screw which serves to retain the float needle seat retaining plate in position between the two float pivot pin columns. Withdraw the plate and, using a pair of long-nose pliers, pull the needle seat out of its location in the carburettor body. In order to avoid causing any damage to the seat, it is best to pad the jaws of these pliers with tape.

6 On all models, unscrew and remove the main jet and carefully hook its washer out of its location in the carburettor body. Note that, when unscrewing any jet from the carburettor, a close fitting screwdriver of the correct type must be used to prevent damage occurring to the soft material from which the jet is constructed. With the main jet and its washer removed, the needle jet may be pushed out of its location so that it leaves the carburettor body through the top of the mixing chamber. Take note of the alignment pin cast in the needle jet location for reference when refitting.

7 Unscrew and remove the pilot jet (with US models, take careful note of the information given in the following Section). In practice, it was found that this jet was very tight and great care combined with some effort was needed to free it. Note the setting of the throttle stop screw by counting the number of turns required to screw it fully in. Remove the throttle stop screw, taking care to retain its spring.

8 Note the setting of and remove the pilot air screw with its spring. Failure to note the settings of the aforementioned screws will make it less easy to 'retune' the carburettor after it has been reassembled and refitted to the machine.

9 The only removable component parts now left in the body of the carburettor are the throttle valve sighting plug with its sealing washer and those components which make up the choke assembly. Unless the condition of its sealing washer is suspect, the sighting plug may now be left fitted. Reference to the figures accompanying this test will show that the type of carburettor fitted to the TS185 and TS250 models incorporates a choke assembly but, unlike the TS100 and TS125 type, does not show the component parts of this assembly as separate obtainable items. Nevertheless, it is worth removing the complete choke assembly from the carburettor body, not only for examination but to allow proper cleaning of the body casting. The main body of the carburettor is now devoid of all removable components and should be placed to one side in readiness for cleaning.

10 The only removable component fitted to the float chamber of the carburettor is the drain plug, which takes the form of a single slotted screw with a sealing washer located beneath its head. It is not necessary to remove this plug except for renewal or replacement of the screw itself.

11 Prior to examination of the carburettor component parts, clean each part thoroughly in clean fuel before placing it on a piece of clean rag or paper. Use a soft nylon-bristled brush to remove any stubborn contamination on the castings and blow dry each part with a jet of compressed air. Avoid using a piece of rag for cleaning since there is always a risk of particles of lint obstructing the airways or jet orifices. Never use a piece of wire or any pointed metal object to clear a blocked jet, it is only too easy to enlarge a jet under these circumstances and increase the rate of petrol consumption. If an air line is not available, a blast of air from a tyre pump will usually suffice. If all else fails to clear a blocked jet, remove a bristle from the soft-bristled

7.11a Use compressed air to blow clear blocked airways

7.11b Never use a piece of wire to clear a blocked jet

brush and carefully pass it through the jet to clear the blockage. Remember to observe the necessary fire precautions during the cleaning procedure and take care to guard against any blow-back of fuel by wearing eye protection.

12 Check each casting for cracks or damage and check that each mating surface is flat by laying a straight-edge along its length. Any distorted casting must be replaced with a serviceable item.

13 Remove all O-rings and sealing gaskets from the component parts and replace them with new items. Ensure that, where applicable, they are correctly seated in their retaining grooves. Any spring washers that have become flattened should now be renewed.

14 The springs on the throttle stop and pilot air screws should now be carefully inspected for signs of corrosion and fatigue and renewed if necessary.

15 The seating area of the float needle will wear after lengthy service and should be closely examined with a magnifying glass. Wear usually takes the form of a ridge or groove, which will cause the float needle to seat imperfectly. If the needle has to be renewed, remember that the needle seat will have worn in unison and in extreme cases, will also need renewing. Check also that the small pin which protrudes from the end of the needle is free to move and is returned to its extended position by the action of the spring fitted beneath it. The correct

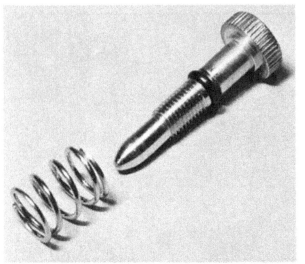

7.14a Inspect the spring of the throttle stop assembly ...

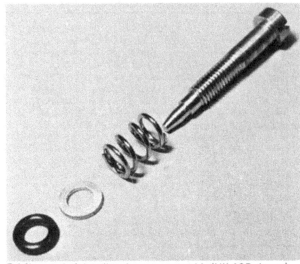

7.14b ... and of the pilot air screw assembly (UK 125 shown)

7.15a Inspect the seating area of the float needle

7.15b Examine the float needle seat (100 and 125)

7.15c Renew the O-ring of the float needle seat (185 and 250)

7.16 Examine the float assembly for signs of damage

7.17a Compress the return spring before releasing the throttle cable from the valve (125 shown)

7.17b US TS100 and 125 models and all 185 and 250 models have the jet needle retaining plate held in position by a single screw

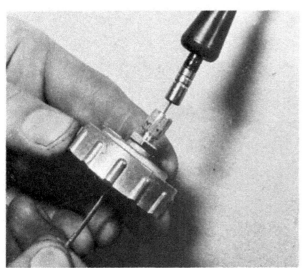

7.17c Remove the carburettor top from the throttle cable

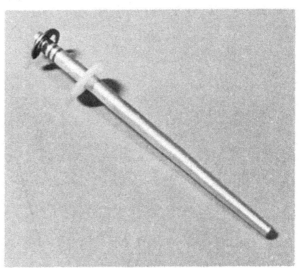

7.18a The jet needle assembly (UK 100 and 125)

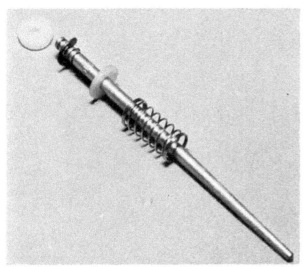

7.18b The jet needle assembly (US 100 and 125 and all 185 and 250)

7.18c The sealing ring within the top of the mixing chamber must be serviceable

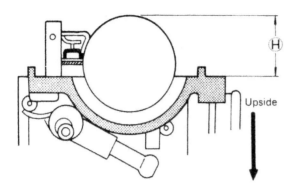

Fig. 2.4 Carburettor float level measurement

H – Float height (see specifications)

action of this pin is essential in cushioning the movement of the float against the needle which in turn acts to reduce the amount of wear on the seating area of the needle.

16 Closely examine the twin float assembly for signs of damage, especially around the soldered joints. Shake the float to establish if a leak is present. Where a brass float assembly is fitted, it is theoretically possible to repair a float by soldering. Unfortunately, any attempt to do so is likely to cause a small but dramatic explosion, having a detrimental effect on both the float and the operator. It is far preferable to renew the float, although a temporary repair may be made with Araldite or Petseal. Where an assembly incorporating plastic floats is fitted, it is not possible to effect a permanent repair. In consequence, a new item must always be fitted if damage is found.

17 Move to the machine and inspect the throttle valve for wear. Any wear will be denoted by polished areas on the external diameter of the valve. Excessive wear will allow air leaks, which in turn will weaken the mixture and produce erratic slow running. Many mysterious carburation maladies may be attributed to this defect, the only cure being to renew the valve, and if worn badly in corresponding areas, the carburettor body. If removal of the valve is necessary, grasp the valve firmly in one hand whilst compressing the return spring against the carburettor top with the other. On US TS100 and 125 models and on all TS185 and 250 models, the jet needle retaining plate is held in position by a single crosshead screw; this screw should now be removed. On all models, disengage the throttle cable from its retaining slot in the valve and withdraw the retaining plate, followed by the jet needle and its clip, from the valve. Before renewing the valve, take care to inspect each individual component part, commencing with the needle.

18 Examine the needle carefully for scratches or wear along its length. Ensure that the needle is not bent by rolling it on a flat surface, such as a sheet of plate glass. If in doubt as to the condition of the needle, return it to an official Suzuki service agent who will be able to give further advice and, if necessary, provide a new component. Check the condition of the sealing ring which is located within the top of the mixing chamber. This seal must be replaced with a serviceable item if it is suspected that it is no longer capable of being effective. Note that the seal is not listed as a separate component; nevertheless, it should not prove difficult to cut a perfectly adequate replacement from a piece of rubber of similar thickness and material composition.

19 Inspect the return spring for signs of fatigue, failure or severe corrosion and renew it if found necessary. The procedure adopted for reassembly of the throttle valve component parts should be a direct reversal of that used for dismantling.

20 Inspect the surface of the choke valve for signs of scratches or excessive wear and the return spring for signs of fatigue, failure or corrosion. Renew each component as necessary. The rubber seal which prevents ingress of moisture and dirt into the valve assembly should also be renewed if found to be split or perished.

21 Prior to reassembly of the carburettor, check that all the

component parts, both new and old, are clean and laid out on a piece of clean rag or paper in a logical order. On no account use excessive force when reassembling the carburettor because it is easy to shear a jet or some of the smaller screws. Furthermore, the carburettor is cast in a zinc based alloy which itself does not have a high tensile strength. If any of the castings are damaged during reassembly, they will almost certainly have to be renewed.

22 Reassembly is basically a reversal of the dismantling procedure, whilst noting the following points. If in doubt as to the correct fitted position of a component part, refer either to the figure accompanying this text or to the appropriate photograph. When fitting the throttle stop and pilot air screws, ensure that each screw is first screwed fully in, until it seats lightly and then set to its previously noted position. Alternatively, set the pilot air screw to the setting given in the Specifications Section of this Chapter (UK models only). The setting of the throttle stop screw will then have to be determined by following the adjustment procedure listed in Section 9 of this Chapter. With US models, take careful note of the information given in the following Section.

23 Do not omit to align the slot in the top of the needle jet with the alignment pin cast in the jet location before pushing the jet fully home. Finally, smear the external diameter of the throttle valve with light machine oil before refitting the carburettor to the machine.

7.20a The choke valve assembly (UK 125)

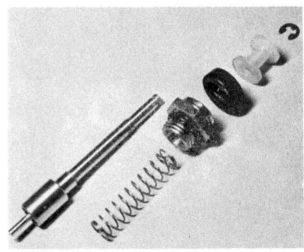

7.20b The choke valve assembly (UK 185)

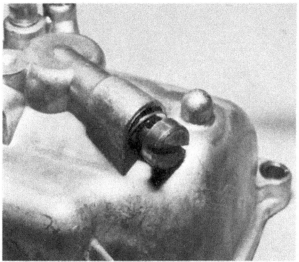

7.22a On 100 and 125 models, commence reassembly of the carburettor by fitting the drain plug, with its sealing washer, to the float chamber.

7.22b Clean the throttle valve sighting plug and renew its sealing washer ...

7.22c ... before fitting the plug in the carburettor body

7.22d The needle jet must be cleaned of all contamination ...

7.22e ... prior to fitting the jet over its alignment pin

7.22f Clean the main jet and its seating washer ...

7.22g ... before fitting the washer and screwing the main jet into the needle jet

7.22h Fit the first sealing washer over the float needle seat location ...

7.22i ... followed by the baffle plate with the second sealing washer ...

7.22j ... and then screw the float needle seat into position

7.22k Fit the float needle into its seat ...

7.22l ... and retain the float assembly in position with its pivot pin

7.22m On 185 and 250 models, commence reassembly of the carburettor by checking that the pilot jet is clean ...

7.22n ... before fitting the jet into the carburettor body

7.22o Clean the needle jet ...

7.22p ... and push the jet into position over its alignment pin

7.22q Clean the main jet and its seating washer ...

7.22r ... and position the washer over the needle jet before fitting the main jet

7.22s Push the float needle seat into its housing ...

7.22t ... and place the float needle in its seat

7.22u Fit the float needle seat retaining plate and screw ...

7.22v ... before retaining the float assembly in position with its pivot pin

7.22w On all carburettor types, secure the float chamber in position with the four screws with spring washers

7.22x Refit the throttle stop screw ...

7.22y ... followed by the pilot air screw (UK models only)

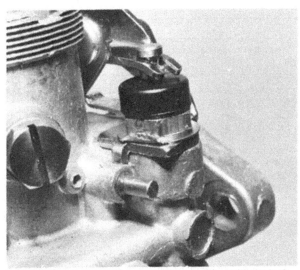

7.22z Lock the choke valve assembly in position with the lock washer ...

7.22aa ... and align the valve lever (125 shown)

7.22bb Ensure the circlip is located correctly in the top of the choke valve needle ..

7.22cc ... before fitting the valve lever (185 shown)

7.22dd The end nipple of the throttle cable must be correctly located in the throttle valve (125 shown)

8   Carburettor adjustment and exhaust emissions: general note

In some countries legal provision is made for describing and controlling the types and levels of toxic emissions from motor vehicles.

In the USA exhaust emission legislation is administered by the Environmental Protection Agency (EPA) which has introduced stringent regulations relating to motor vehicles. The Federal law entitled the Clean Air Act, specifically prohibits the removal (other than temporary) or modification of any component incorporated by the vehicle manufacturer to comply with the requirements of the law. The law extends the prohibition to any tampering which includes the addition of components, use of unsuitable replacement parts or maladjustment of components which allows the exhaust emissions to exceed the prescribed levels. Violations of the provisions of this law may result in penalties of up to $10 000 for each violation. It is strongly recommended that appropriate requirements are determined and understood prior to making any change to or adjustments of components in the fuel, ignition, crankcase breather or exhaust systems.

To help ensure compliance with the emission standards some manufacturers have fitted to the relevant systems fixed or pre-set adjustment screws as anti-tamper devices. In most cases this is restricted to plastic or metal limiter caps fitted to the carburettor pilot adjustment screws, which allow normal adjustment only within narrow limits. Occasionally the pilot screw may be recessed and sealed behind a small metal blanking plug, or locked in position with a thread-locking compound, which prevents normal adjustment.

It should be understood that none of the various methods of discouraging tampering actually prevents adjustment, nor, in itself, is re-adjustment an infringement of the current regulations. Maladjustment, however, which results in the emission levels exceeding those laid down, is a violation. It follows that no adjustments should be made unless the owner feels confident that he can make those adjustments in such a way that the resulting emissions comply with the limits. For all practical purposes a gas analyser will be required to monitor the exhaust gases during adjustment, together with EPA data of the permissible Hydrocarbon and CO levels. Obiouviously, the home mechanic is unlikely to have access to this type of equipment or the expertise required for its use, and, therefore, it will be necessary to place the machine in the hands of a component motorcycle dealer who has the equipment and skill to check the exhaust gas content.

For those owners who feel competent to carry out correctly the various adjustments, specific information relating to the anti-tamper components fitted to the machines covered in this manual is given in the relevant Sections of this Chapter.

The main jet, pilot jet and needle fitted to emission controlled carburettors are manufactured to particularly close tolerances. To identify these close tolerance items Arabic numerals of a different style than usually used are embossed on the components. When renewing these parts ensure that the numeral style of the new items matches that of the old. A comparison of the two types of identifying numeral is shown in the accompanying figure.

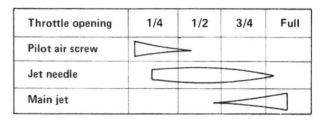

Conventional Figures Used on Standard Tolerance Jet Components	1 2 3 4 5 6 7 8 9 0	
Emission Type Figures Used on Close Tolerance Jet Components	1 2 3 4 5 6 7 8 9 0	

Fig. 2.5 Carburettor component identification numeral types

Throttle opening	1/4	1/2	3/4	Full
Pilot air screw				
Jet needle				
Main jet				

Fig. 2.6 Carburettor mixture strength chart

9   Carburettor: adjustment

1   If flooding of the carburettor or excessive mixture weakness has been experienced, it is wise to start operations by checking the float level, which will involve detaching the carburettor, if not already removed, inverting it and removing the float chamber followed by the chamber to carburettor body sealing gasket. Take care to avoid tearing this gasket during removal. In practice, it was found impossible to free the gasket without tearing it as it was firmly stuck to the mating surface of the carburettor body. It is, therefore, worth anticipating the occurrence of this problem by purchasing a new gasket, or at least carrying out research as to its availability, before proceeding with the following operation.

2   To measure the float level, stand the inverted carburettor on a flat and level work surface and measure the distance between the gasket surface of the carburettor body and the furthest point of the float as shown in the photograph accompanying this text. The float level is correct when the measurement obtained is within the limits given in the Specifications Section of this Chapter. If necessary, bend the small tongue, sited between the two floats, to obtain the correct setting.

9.2 Remove the float chamber base gasket and measure the float level (185 shown)

3   With the carburettor correctly refitted and with the engine running at its normal operating temperature, set the throttle stop screw to give the slowest possible idle speed. Turn the pilot air screw in by a fraction of a turn at a time until the engine starts to falter. Now back the screw off progressively whilst noting the number of turns required to reach the point where the engine starts to run erratically. The correct position for the pilot air screw is mid-way between these two extremes when it will be found that the engine is idling at its fastest. This should be close to the specified setting. At this point, the engine should be idling at the recommended speed. If the reading on the tachometer (where fitted) indicates an idle speed which is

slightly outside that recommended, then the throttle stop screw should be turned until the indicated speed is correct.

4   Always guard against the possibility of incorrect carburettor adjustment which will result in a weak mixture. Two-stroke engines are very susceptible to this type of fault, causing rapid overheating and often subsequent engine seizure. Changes in carburation leading to a weak mixture will occur if the air cleaner is removed or disconnected, or if the silencer is tampered with in any way. Above all, do not add oil to the petrol, in the mistaken belief that it will aid lubrication. Adequate lubrication is provided by the throttle controlled oil pump.

10 Carburettor: checking and settings

1   The various jet sizes, throttle valve cutaway (on slide type carburettors) and needle position are predetermined by the manufacturer and should not require modification. Check with the Specifications list at the beginning of this Chapter if there is any doubt about the types fitted. If a change appears necessary it can often be attributed to a developing engine fault unconnected with the carburettor. Although carburettors do wear in service, this process occurs slowly over an extended length of time and hence wear of the carburettor is unlikely to cause sudden or extreme malfunction. If a fault does occur check first other main systems, in which a fault may give similar symptoms before proceeding with carburettor examination or modification.
2   Note that on all UK models the fitted position of the jet needle clip is indicated by the suffix number of the jet needle identification. For example, on the UK TS100 ERX model, the jet needle identification is 4DH7-2. In this case, the number 2 indicates that the fitted position of the clip is in the 2nd groove down from the top of the needle. US models are fitted with a jet needle which has only a single groove; adjustment is, therefore, not possible.
3   If the machine is found to perform badly on a particular throttle setting, check that the float level is correct, that all pipes to the carburettor are in sound condition and that the air filter assembly and the exhaust system are both in good order before riding the machine on the throttle setting concerned for a distance of approximately 6 miles.
4   On completion of the ride, remove the spark plug and inspect the insulator and electrodes for condition and colour in accordance with the instructions given in Chapter 3 of this Manual. Upon deciding the mixture strength at the throttle setting concerned, refer to the chart accompanying this text and note which of the three components listed (ie, main jet, jet needle and pilot air screw) need to be reset or replaced in order to weaken or richen the mixture strength over the throttle setting as desired. Note that the divisions shown in the chart indicate a certain amount of overlap between the various stages. Follow the instructions given in the following table when resetting or replacing the component concerned, taking care to err slightly on the side of a rich mixture, since a weak mixture will cause the engine to overheat with the subsequent risk of seizure.

Component	Weaker mixture	Richer mixture
Main jet	Fit smaller No jet	Fit larger No jet
Jet needle	Move clip towards top of needle to lower needle	Move clip towards tip of needle to raise needle
Pilot air screw	Turn screw anti clockwise (out)	Turn screw clockwise (in)

5   Note that if the machine is found to perform badly when on $\frac{1}{2}$ throttle setting, it is advisable to make any necessary adjustment to mixture strength by altering the settings of the pilot air screw and jet needle as opposed to changing the main jet.
6   Where non-standard items, such as exhaust systems and air filters, have been fitted to a machine, some alterations to

carburation may be required. Arriving at the correct settings often requires trial and error, a method which demands skill born of previous experience. In many cases the manufacturer of the non-standard equipment will be able to advise on correct carburation changes.

11 Reed valve induction system: mode of operation

1   The various systems of controlling the induction cycle of a two-stroke engine, Suzuki has chosen to adopt the reed valve, a device which permits precise control of the incoming mixture, allowing more favourable port timing to give improved torque and power outputs. The reed valve assembly comprises a wedge-shaped die-cast aluminium alloy valve case which is in the inlet tract. The valve case has rectangular ports which are closed off by flexible stainless steel reeds. The reeds seal against a heat and oil resistant synthetic rubber gasket which is bonded to the valve case. A specially shaped valve stopper, made from cold rolled stainless steel plate, controls the extent of movement of the valve reeds.
2   As the piston ascends in the cylinder, a partial vacuum is formed beneath the cylinder in the crankcase. This allows a charge of fuel/air mixture to flow past the valve and into the crankcase. As the pressure differential becomes equalised, the valves close, and the incoming charge is then trapped. The charge of mixture in the cylinder is by this time fully compressed, the ignition takes place, thus driving the piston downwards. The descending piston eventually uncovers the exhaust port, and the hot exhaust gases, still under a certain amount of pressure, are discharged into the exhaust system. At this stage, the charge of combustion mixture which has been compressed in the crankcase is released into the cylinder via the transfer ports, and the piston again ascends to close the various ports and begin compression. The reed valves open once more and another partial vacuum is created in the crankcase; the cycle of induction thus repeats. It will be noted that no direct mechanical operation of the valve takes place, the pressure differential being the sole controlling factor.

12 Reed valve: removal, examination and fitting

1   The reed valve assembly is mounted at the base of the cylinder barrel, within the barrel to crankcase mating surface. It is therefore necessary to remove the cylinder head and barrel in order to gain access to this assembly. In effect, the reed valve is a flap valve which relies on atmospheric pressure to open it during the induction stroke, allowing the incoming mixture to flow into the crankcase, and to close it when the crankcase pressure increases, to help prevent mixture blown back through the carburettor.
2   Failure of the reed valve in service is not usual, although after a considerable amount of mileage has been covered, the valve reeds may lose some springiness and their performance will suffer accordingly. If it is found necessary to gain access to the valve, then refer to the appropriate Sections of Chapter 1 of this Manual for removal of the cylinder head and barrel and their associated components. With the barrel removed from the machine, position it, cylinder head mating face downwards, on a clean work surface.
3   Check the valve unit for any obvious signs of damage, such as cracked reeds. After an extended period of use, the valve may wear, leaving a permanent gap between the valve reeds and their seat. If this gap exceeds 0.2 mm (0.008 in), then the valve assembly must be renewed. Refer to the figure accompanying this text and check the distance between the edge of the valve reeds and the inlet tract of the valve. This distance should measure no less than 1.0 mm (0.04 in). If the valve is found to be unserviceable, then proceed as follows.
4   On early versions of all the model types covered in this

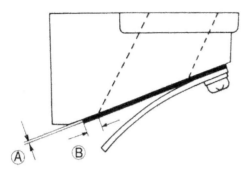

Fig. 2.7 Reed valve unit measuring points

A – Gap between valve reeds and seat – not to exceed 0.2 mm
(0.008 in)

B – Distance between the edge of valve reeds and inlet tract –
no less than 1.0 mm (0.04 in)

Manual, the reed valve assembly is listed as one complete item which indicates that Suzuki intended it to be replaced as such. For later models, however, Suzuki list the valve body, valve reeds and stopper plate as separate items, yet retain the same part number for the assembly as a whole as that given for the earlier models. It is therefore worth checking with an official Suzuki service agent as to whether valve reeds obtained for a late model can be fitted to the valve body of an earlier model of the same type. Doing this will save considerable cost.

5   In the unlikely event of the valve body being damaged, it can be detached from the cylinder barrel by removal of its two retaining screws. Discard the sealing gasket located between the valve and barrel and replace it with a new item. Before fitting the replacement valve body, degrease the threads of its retaining screws and coat them with a thread locking compound. Tighten these screws evenly and in small increments to avoid distorting the valve body.

6   Where the valve reeds and/or stopper plate are found to be damaged, detach both components from the valve body by removing their two retaining screws. Renew each component as necessary and position them on the valve body. Degrease the threads of each retaining screw and coat them with a thread locking compound. Fit and tighten each screw to its recommended torque loading of 0.07 – 0.09 kgf m (0.50 – 0.65 lbf ft). The cylinder barrel, cylinder head and their associated components can now be refitted by referring to the appropriate Sections of Chapter 1.

12.1 The reed valve is mounted within the cylinder barrel to crankcase mating surface

13 Air filter element: removal, examination, cleaning and fitting

1   The air filter assembly consists of a large oil-impregnated polyurethane foam filter element which is housed in a frame mounted container located just to the rear of the carburettor. This container is connected to the carburettor mouth by means of a large-diameter rubber hose and is attached to the frame by means of a mounting bracket with two bolts and washers. TS185 and 250 models also incorporate two mounting rubbers with spacers in this assembly. The element itself must be removed from its container at the intervals specified in the Routine Maintenance Chapter of this Manual for the purposes of examination and cleaning.

2   To remove the filter element, expose the side cover of the filter housing by detaching the left-hand sidepanel from the machine. Remove the filter element as follows.

US TS250 and all TS100, 125 and 185 models

3   The filter housing cover of these machines is retained in position by a single, centrally-located nut with washer. With this nut removed, the cover can be pulled clear of its mounting stud. Note the fitted position of the spacer (and seal on TS185 models) which is located between the cover and filter frame before removing both it and the frame from the mounting stud. The element can now be detached from the frame by unhooking its retaining clip and easing it from position.

UK TS250 models

4   The filter housing cover of these machines is retained in position by two clips which slide upwards off projections on both the cover and housing body. In practice, it was found that these clips were difficult to dislodge, the best method of removal being to push one of the clips completely off its location with the flat of a large screwdriver, thus taking the tension off the remaining clip which may then be removed without much effort. The element should now be carefully drawn clear of its mounting frame.

All models

5   Carry out a close inspection of the filter element. If the foam of the element shows signs of having become hardened with age, or is seen to be very badly clogged, then it must be renewed. If the element is found to be serviceable then it should be cleaned as follows.

6   Immerse the element in a non-flammable solvent such as white spirit (available as Stoddard solvent in the US), whilst gently squeezing it to remove any oil and dust. After cleaning, squeeze out the foam by pressing it between the palms of both hands and then allow a short time for any solvent remaining in the form to evaporate. Do not wring out the foam as this will cause damage and thus lead to the need for early renewal.

7   Reimpregnate the foam with clean 2-stroke oil and gently squeeze out any excess. Refit the element onto its mounting frame and reassemble the component parts of the filter housing by reversing the procedure used for their removal. Check the condition of the spacer seal fitted to the TS185 models and renew it if necessary. Take great care when positioning each component to ensure that no incoming air is allowed to bypass the element. If this is allowed to happen, it will allow any dirt or dust that is normally retained by the element to find its way into the carburettor and crankcase assemblies; it will also effectively weaken the fuel/air mixture.

8   Finally, note that if the machine is being run in a particularly dusty or moist atmosphere, then it is advisable to increase the frequency of cleaning and reimpregnating the element. Never run the engine without the element fitted. This is because the carburettor is specially jetted to compensate for the addition of this component and the resulting weak mixture will cause overheating of the engine with the probable risk of severe engine damage.

13.7a Relocate the air filter element over its mounting frame ...

13.7b ... and retain the element in position with its spring clip

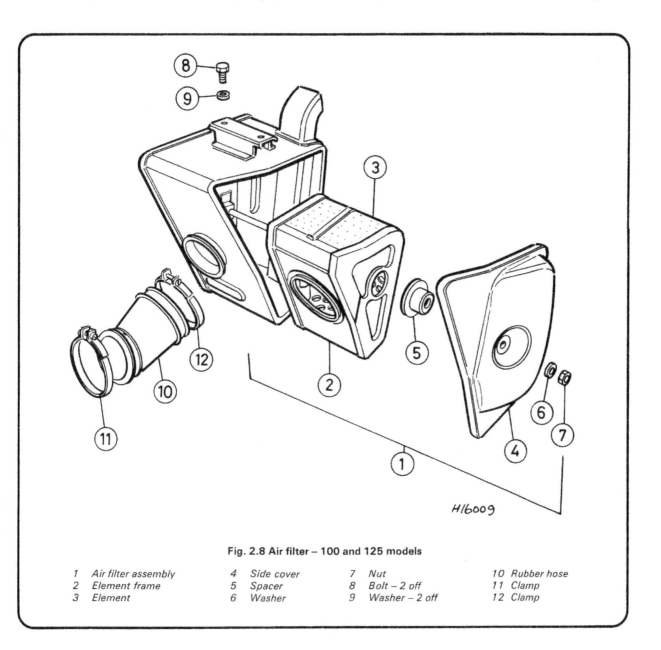

H16009

Fig. 2.8 Air filter – 100 and 125 models

1	Air filter assembly	4	Side cover	7	Nut	10	Rubber hose
2	Element frame	5	Spacer	8	Bolt – 2 off	11	Clamp
3	Element	6	Washer	9	Washer – 2 off	12	Clamp

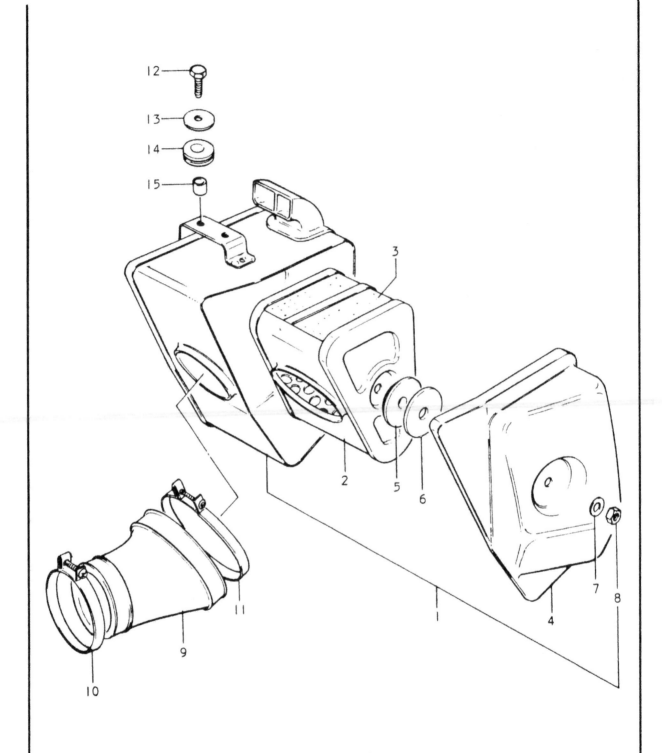

Fig. 2.9 Air filter – 185 model

1	Air filter assembly	5	Seal	9	Rubber hose	13	Washer – 2 off
2	Element frame	6	Washer	10	Clamp	14	Grommet – 2 off
3	Element	7	Washer	11	Clamp	15	Spacer – 2 off
4	Side cover	8	Nut	12	Bolt – 2 off		

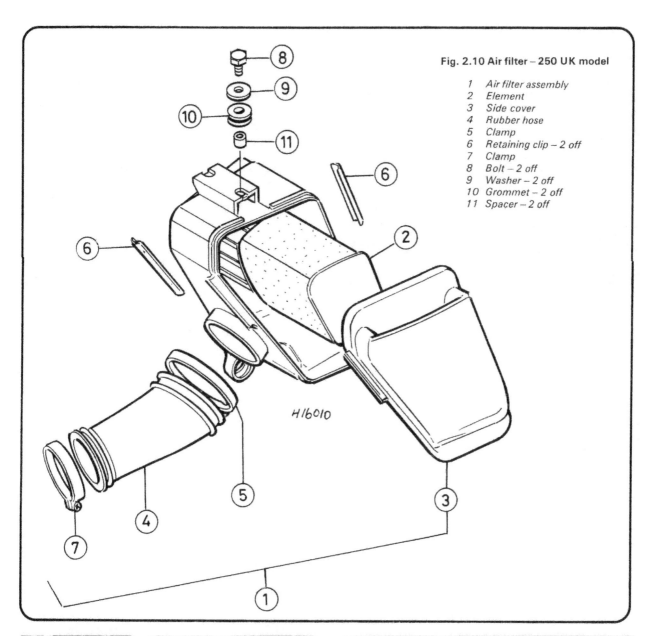

Fig. 2.10 Air filter – 250 UK model

1 Air filter assembly
2 Element
3 Side cover
4 Rubber hose
5 Clamp
6 Retaining clip – 2 off
7 Clamp
8 Bolt – 2 off
9 Washer – 2 off
10 Grommet – 2 off
11 Spacer – 2 off

H16010

13.7c Check that all seals are in good condition and that the filter housing is clean ...

13.7d ... before inserting the filter element into position over its mounting stud ...

13.7e ... refitting the spacer piece ...

13.7f ... followed by the filter housing cover, complete with a serviceable seal (125 shown)

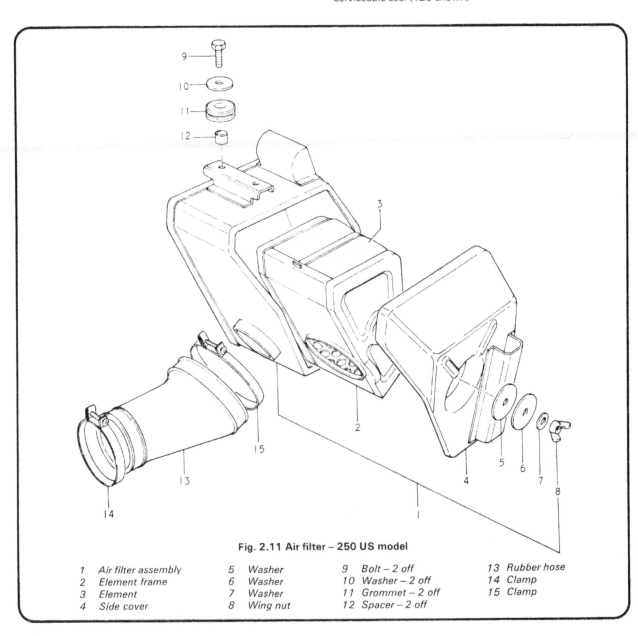

Fig. 2.11 Air filter – 250 US model

1	Air filter assembly	5	Washer	9	Bolt – 2 off	13	Rubber hose
2	Element frame	6	Washer	10	Washer – 2 off	14	Clamp
3	Element	7	Washer	11	Grommet – 2 off	15	Clamp
4	Side cover	8	Wing nut	12	Spacer – 2 off		

14 Engine lubrication system: general maintenance

1   As mentioned in Section 1 of this Chapter, the engine is pressure lubricated by an oil pump which is mounted to the rear of the left-hand crankcase half. It is recommended that adjustment to the control cable of this pump is checked every 2000 miles (3000 km), or on any occasion where the pump has been removed or where over or under lubrication is suspected. Full details of adjustment are contained in the following Section of this Chapter.
2   It is most important that an adequate level of oil is maintained in the frame-mounted oil tank. A sight glass is provided as a means of indicating that the oil level is low. Only the oil types listed in the Specifications Section of this Chapter should be used.
3   whenever removing or carrying out adjustment to the pump, check the oil feed and delivery pipes for any signs of splitting or perishing and ensure that all connections and unions in the lubrication system are tight and free from leaks. Any fault found must be rectified immediately, as leakage will cause a loss of engine lubriciation which will result in rapid wear of the engine components concerned or, at the worse, complete engine seizure.

15 Oil pump: removal and fitting

1   The oil pump can be expected to give long service, requiring no maintenance, but in the event of failure it must be renewed. No replacement parts are obtainable and the pump is, therefore effectively a sealed unit.
2   To gain access to the pump unit, first remove the gearchange lever by unscrewing its retaining bolt and then pulling it clear of the gearchange shaft end. Remove the gearbox sprocket cover. On all but the TS250 models this cover is combined with the flywheel generator rotor cover and is held in position by four retaining screws. On the TS250 models, this cover is separate and is retained by three screws with lock washers.
3   With the gearbox sprocket thus exposed, rotate the rear wheel until the split link in the final drive chain appears between the chain guard and the gearbox sprocket whilst on its upper run. Removal of the chain may now be achieved by removing the spring clip of the split link with a pair of flat-nose pliers and then withdrawing the link to allow the ends of the chain to separate. Lift the chain off the teeth of the sprocket and allow its ends to rest on a piece of clean rag or paper placed beneath the machine.
4   Remove the two screws that serve to retain the oil pump cover plate in position and manoeuvre the cover clear of the machine. Disconnect the pump control cable from the pump lever by pushing the end of the lever up so that tension is taken off the cable inner and then detaching the cable nipple from its operating arm. With the rubber sealing cap detached from the cable adjuster, the cable may now be pulled through the adjuster so that it is clear of the pump.
5   It is now necessary to make provision for catching any oil that will issue from both the feed and delivery pipes once they are disconnected from the pump. To prevent complete draining of the oil tank, the feed pipe should be plugged as soon as it is disconnected; a clean screw or bolt of the appropriate thread diameter is ideal for this purpose. Slacken and remove both of the cross-head retaining screws and lift the pump unit clear of its drive shaft end. Discard the pump base gasket and replace it with a new item. Note the condition of the spring washers fitted beneath the heads of the pump retaining screws and renew them if they are flattened. On TS185 models, ensure that the keyed drive attachment which connects the pump to the kickstart drive pinion assembly remains in its location. On TS250 models, note the fitted position of the pump to

crankcase spacer washer.
6   To fit the replacement pump unit, clean both the pump and crankcase mating surfaces, place the new gasket onto the pump mating surface and align the central driven spigot of the pump with its drive slot before fitting the pump into its crankcase housing. With the pump correctly seated, fit and tighten its two retaining screws. Note that each one of these screws must have a serviceable spring washer located beneath its head.
7   Unplug and reconnect the tank to pump feed pipe. Ensure that this pipe is a good push fit on its retaining stub and, where applicable, is securely held by its spring clip. Check that all pipes are routed correctly and are neither twisted nor crimped between any component parts. The oil pump must now be bled of air by following the instructions listed in the following Section of this Chapter.
8   With the oil pump thus primed with oil, reconnect the pump control cable to the nylon holder in the pump lever end and proceed to check the oil pump for correct adjustment in accordance with the following instructions.
9   Remove the single screw with sealing washer from the wall of the carburettor mixing chamber. Check the condition of the sealing washer and renew it if necessary. On TS185 and 250 models, this screw is located on the left-hand side of the carburettor, which makes a sighting comparison between the throttle valve and the oil pump relatively easy. Unfortunately, this is not so with TS100 and 125 models, because the screw is located on the right-hand side of the carburettor and therefore faces away from the oil pump. In this instance, it is best to recruit the help of an assistant to sight the position of the throttle valve whilst the oil pump control cable is adjusted.
10   Rotate the throttle twistgrip until the circular indicator mark on the side of the throttle valve comes into alignment with the upper edge of the screw hole in the wall of the mixing chamber (see accompanying figure). With the throttle set in this position, check that the mark scribed on the pump lever boss is in exact alignment with the mark cast in the pump body. If this is not the case, then the marks should be made to align by rotating the control cable adjuster, after having released the locknut. On completion of the adjustment procedure, retighten the locknut whilst holding the cable adjuster in position and slide the rubber sealing cap down the cable to cover the adjuster. It should be noted that any adjustment of the oil pump control cable may well affect the adjustment of the throttle cable. It is, therefore, necessary to refer to the Routine Maintenance Chapter of this Manual and check the throttle cable for correct adjustment before proceeding further. Refit and tighten the screw, with its sealing washer, to the carburettor.
11   Refit the oil pump cover plate in position and fit and tighten its two retaining screws. Loop the final drive chain around the gearbox sprocket and reconnect its two ends with the split link. It is most important that the spring clip of this link is correctly fitted with the closed end facing the normal direction of chain travel. It is quite possible that the rear wheel will have to be moved forward in order to place enough slack on the chain to allow insertion of the link. In either case, the chain must be checked for correct tension and adjusted accordingly by referring to the instructions given in Chapter 5 of this Manual.
12   Wipe clean the mating surfaces of the gearbox sprocket cover and the crankcase and refit the cover. Tighten the cover retaining screws evenly and in a diagonal sequence in order to lessen the risk of the cover becoming distorted. On TS250 models, check the condition of the spring washers fitted beneath the heads of the cover retaining screws. If any one of these washers is flattened then it should be renewed.
13   The gearchange lever can now be slid into position on the gearchange shaft. Ensure that the lever is positioned correctly in relation to the footrest and that its bolt hole is aligned with the channel in the shaft spline before inserting and tightening its retaining bolt.
14   Finally, check that the oil tank is topped-up with oil of the specified type before starting the engine and allowing it to idle for a few minutes.

15.3 Split the final drive chain to allow removal of the oil pump cover plate and the pump itself

15.5 The keyed drive attachment must remain with the pump (185 only)

15.9a Remove the screw from the wall of the carburettor mixing chamber ...

15.9b ... to allow sighting of the throttle valve alignment mark (185 shown)

15.10 Check for alignment of the oil pump lever with the mark cast in the pump body

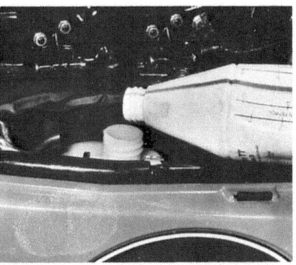

15.14 The oil tank must be topped-up with oil of the specified type

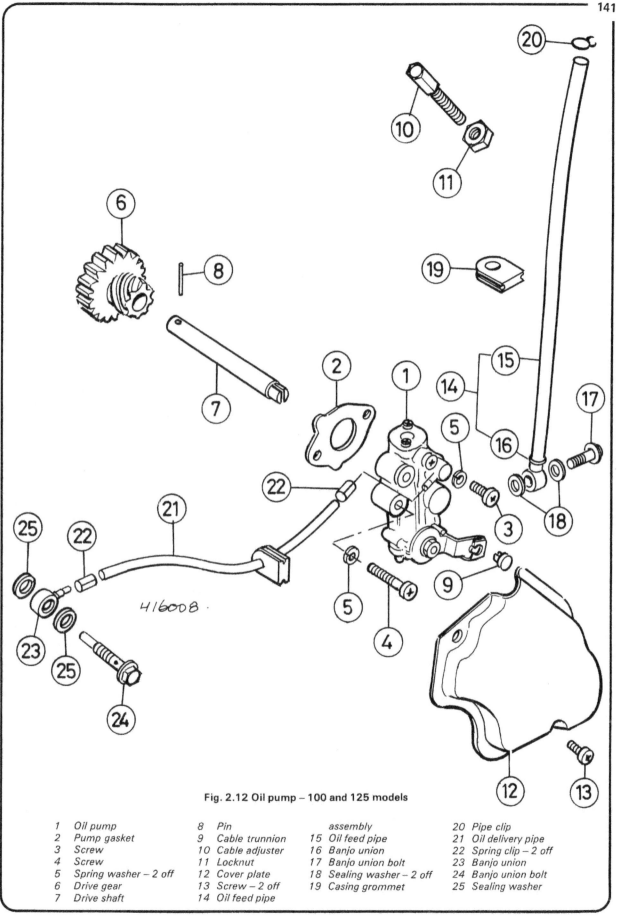

Fig. 2.12 Oil pump – 100 and 125 models

1	Oil pump	8	Pin		assembly	20	Pipe clip
2	Pump gasket	9	Cable trunnion	15	Oil feed pipe	21	Oil delivery pipe
3	Screw	10	Cable adjuster	16	Banjo union	22	Spring clip – 2 off
4	Screw	11	Locknut	17	Banjo union bolt	23	Banjo union
5	Spring washer – 2 off	12	Cover plate	18	Sealing washer – 2 off	24	Banjo union bolt
6	Drive gear	13	Screw – 2 off	19	Casing grommet	25	Sealing washer
7	Drive shaft	14	Oil feed pipe				

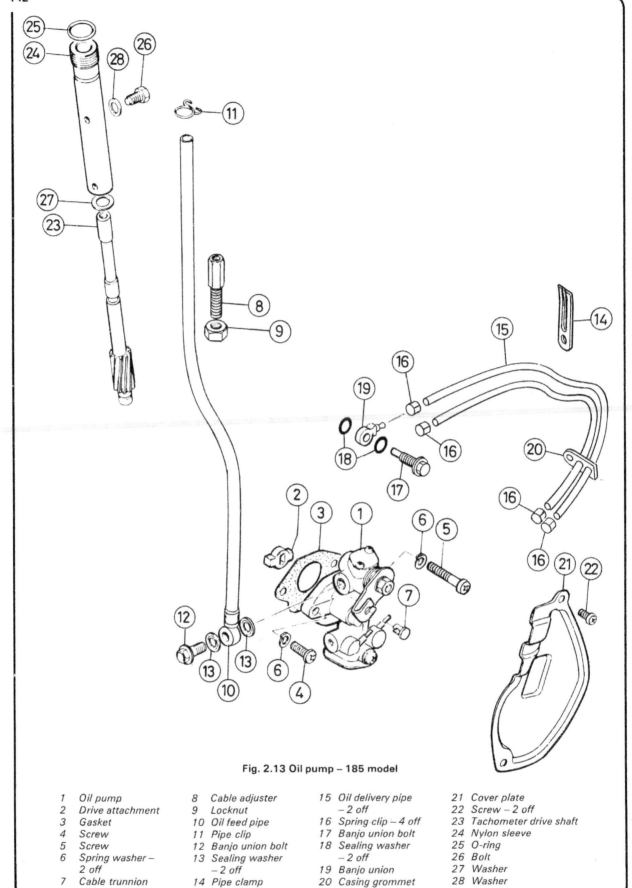

Fig. 2.13 Oil pump – 185 model

1	Oil pump	8	Cable adjuster	15	Oil delivery pipe – 2 off	21	Cover plate
2	Drive attachment	9	Locknut			22	Screw – 2 off
3	Gasket	10	Oil feed pipe	16	Spring clip – 4 off	23	Tachometer drive shaft
4	Screw	11	Pipe clip	17	Banjo union bolt	24	Nylon sleeve
5	Screw	12	Banjo union bolt	18	Sealing washer – 2 off	25	O-ring
6	Spring washer – 2 off	13	Sealing washer – 2 off	19	Banjo union	26	Bolt
7	Cable trunnion	14	Pipe clamp	20	Casing grommet	27	Washer
						28	Washer

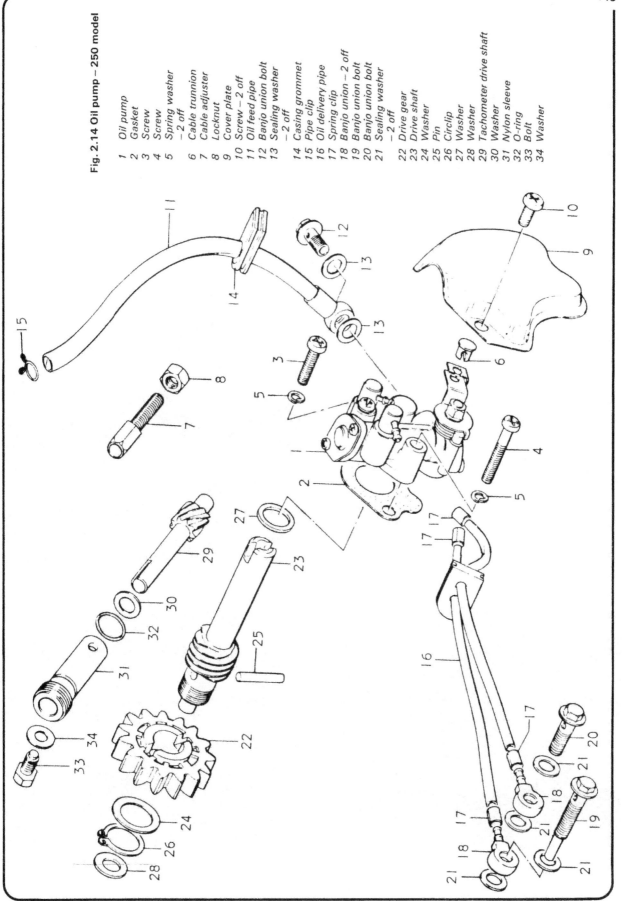

Fig. 2.14 Oil pump – 250 model

1 Oil pump
2 Gasket
3 Screw
4 Screw
5 Spring washer
   – 2 off
6 Cable trunnion
7 Cable adjuster
8 Locknut
9 Cover plate
10 Screw – 2 off
11 Oil feed pipe
12 Banjo union bolt
13 Sealing washer
   – 2 off
14 Casing grommet
15 Pipe clip
16 Oil delivery pipe
17 Spring clip
18 Banjo union – 2 off
19 Banjo union bolt
20 Banjo union bolt
21 Sealing washer
   – 2 off
22 Drive gear
23 Drive shaft
24 Washer
25 Pin
26 Circlip
27 Washer
28 Washer
29 Tachometer drive shaft
30 Washer
31 Nylon sleeve
32 O-ring
33 Bolt
34 Washer

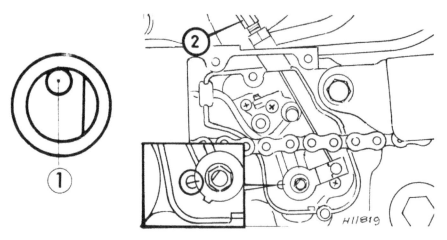

Fig. 2.15 Oil pump cable adjustment marks

1   Throttle valve punch   2   Cable adjuster
    mark

16 Oil pump: bleeding of air

1   If the oil pump has been removed, or the oil system drained, then it is most important to bleed any air from the system before the engine is started. Failure to do this will result in the engine seizing, with the resulting expense of a complete engine rebuild and the potential danger to the rider of the machine should the rear wheel lock unexpectedly whilst the machine is in motion.

2   Bleeding of the oil pump and the oil feed pipe between the pump and oil tank is effected by loosening the cross-headed screw on the side of the pump body and waiting for a steady stream of air-free oil to emerge before retightening the screw. It is also necessary to prime the oil feed pipe(s) running from the pump to the base of the cylinder barrel. To do this, use a syringe or similar container full of clean engine oil to inject oil into the pipe until it is full. Prevent any loss of oil from the pipe by pinching its end between two fingers and then push it over its retaining stub. Ensure that the pipe is retained in position by its spring clip. The priming operation is now complete.

16.2a Commence bleeding of the oil pump and its feed pipe by loosening the screw (arrowed) ...

16.2b ... with the bleed screw tightened, prime the pump to cylinder barrel feed pipe with oil ...

16.2c ... and then reattach each pipe in position with its spring clip

17 Gearbox lubrication: general maintenance

1   General maintenance for the gearbox lubrication system consists solely of checking the level of oil within the gearbox casing at frequent intervals and changing the oil at the intervals recommended in the Routine Maintenance Chapter of this Manual. Carrying out these two service procedures will preclude any risk of the gearbox components becoming starved of oil or having to run in oil that has deteriorated to the point where it is ineffective in its prime functions as a lubricating medium.
2   Full details of checking the oil level and of carrying out an oil change can be found in Routine Maintenance at the front of this Manual.

18 Exhaust system: decarbonisation

1   After a set mileage has been covered, it will be necessary to decarbonise the complete exhaust system. TS100, 125 and 185 models have a detachable silencer baffle fitted into the tail end of their exhaust systems and it is this component that is most likely to require attention as it will tend to become choked if not cleaned at frequent intervals. A two-stroke engine is very susceptible to this fault which is caused by the oily nature of the exhaust gases. As the sludge builds up back pressure will increase with a resulting fall off in performance.
2   It is not necessary to remove the exhaust system in order to gain access to this baffle. It is retained in the end of the silencer by a screw with lock washer. With this screw removed, grip the end of the baffle with a pair of pliers or mole wrench and withdraw it from position. It will be found that, more often than not, both the baffle retaining screw and the baffle itself will be stuck fast in position due to the build-up of carbon in the silencer assembly. In practice, the most satisfactory way found of freeing these components was to use a close-fitting socket on the head of the screw and a mole wrench, in conjunction with a tommy bar and hammer, on the end of the baffle. The baffle proved to be by far the most stubborn of the two components to free and was eventually pulled clear of the silencer by passing the tommy bar through the fitted mole wrench and striking the bar with the hammer, to shock the baffle from position.
3   If the build up of carbon and oil on the baffle is not too great, a wash with a petrol/paraffin mix will suffice to clean the component. Remember to take the usual precautions against fire when employing this cleaning method. When removing this component for the first time, it will be found that the rearmost end of the baffle is wrapped in an asbestos tape. This tape is not essential to the efficient operation of the silencer and should be cut away from the baffle and discarded.
4   If the build up of carbon and oil on the baffle is heavy, then more drastic action will be needed to clear away the accumulated deposits. It should be noted that a heavily con-taminated baffle may well indicate the presence of similar contamination within the silencer casing and exhaust pipe assembly, in which case the complete exhaust system should be removed from the machine for inspection and cleaning as described in the following paragraphs of this Section. The most efficient method of removing heavy contamination from the baffle is to burn away the deposits by running the flame of·a blowlamp or welding torch along the length of the baffle. On completion of this burning process, and with the baffle fully cooled, tap the baffle along its length with a piece of hard wood to dislodge any remaining deposits of carbon before giving it a final clean with a wire-bristled brush. Check that all the baffle holes are unobstructed.
5   To clean the silencer casing and exhaust pipe assembly, remove the system from the machine and suspend it from its silencer end. Block up the end of the exhaust pipe with a cork or wooden bung. If wood is used, allow an outside projection of three or four inches with which to grasp the bung for removal.

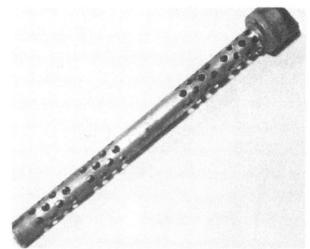

18.4 Clean the exhaust baffle of all carbon deposits

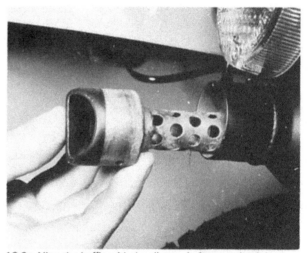

18.9a Align the baffle with the silencer before tapping it into position ...

18.9b ... and securing the baffle with the single screw and spring washer

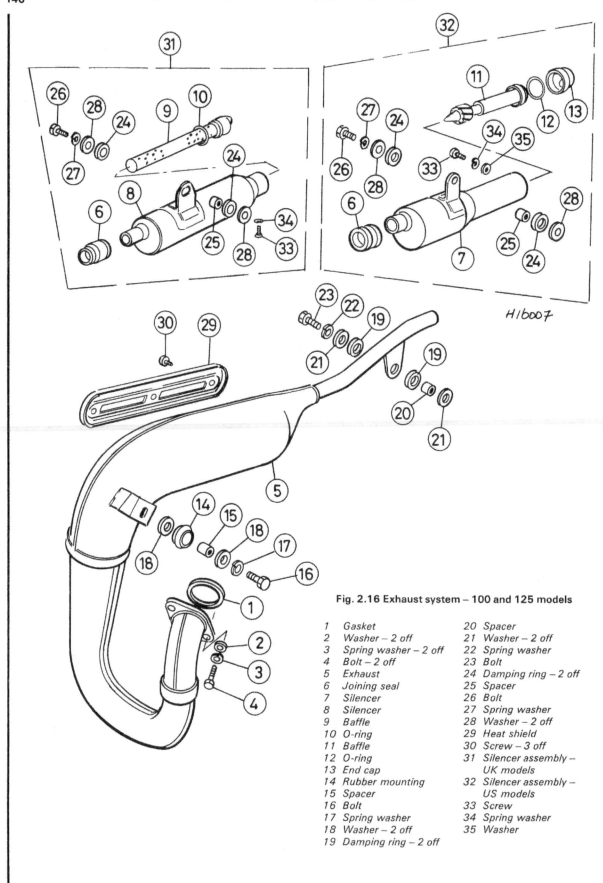

H16007

Fig. 2.16 Exhaust system – 100 and 125 models

1	Gasket	20	Spacer
2	Washer – 2 off	21	Washer – 2 off
3	Spring washer – 2 off	22	Spring washer
4	Bolt – 2 off	23	Bolt
5	Exhaust	24	Damping ring – 2 off
6	Joining seal	25	Spacer
7	Silencer	26	Bolt
8	Silencer	27	Spring washer
9	Baffle	28	Washer – 2 off
10	O-ring	29	Heat shield
11	Baffle	30	Screw – 3 off
12	O-ring	31	Silencer assembly – UK models
13	End cap	32	Silencer assembly – US models
14	Rubber mounting	33	Screw
15	Spacer	34	Spring washer
16	Bolt	35	Washer
17	Spring washer		
18	Washer – 2 off		
19	Damping ring – 2 off		

1   Gasket
2   Front clamp
3   Ring
4   Bolt – 2 off
5   Spring washer – 2 off
6   Retaining spring
7   Exhaust
8   Silencer assembly –
    US models
9   O-ring
10 Baffle
11 End cap
12 Silencer assembly –
    UK models
13 Bolt
14 Washer
15 Spacer
16 Mounting rubber
17 Spring washer
18 Washer
19 Spring washer
20 Bolt
21 Joining seal
22 Washer – 2 off
23 Spacer
24 Damping ring
25 Bolt
26 Spring washer
27 Washer
28 Heat shield
29 Spring washer – 3 off
30 Screw – 3 off
31 Screw
32 Spring washer
33 Washer
34 Clamp
35 Screw
36 Sealing washer
37 Drain plug – 185
    model only

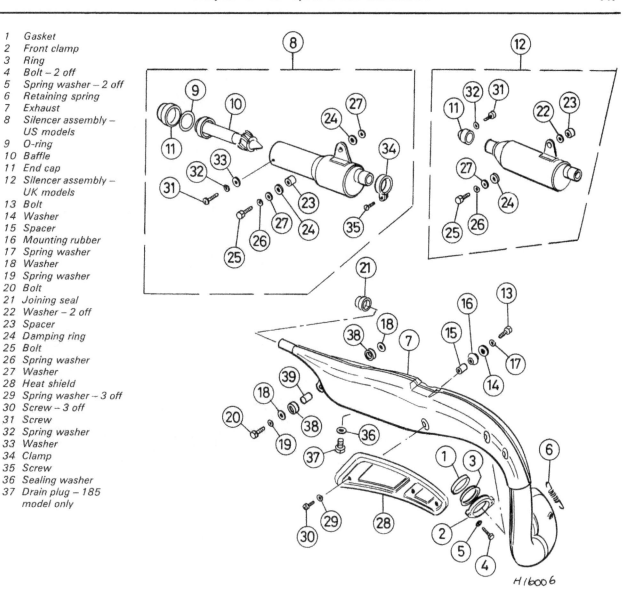

Fig. 2.17 Exhaust system – 185 and 250 models

The system can of course be separated into its two component parts for cleaning by undoing the seal retaining clamp(s) (where fitted) and pulling the silencer away from the end of the pipe assembly.

6   The method used to remove heavy carbon deposits from within the system is to dissolve them by using a chemical solution. Caustic soda dissolved in water is the solution most usually utilised as it is highly effective and relatively cheap. The mixture used is a ratio of 3 lbs caustic soda to over a gallon of fresh water. This is the strongest solution ever likely to be required. Obviously the weaker the mixture the longer the time required for the carbon to be dissolved. Note, whilst mixing the solution, that the caustic soda should be added to the water gradually, whilst stirring. Never pour water into a container of caustic soda powder or crystals; this will cause a violent reaction to take place which will result in great danger to one's person.

7   Bear in mind that it is very important to take great care when using caustic soda as it is a very dangerous chemical. Always wear protective clothing, this must include proper eye protection. If the solution does come into contact with the eyes,

or skin it must be washed clear immediately with clean, fresh running water. In the case of an eye becoming contaminated, seek expert medical advice immediately. Also, the solution must not be allowed to come into contact with aluminium alloy – especially at the above recommended strength – caustic soda reacts violently with aluminium and will cause severe damage to the component.

8   Commence the cleaning operation by pouring the solution into the system until it is quite full. Do not plug the open end of the system. The solution should now be left overnight for its dissolving action to take place. Note that the solution will continue to give off noxious fumes throughout its dissolving process; the system must therefore be placed in a well ventilated area. After the required time has passed, carefully pour out the solution and flush the system through with clean, fresh water. The cleaning operation is now complete.

9   Note the information contained in the following Section of this Chapter before refitting the exhaust system to the machine. When refitting the baffle in the silencer, ensure that the threaded hole in the baffle is aligned with the corresponding hole in the silencer before tapping the baffle into place. Check

the condition of, and if necessary renew, the spring washer fitted beneath the head of the baffle retaining screw before fitting and tightening the screw. It is a good idea to prevent further seizure of this screw by lightly smearing its threads with graphite grease before insertion. Note that if this screw is not fitted or tightened properly and, as a result of this falls out, then the baffle will work loose, creating excessive exhaust noise accompanied by a marked fall off in performance.

10   Do not run the machine without the baffles in the silencer or modify the baffles in any way. Although the changed exhaust note may give the illusion of greater power, the chances are that the performance will fall off, accompanied by a noticeable lack of acceleration. The carburettor is jetted to take into account the fitting of silencers of a certain design and if this balance is disturbed the carburation will suffer accordingly.

11   Finally, it should be noted that the type of system fitted to TS185 models has a drain point incorporated at the base of the expansion chamber. Access to this point can be gained by removal of the right-hand sidepanel from the machine. It is recommended that the drain plug and its sealing washer be removed at frequent intervals to allow any moisture contained in the expansion chamber to drain away. Carrying out this simple procedure will greatly reduce the likelihood of the system becoming corroded from within by the acids contained in the moisture. Note that the sealing washer must be in good condition at all times to prevent the escape of exhaust gases whilst the machine is running.

18.11 A drain plug may be incorporated at the base of the expansion chamber (185 shown)

19   Exhaust system: removal and fitting

1   The exhaust system can be removed from the machine as a complete assembly after full access has been gained to its frame attachment points. This will require removal of the right-hand sidepanel. If working on TS185 and 250 models, access to the middle mounting point of the pipe assembly will be made easier by removal of the left-hand sidepanel also.

2   Reference to the figures contained in this Chapter will show that the design of the exhaust system fitted to TS100 and 125 models differs from that of the system fitted to TS185 and 250 models. Nevertheless, the same basic procedure for removal, and the precautions involved, apply equally to each system type.

3   Commence removal of the exhaust system by unscrewing the two bolts which serve to retain the exhaust pipe to the cylinder head. If working on a TS185 or 250 model, unhook the pipe retaining spring by using a pair of pliers. Note the condition of the spring washer fitted beneath the head of each retaining bolt and, if it is flattened or broken, replace it wth a new item.

4   Where necessary, unclip the clutch cable from its guide on the exhaust system. It is now necessary to move along the complete system, undoing each mounting bolt whilst doing so and making a careful note of the fitted position of each washer, spacer and mounting rubber. With all the mounting bolts removed, the system can be lifted away from the machine. Note that under no circumstances should the system be allowed to hang unsupported from the cylinder barrel mounting as this will impose an unacceptable strain on the threads of the two mounting bolts.

5   With the exhaust system removed from the machine, take this opportunity to carefully inspect the system itself and all of its mounting components. Slight damage to the system, such as small cracks or rust holes, can be repaired by brazing a patch over the affected area. The system is finished with a matt black heat-dispersant finish which will flake off in use. This can be restored by wire-brushing and degreasing the system, and spraying it with a suitable aerosol paint finish. It follows that this should be of a type suitable for use on exhaust systems, such as Sperex VHT or similar.

6   If the system is found to be cracked, especially around its mounting points, then suspect the mounting rubbers. These

19.7a Before fitting the exhaust system, fit a new sealing ring into the recess of the cylinder barrel

19.7b Inspect the spacer ring located in the exhaust pipe flange (125 shown)

19.7c The exhaust pipe spacer ring must be correctly fitted (185 shown)

19.7d Renew all damaged system mounting components before attaching the system to the machine

19.7e All exhaust seals must be in good condition and correctly fitted

19.7f Tighten the exhaust pipe to cylinder barrel securing bolts to the correct torque loading

19.7g Fit a serviceable pipe retaining spring (185 and 250)

19.7h Route the clutch cable clear of the exhaust system (125 shown)

rubbers serve to isolate the system from engine vibration passed through the frame structure; if they are perished, have become hardened or have simply disintegrated then they are no longer effectively carrying out this function and must be renewed. Renew any spacers or washers that have also become damaged.

7   Refitting of the exhaust system is a straightforward reversal of the removal procedure. If in doubt as to the fitted position of any one component part of the system, then refer to the figure of the system type which accompanies this text. Do not tighten any one fixing bolt until the complete system is fitted and tighten the exhaust pipe to cylinder barrel securing bolts to the recommended torque loading. Remember to renew any spring washers that have become flattened during use and also to fit a new sealing ring into the recess provided in the cylinder barrel.

Before fitting the pipe retaining spring to TS185 and 250 models, check that it has not become fatigued or weakened through corrosion. Renew the spring if thought necessary. Check also that the clutch operating cable is routed so as not to contact the exhaust system or cylinder assembly.

8   After-market exhaust systems where available, must be treated with a degree of caution. Although they can effect some improvement at particular engine speeds, there is inevitably a drop in performance elsewhere in the engine speed range. It is essential to ensure that the system is of well-known and reputable manufacture. It should be noted that some adjustment of the carburettor jetting will probably be required. This should be checked with the supplier when the system is purchased.

Chapter 3 Ignition system

For modifications and information relating to later models, see Chapter 7

Contents

General description 1
Contact breaker points: adjustment – contact
breaker models 2
Ignition timing (static): checking and resetting –
contact breaker models 3
Ignition timing (dynamic): checking and resetting 4
Contact breaker assembly: removal, renovation and
refitting – contact breaker models 5
Condenser: testing, removal and fitting –
contact breaker models 6

Pulser coil: testing, removal and fitting – PEI TS100
and 125 models 7
CDI unit: removal, testing and refitting – TS185
and 250 models 8
Flywheel generator: checking the output 9
Ignition coil: location and testing 10
Spark plug: cleaning and resetting the gap 11
High tension lead: examination 12

Specifications

Ignition system
Type:
    UK TS100 and 125 models .... Coil and contact breaker
    All other models .... Suzuki PEI Electronic

Ignition timing
    UK TS100 and 125 models .... $20° \pm 2°$ BTDC (1.90 mm/0.075 in piston position)
    US TS100 and 125 models .... $20°$ BTDC at 6000 rpm
    TS185 models .... $21.5°$ BTDC at 6000 rpm
    TS250 models .... $24°$ BTDC at 6000 rpm

Contact breaker
Gap:
    All contact breaker models .... 0.3 – 0.4 mm (0.012 – 0.016 in)

Ignition coil
Winding resistance:

	Primary	Secondary
All UK TS100 and 125 models	0.75 ohm	5.7 K ohm
US TS100 and 125 models	See table in Section 10	
UK TS185 N model	0.4 – 1.0 ohm	9.0 – 15.0 K ohm
TS185 models	0 – 1.0 ohm	20.0 – 21.0 K ohm
TS250 models	0 – 1.0 ohm	20.0 – 25.0 K ohm

Condensor
Capacity:
    All contact breaker models .... 0.16 – 0.20 microfarad

Spark plug
Make .... NGK or ND
Type:
    TS100 models US
    TS125 models UK TS
    125 T and X models .... B8ES or W24ES
    UK TS125 N models .... BP8ES or W24EP
    TS185 and 250 models .... BP7ES or W22EP
Gap .... 0.6 – 0.8 mm (0.024 – 0.031 in)

Flywheel generator
Ignition source coil resistance:
    All contact breaker models .... 2.3 ohm
Pulser coil/Exciter coil resistances:
    TS100 and 125 PEI models .... 25 – 30 ohm/1.5 – 2.0 ohm
L_1/L_2 coil resistances:
    TS185 X models .... 160 – 240 ohm/30 – 40 ohm
    TS250 X models .... 1.0 – 2.0 ohm/120 – 160 ohm

1 General description

Three types of ignition are featured in the text of this Chapter. They are the conventional type of contact breaker system and two versions of Suzuki's own pointless electronic ignition (PEI) system. Reference to the Specifications Section of this Chapter will show which machines have the PEI system fitted. Further to this, it should be noted that the PEI system fitted to those TS100 and 125 models that are not equipped with the contact braker system differs appreciably from the PEI system fitted to TS185 and 250 models.

The conventional contact breaker system functions as follows. As the flywheel generator rotor operates, alternating current (ac) is generated in the ignition source coil mounted on the generator stator. Because the contact breaker is closed, the power runs to earth. When the contact breaker points open, the current is transferred to the primary windings of the ignition coil. In doing this, a high voltage is produced in the ignition coil secondary windings (by means of mutual induction) which is fed to the spark plug via the HT lead. As the energy flows to earth across the spark plug electrodes, a spark is produced and the combustible gases in the cylinder ignited. A capacitor (condenser) is fitted into the system to prevent arcing across the points; this helps reduce erosion due to burning.

With the type of PEI system fitted to TS100 and 125 models, Suzuki has acted to replace the mechanical switch of the contact breaker system with an electronic switch which is known as a thyristor (silicone controlled rectifier – SCR).

The system operates as follows. As the flywheel generator begins to rotate, alternating current (ac) is generated in the exciter coil. This current is fed to the base of the transistor in the system via the primary windings of the ignition coil and through the system control unit, so that it reaches the transistor as a small bleed current. The current then acts to hold the transistor, which should be regarded as a switch, in its closed position, thereby allowing the inducing voltage from the exciter coil to flow through the transistor and direct to earth.

Alternating current generated by the pulser coil is fed direct to the thyristor. The alternating wave forms of this current open and close the thyristor. As the thyristor opens, the current flowing through the primary windings of the ignition coil takes the line of least resistance and passes through the thyristor direct to earth instead of passing through the control unit and to the base of the transistor. Thus the small bleed current holding the transistor in its closed position is cut which allows the transistor to open. This has the effect of increasing the supply of current to the primary windings of the ignition coil which in turn induces a high tension pulse in the secondary windings of the coil, the end result being a spark across the electrodes of the spark plug. Closing of the thyristor will return the current to the base of the transistor, thereby returning the supply of current being fed to the primary winding of the ignition coil to its previous rating. The output of the pulser coil will vary slightly with engine speed and because of this, the ignition timing will advance as the engine speed rises.

It should be noted that the control unit of the PEI system fitted to TS100 and 125 models is incorporated in the ignition coil assembly, whereas the control (CDI) unit fitted to TS185 and 250 models is a separate item which takes the form of a clearly labelled box. The PEI system fitted to TS185 and 250 models functions as follows.

There are two coils incorporated in the ignition side of the flywheel generator assembly, both of which supply a low tension alternating current (ac) to the frame mounted CDI unit. The alternating wave forms of this current control the timing of the spark to the plug electrodes by the following method. As the flywheel generator rotor passes through one half of a revolution, two pulses of energy are supplied from the coils (L_1 and L_2). One pulse is positive and it is this pulse which charges the capacitor on the CDI unit. The other pulse is negative and it is this pulse which passes to the timing circuit within the CDI unit

and which triggers the thyristor (silicone controlled rectifier – SCR) into a conductive state. As the thyristor becomes conductive, that is open, the capacitor discharges instantly through it and into the primary windings of the ignition coil which in turn induces a high tension pulse in the secondary windings of the coil, the end result being a spark across the electrodes of the spark plug. This spark will occur twice during each complete revolution of the crankshaft.

The two coil arrangement in the ignition side of the flywheel generator assembly also provides a method of progressively advancing the ignition timing as the engine speed rises. With the engine running in its low speed range, the coil L_1, supplies the greater part of the indicating voltage to the capacitor. As the engine speed rises into its high speed range, the supply from the coil L_1, falls and the coil L_2 then begins to supply the greater part of the voltage required.

Finally, there are several advantages that both types of PEI system hold over the traditional contact breaker points system. The most important advantage is that it removes the majority of mechanical components from the system. Because there are no contact breakers to wear, the owner is freed from the task of periodically adjusting or renewing them. Once the transistorised system has been set up, it need not be attended to unless it has been disturbed in the course of dismantling or failure occurs somewhere in the system. Other advantages of the transistorised ignition system include a greater resistance to the effects of vibration, dirt and moisture, and a constantly 'strong' spark at exactly the correct moment, with no wastage of electrical energy due to arcing etc.

2 Contact breaker points: adjustment – contact breaker models

1 Access to the contact breaker assembly can be gained by removing the cover from the left-hand crankcase. To avoid distorting the cover loosen its four retaining screws evenly and in a diagonal sequence and take care to avoid tearing its gasket as the cover is lifted from position. This will reveal the flywheel generator, which has four elongated holes in its outer face. The larger two of these holes are provided to permit inspection and adjustment of the contact breaker points.

2 Using the flat of a small screwdriver, open the contact breaker points against the pressure of the return spring so that the condition of the point contact faces may be checked. A piece of stiff card or crokus paper may be used to remove any light surface deposits, but if burnt or pitted, the complete contact breaker assembly must be removed to facilitate further examination of the point contact faces, and if necessary, renewal of the assembly. Refer to Section 5 of this Chapter for

2.3 Use a feeler gauge to measure the contact breaker gap

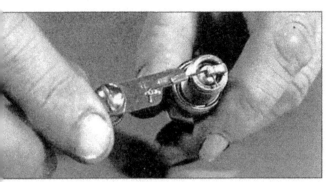

Electrode gap check - use a wire type gauge for best results

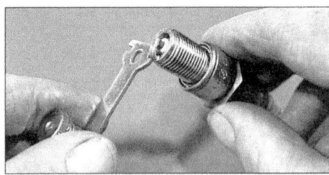

Electrode gap adjustment - bend the side electrode using the correct tool

Normal condition - A brown, tan or grey firing end indicates that the engine is in good condition and that the plug type is correct

Ash deposits - Light brown deposits encrusted on the electrodes and insulator, leading to misfire and hesitation. Caused by excessive amounts of oil in the combustion chamber or poor quality fuel/oil

Carbon fouling - Dry, black sooty deposits leading to misfire and weak spark. Caused by an over-rich fuel/air mixture, faulty choke operation or blocked air filter

Oil fouling - Wet oily deposits leading to misfire and weak spark. Caused by oil leakage past piston rings or valve guides (4-stroke engine), or excess lubricant (2-stroke engine)

Overheating - A blistered white insulator and glazed electrodes. Caused by ignition system fault, incorrect fuel, or cooling system fault

Worn plug - Worn electrodes will cause poor starting in damp or cold weather and will also waste fuel

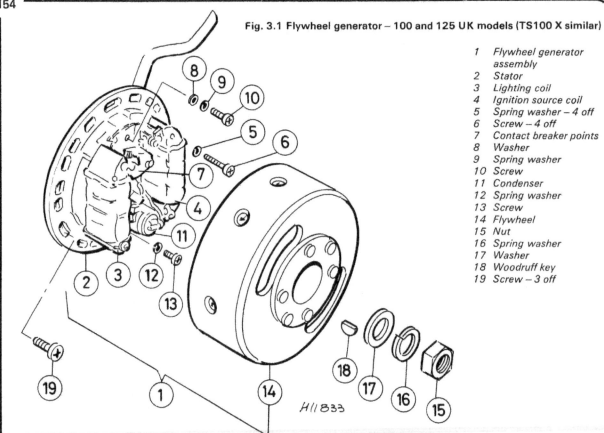

Fig. 3.1 Flywheel generator – 100 and 125 UK models (TS100 X similar)

1 Flywheel generator assembly
2 Stator
3 Lighting coil
4 Ignition source coil
5 Spring washer – 4 off
6 Screw – 4 off
7 Contact breaker points
8 Washer
9 Spring washer
10 Screw
11 Condenser
12 Spring washer
13 Screw
14 Flywheel
15 Nut
16 Spring washer
17 Washer
18 Woodruff key
19 Screw – 3 off

H11833

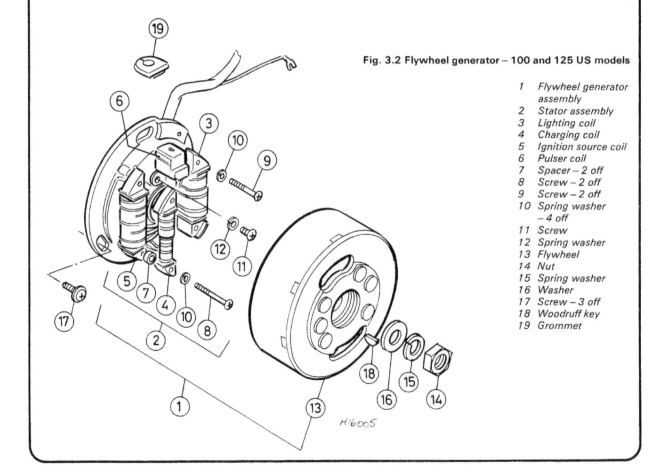

Fig. 3.2 Flywheel generator – 100 and 125 US models

1 Flywheel generator assembly
2 Stator assembly
3 Lighting coil
4 Charging coil
5 Ignition source coil
6 Pulser coil
7 Spacer – 2 off
8 Screw – 2 off
9 Screw – 2 off
10 Spring washer – 4 off
11 Screw
12 Spring washer
13 Flywheel
14 Nut
15 Spring washer
16 Washer
17 Screw – 3 off
18 Woodruff key
19 Grommet

H16005

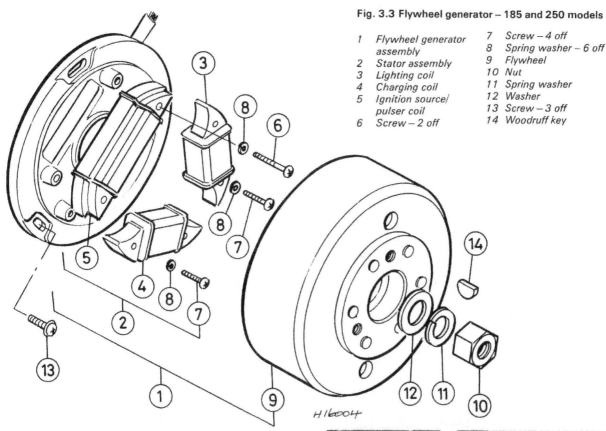

Fig. 3.3 Flywheel generator – 185 and 250 models

1   Flywheel generator assembly
2   Stator assembly
3   Lighting coil
4   Charging coil
5   Ignition source/pulser coil
6   Screw – 2 off
7   Screw – 4 off
8   Spring washer – 6 off
9   Flywheel
10 Nut
11 Spring washer
12 Washer
13 Screw – 3 off
14 Woodruff key

H16004

full details on removal, renovation and fitting of the assembly. If the points are found to be in sound condition, then proceed with adjustment as follows.

3   Rotate the crankshaft slowly until the contact breaker points are seen to be in their fully open position. Measure the gap between the points with the feeler gauge. If the gap is correct, a gauge of 0.35 mm (0.014 in) thickness will be a light sliding fit between the point faces. If this is not the case, slacken the single crosshead screw which serves to retain the fixed contact in position, just enough to allow movement of the contact. With the flat of a screwdriver placed in the indentation provided in the edge of the fixed contact plate, move the plate in the appropriate direction until the gap is correct. Retighten the retaining screw and recheck the gap setting; it is not unknown for this setting to alter slightly upon retightening of the screw.

4   Prior to refitting the crankcase cover, apply one or two drops of light machine oil to the cam lubricating wick whilst taking care not to allow excess oil to foul the point contact surfaces. Renew the cover sealing gasket if it is in any way damaged.

3   Ignition timing (static): checking and resetting – contact breaker models

1   It cannot be overstressed that optimum performance of the engine depends on the accuracy with which the ignition timing is set. Even a small error can cause a marked reduction in performance and the possibility of engine damage as the result of overheating.

2   Static timing of the ignition can be carried out simply by aligning the timing mark scribed on the wall of the flywheel generator rotor with the corresponding mark on the crankcase and then checking that the contact breaker points are just on the point of separation.

3.2 Align the mark on the rotor wall with that on the crankcase (125 shown)

3   Before commencing a check of the ignition timing, refer to the previous Section of this Chapter and check that the contact breaker points are both clean and correctly gapped. In order to provide an accurate indication as to when the contact breaker points begin to separate, it will be necessary to obtain certain items of electrical equipment. This equipment may take the form of a multimeter or ohmmeter, or a high wattage 6 volt bulb, complete with three lengths of electrical lead, which will be used in conjunction with the battery of the motorcycle.

4   When carrying out the above described method of static timing and the method of dynamic timing described in the following Section of this Chapter, note that the accuracy of both methods of timing depends very much on whether the flywheel

generator rotor is set correctly on the crankshaft. Any amount of wear between the keyways in both the crankshaft and rotor bore and the Woodruff key will cause some amount of variation between the timing marks which will, in turn, lead to inaccurate timing. Inaccuracy in the timing mark position may also be a result of manufacturing error. The only means of overcoming this is to remove any movement between the two components and then to set the piston at a certain position within the cylinder bore before checking that the timing marks have remained in correct alignment. In order to accurately position the piston, it will be necessary to remove the spark plug and replace it with a dial gauge or a slide gauge either one of which is adapted to fit into the spark plug hole of the cylinder head.

5   Position the piston in the cylinder bore by first rotating the crankshaft until the piston is set in the top dead centre (TDC) position. Set the gauge at zero on that position and then rotate the crankshaft backwards (clockwise) until the piston has passed down the cylinder bore a distance of at least 4 mm (0.16 in). Reverse the direction of rotation of the crankshaft until the piston is exactly 1.90 mm (0.075 in) from TDC. The timing marks should now be in exact alignment. If this is not the case, then new marks will have to be made. All subsequent adjustment of the timing may be made using these marks.

6   Commence a check of the ignition timing by tracing the electrical lead from the fixed contact point (colour code, Black/Yellow) to a point where it can be disconnected. To set up a multimeter, set it to its resistance function, connect one of its probes to the lead end and the other probe to a good earthing point on the crankcase. Use a similar method to set up an ohmmeter. In each case, opening of the contact breaker points will be indicated by a deflection of the instrument needle from one reading of resistance to another.

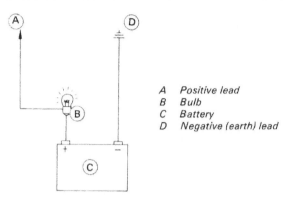

A   Positive lead
B   Bulb
C   Battery
D   Negative (earth) lead

Fig. 3.5 Battery and bulb method of checking the ignition timing – contact breaker models

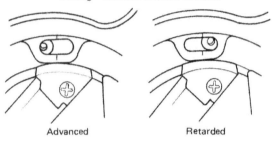

Advanced   Retarded

Fig. 3.6 Ignition timing stator plate positions

If this is not the case, then the points will have to be adjusted by moving the stator plate clockwise or anti-clockwise, depending on whether the opening point needs to be advanced or retarded. This is accomplished by removing the flywheel generator rotor in order to gain full access to the three stator plate retaining screws, each one of which passes through an elongated slot cut in the plate.

9   Removal of the flywheel rotor may be achieved by carrying out the following procedure. Prevent the crankshaft from rotating by selecting top gear and then applying the rear brake. Alternatively, fit a strap wrench around the rotor. Remove the rotor retaining nut, spring washer and plain washer. The rotor is a tapered fit on the crankshaft end and is located by a Woodruff key; it therefore requires pulling from position. The rotor boss is threaded internally to take the special Suzuki service tool No 09930 – 30102 with an attachment, No 09930 – 30161. If this tool cannot be acquired, then it is possible to remove the rotor by careful use of a two-legged puller.

10   If it is decided to make use of a two-legged puller, great care must be taken to ensure that both the threaded end of the crankshaft and the rotor itself remain free from damage. To protect the crankshaft end, refit the rotor retaining nut and screw it on until its outer face lies flush with the end of the crankshaft. Assemble the puller, checking that its feet are fitted through the two larger slots in the rotor face and are resting securely on the strengthened hub. Remember that the closer the feet of the puller are to the centre of the rotor then the less chance there is of the rotor becoming distorted. Gradually tighten the centre bolt of the puller to apply pressure to the rotor. Do not overtighten this bolt but apply reasonable pressure and then stroke the end of the bolt with a hammer to break the taper joint. If this fails at first, tighten the centre bolt to apply a little more pressure and then try shocking the rotor free again.

11   It will be appreciated that resetting of the timing is a repetitive process of loosening the three stator plate retaining screws, rotating the stator plate a small amount in the required direction, retightening the screws, pushing the rotor back onto the crankshaft taper and then rechecking the timing. With the ignition timing correctly set, check tighten the stator plate retaining screws and then refit the rotor by first cleaning and

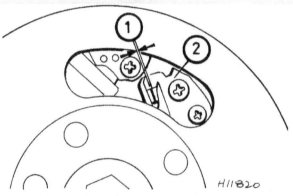

Fig. 3.4 Measuring the contact breaker gap

1   0.3 – 0.4 mm   2   Adjusting plate
    (0.012 – 0.016 in)

7   When using the battery and bulb mentioned, remove the battery from the machine and position it at a convenient point next to the left-hand side of the engine. Connect one end of a wire to the postive (+) terminal of the battery and one end of another wire to the negative (–) terminal of the battery. The negative lead may now be earthed to a point on the engine casing. Ensure the earth point on the casing is clean and that the wire is positively connected; a crocodile clip fastened to the wire is ideal. Take the free end of the positive lead and connect it to the bulb. The third length of wire may now be connected between the bulb and the electrical lead of the fixed contact point. As the final connection is made, the bulb should light with the points closed. Opening of the contact breaker points will be indicated by a dimming of the bulb. The reason for recommending the use of a high voltage bulb is that this dimming will be more obvious to the eye.

8   Ignition timing is correct when the contact breaker points are seen to open just as the timing marks come into alignment.

3.11a Note that Suzuki provide a reference mark for timing which should be aligned with the centre of the stator securing screw

3.11b Degrease the crankshaft taper and fit the Woodruff key ...

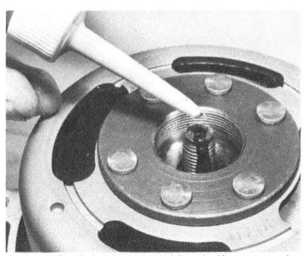

3.11c ... before fitting the rotor, applying a locking compound to the threads and securing the rotor in position

degreasing both the taper of the crankshaft and the bore of the rotor where the two components come into contact. Check that the Woodruff key is correctly fitted in the crankshaft keyway and then push the rotor onto the crankshaft. Gently tap the centre of the rotor with a soft-faced hammer to set it on the crankshaft taper and then fit the plain washer followed by the spring washer. Clean and degrease the threads of the rotor retaining nut and of the crankshaft end. Apply a thread locking compound to these threads and fit and tighten the nut, finger-tight. Lock the crankshaft in position by employing the method used for rotor removal and tighten the rotor retaining nut to the specified torque loading.

12 Finally, on completion of the timing procedure, remove all test equipment from the machine, reconnect any disturbed electrical connections and then refit the rotor cover.

4 Ignition timing (dynamic): checking and resetting

1 On those models equipped with a contact breaker assembly, checking the ignition timing by the following method provides an alternative to the method of static timing described in the previous Section of this Chapter. On those models equipped with a PEI system, this method provides the only means of checking the ignition timing.

2 Ignition timing by the dynamic method is carried out with the engine running whilst using a stroboscopic lamp. This will entail gaining access to the wall of the flywheel generator rotor by removing its cover. When the light from the lamp is aimed at the timing marks on the crankcase and rotor wall, it has the effect of 'freezing' the moving mark on the rotor in one position and thus the accuracy of the timing can be seen. It cannot be overstressed that optimum performances of the engine depends on the accuracy with which the ignition timing is set. Even a small error can cause a marked reduction in performance and the possibility of engine damage as the result of overheating.

4.2 Use a stroboscopic lamp to 'freeze' the moving marks on the rotor (185 shown)

3 Prepare the timing marks by degreasing them and then coating each one with a trace of white paint. This is not absolutely necessary but will make the position of each mark far easier to observe if the light from the 'strobe' is weak or if the timing operation is carried out in bright conditions.

4 Two basic types of stroboscopic lamp are available, namely the neon and xenon tube types. Of the two, the neon type is much cheaper and will usually suffice if used in a shaded position, its light output being rather limited. The brighter but more expensive xenon types are preferable, if funds permit, because they produce a much clearer image.

5   Connect the 'strobe' to the HT lead, following the maker's instructions. If an external 6 volt power source is required, use the battery from the machine but make sure that when the leads from the machine's electrical system are disconnected from the terminals of the battery, they have their ends properly insulated with tape in order to prevent them from shorting on the cycle components.

6   Start the engine and aim the 'strobe' at the timing marks. The models equipped with a PEI system incorporate a means of automatically advancing and retarding the ignition timing throughout the engine speed range. Because of this, it is necessary to bring the ignition timing to its fully advanced position before noting the position of the timing marks. The ignition timing is fully advanced with the engine speed set at 6000 rpm. Models equipped with a contact breaker system have fixed ignition timing which means that the position of the timing marks can be noted at any given engine speed. On all models, ignition timing is correct when the mark on the crankcase wall is seen to be in exact alignment with the mark on the rotor wall (the centre mark where three marks are present).

7   If it is found necessary to adjust the ignition timing, then it is first necessary to remove the flywheel generator rotor in order to gain full access to the three stator plate retaining screws, each one of which passes through an elongated slot cut in the plate. Before commencing this task, take careful note of the information given in paragraphs 4 and 5 of the previous Section of this Chapter. Remove the flywheel rotor as follows.

8   Prevent the crankshaft from rotating by selecting top gear and then applying the rear brake. Alternatively, fit a strap wrench around the rotor. Remove the rotor retaining nut, spring washer and plain washer. The rotor is a tapered fit in the crankshaft end and is located by a Woodruff key; it therefore requires pulling from position. Because the type of rotor fitted varies between the model types covered in this Manual, then the method of rotor removal must also vary.

TS100 and 125 models

9   The rotor fitted to these models has a boss which is threaded internally to take the special Suzuki service tool No 09930-30102 with an attachment. No 09930-30161. If this tool cannot be acquired, then it is possible to remove the rotor by careful use of a two legged puller.

10 If it is decided to make use of a two-legged puller, great care must be taken to ensure that both the threaded end of the crankshaft and the rotor itself remain free from damage. To protect the crankshaft end, refit the rotor retaining nut and screw it on until its outer face lies flush with the end of the crankshaft. Assemble the puller, checking that its feet are lifted through the two larger slots in the rotor face and are resting securely on the strengthened hub. Remember that the closer the feet of the puller are to the centre of the rotor then the less chance there is of the rotor becoming distorted. Gradually tighten the centre bolt of the puller to apply pressure to the rotor. Do not overtighten this bolt but apply reasonable pressure and then strike the end of the bolt with a hammer to break the taper joint. If this fails at first, tighten the centre bolt to apply a little more pressure and then try shocking the rotor free again.

TS185 and 250 models

11 The rotor fitted to these model types has three threaded holes tapped in its strengthened hub section. These holes are positioned so as to accommodate the special Suzuki service tool No 09930-30713, which is the recommended tool with which to pull the rotor from position. In the absence of this tool, an extractor may be made by using the following procedure.

12 Obtain a piece of mild steel plate which is approximately 7 mm ($\frac{1}{4}$ in) thick. Obtain three high tensile steel bolts which are of the correct thread type and diameter to fit into the holes tapped in the rotor. Finally, obtain a suitably sized bolt and nut to fit through the centre of the extractor as shown in Fig. 1.3 of Chapter 1.

13 Referring to the figure, accurately mark the centres of the four holes shown to correspond with the holes in the flywheel rotor. Drill these holes and fit the bolts through the plate to check that their ends align with their locations in the rotor. To finish the job, tack weld the nut to the centre of the plate and grind a taper on the end of the centre bolt so that it locates in the machined centre of the crankshaft.

14 Arrange the extractor on the rotor whilst ensuring that the three bolts are screwed fully into the rotor hub section. Screw down the centre bolt to draw the rotor off its taper. If the rotor refuses to move, do not overtighten the centre bolt but apply reasonable pressure and then strike the head of the bolt with a hammer to break the taper joint. If this fails at first tighten the bolt a little further and try again.

All models

15 It will be appreciated that resetting of the timing is a repetitive process of loosening the three stator plate retaining screws, rotating the stator plate a small amount in the required direction to advance or retard the timing as required, retightening the screws, setting the rotor back onto the crankshaft taper, starting the engine and then rechecking the timing. With the ignition timing correctly set, check tighten the stator plate retaining screws and then refit the rotor by first cleaning and degreasing both the taper of the crankshaft and the bore of the rotor where the two components come into contact. Check that the Woodruff key is correctly fitted in the crankshaft keyway and then push the rotor into the crankshaft. Gently tap the centre of the rotor with a soft-faced hammer to seat it on the crankshaft taper and then fit the plain washer followed by the spring washer. Clean and degrease the threads of the rotor retaining nut and of the crankshaft end. Apply a thread locking compound to these threads and fit and tighten the nut, finger-tight. Lock the crankshaft in position by employing the method used for rotor removal and tighten the rotor retaining nut to the torque loading given in the Specifications Section of Chapter 1.

16 Finally, on completion of the timing procedure, refit the rotor cover, reconnect any disturbed electrical connections and remove all test equipment from the machine.

4.15 Rotate the stator plate a small amount in the required direction to advance or retard the ignition timing

5   Contact breaker assembly: removal, renovation and fitting – contact breaker models

1   If the contact breaker points are found to be burned, pitted or badly worn, they should be removed for dressing. If, however, it is necessary to remove a substantial amount of material before the faces can be restored, new contacts should be fitted.

2   The contact breaker assembly forms part of the flywheel

generator stator plate assembly. To gain full access to this assembly, it is necessary to remove the crankcase covers surrounding the flywheel generator and then to remove the rotor. Full instructions for the removal and fitting of these components are given in the relevant Sections of Chapter 1.

3   The contact breaker assembly is secured to the stator plate by means of a single crosshead screw with a spring washer beneath its head. Before the assembly can be removed, it is first necessary to detach both the condenser and coil leads from the spring blade of the moving point. They are retained by a single crosshead screw with spring washer, which should be relocated immediately after the leads have been detached in order to avoid loss.

4   Using the flat of a small electrical screwdriver, prise off the clip which retains the moving contact assembly to its pivot pin. Remove the plain washer, followed by the moving contact complete with insulating washers. Make a note of the order in which components are removed, as they are easily assembled incorrectly. Release the fixed contact by unscrewing the single retaining screw which passes through the backplate. Do not, under any circumstances, loosen the three stator mounting screws, otherwise the ignition timing will be lost.

5   The points surfaces may be dressed by rubbing them on an oilstone or fine emery paper, keeping the points square to the abrasive surface. If possible, finish off by using crokus paper to give a polished surface, which is less prone to subsequent pitting. Make sure all traces of abrasive are removed before reassembly.

6   Reassemble the contact breaker assembly by reversing the dismantling sequence, taking care that the insulating washers are replaced correctly. If this precaution is not observed, it is easy to inadvertently earth the assembly thus rendering it inoperative. The pivot pin should be greased sparingly, and a few drops of oil applied to the cam lubricating wick.

7   If the contact breaker is being renewed due to excessive burning of the contacts, this is likely to have been caused by a faulty condenser. Refer to Section 6 of this Chapter if this is suspected.

8   On completion of reassembling and fitting the renovated or renewed contact breaker assembly after having refitted the generator rotor, adjust the contact breaker gap and check the ignition timing, as detailed in the relevant Sections of this Chapter.

6   Condenser: testing, removal and fitting – contact breaker models

1   A condenser is included in the contact breaker circuitry to prevent arcing across the contact breaker points as they separate. The condenser is connected in parallel with the points and if a fault develops, ignition failure is liable to occur.

2   If the engine proves difficult to start, or misfiring occurs, it is possible that the condenser is at fault. To check, separate the contact breaker points by hand when the ignition is switched on. If a spark occurs across the points and they have a blackened and burnt appearance, the condenser can be re-guarded as unserviceable.

3   It is not possible to check the condenser without the appropriate equipment. In view of the low cost involved, it is preferable to fit a new replacement and observe the effect on engine performance. The condenser may be removed and the new item fitted as follows.

4   The condenser forms part of the flywheel generator stator plate assembly. To gain full access to this assembly, it is necessary to remove the crankcase covers surrounding the flywheel generator and then to remove the rotor. Full instructions for the removal and fitting of these components are given in the relevant Sections of Chapter 1.

5   With the stator plate thus exposed, commence removal of the condenser by carefully easing back the lead retaining clip, which is situated on top of the component, until it is clear of the leads. It will now be necessary to place the heated tip of a

5.2 The contact breaker assembly forms part of the stator plate assembly

5.8 Before fitting the rotor, lightly lubricate the cam lubricating wick

6.4 The condenser is secured to the stator plate by a single screw with spring washer

soldering iron on the union of the leads to the condenser in order to melt the soldered joint and thus free the leads. Detach the condenser from the stator plate by removing the single crosshead retaining screw with its spring washer.

6   The procedure adopted for fitting the new condenser should be an exact reversal of that used for removal of the unserviceable component. Take great care to ensure that the soldered joint is properly made and that each lead is properly routed through the retaining clip.

7   Pulser coil: testing, removal and fitting – PEI TS100 and 125 models

1   The pulser coil incorporated in the type of PEI system fitted to TS100 and 125 models is mounted on the stator plate of the flywheel generator assembly and is retained in position by a single screw which has a spring washer fitted beneath its head. It is not necessary to gain access to the coil for the purpose of testing.

2   To test the pulser coil, trace the electrical leads from the flywheel generator and disconnect them at their nearest push connectors. All of these wires are colour coded to avoid confusion whilst reconnecting. The wire to the pulser coil is colour coded Red/white.

3   Closely examine each connection for contamination by dirt moisture or corrosion. If any contamination is found, then the connection in question must be cleaned. Note that the fault which was suspected to be in the coil may well be attributed to this faulty connection. Examine the electrical leads running into the generator housing for signs of damage or deterioration.

4   Set a multimeter to its resistance function and place one of its probes on a good earthing point on the engine crankcase. Connect the other plate to the Red/white wire connection and note the reading on the meter scale. This reading should be within the limits given in the Specifications Section of this Chapter. If this is not the case, then the coil must be removed from the stator plate and replaced with a serviceable item, but before doing this it is well worth having the result of the test confirmed by the Suzuki service agent from whom the new item is to be purchased.

5   Gaining access to the pulser coil will involve removal of the flywheel generator rotor along with its cover. Full details of rotor removal are contained in Section 3 of this Chapter (paragraphs 9 and 10). With the coil thus exposed, remove its retaining screw to free it from the stator plate. It may be necessary to remove the stator plate from its crankcase mountings in order to allow withdrawal of the electrical lead to the coil. If this proves to be the case, then use a sharp scribing tool or a centre punch to carefully mark the fitted position of the stator plate in relation to the crankcase; this will serve as a reference when refitting the plate.

6   Fitting of the replacement coil and the stator plate (if necessary) is a direct reversal of the removal procedure. Take care to ensure that each component is correctly positioned and its retaining screw(s) fully tightened. Refer to Section 3 of this Chapter (paragraph 11) for details of refitting the flywheel generator rotor.

8   CDI unit: removal, testing and refitting – TS185 and 250 models

1   The fully transistorised CDI unit is mounted beneath the frame top tube, directly behind the steering head. The unit is clearly labelled for identification but cannot be detached from its frame mounting until after the fuel tank has been removed.

8.1 The CDI unit is clearly labelled for identification (185 shown)

Refer to Section 2 of the previous Chapter for details of fuel tank removal.

2   With the fuel tank removed, trace the electrical leads from the CDI unit and disconnect each one at its nearest push connector. Note each connection in turn for reference when fitting the unit. Closely examine each connection for contamination by dirt, moisture and corrosion. If any contamination is found, then the connection in question must be cleaned. Note that the fault which was suspected to be in the unit may well be attributed to this faulty connection. Examine the electrical leads from each connection to the unit for signs of damage or deterioration.

3   To test the unit, measure the resistance between the various terminal combinations shown in the appropriate table below. The abbreviated terms used in the table can be interpreted as follows:

ON: Indicates continuity
OFF: Indicates open circuit, or infinitely high resistance
CON: Needle should deflect momentarily, but settle back at the infinite end of the scale

The resistance values, where shown, are approximate, and only large discrepancies will indicate a fault in the unit. Do not omit to short the terminals to be tested before commencing the meter probe leads.

TS185 models

		Positive (+) Probe Pointer to Touch				
		Black/Yellow	Black/White	Black/Red	Red/White	White/Blue
Negative (−) Pointer to Touch	Black/Yellow		CON	Approx. 2MΩ	CON	CON
	Black/White	ON		Approx. 2MΩ	ON	CON
	Black/Red	ON	CON		CON	CON
	Red/White	OFF	OFF	OFF		OFF
	White/Blue	CON	CON	CON	CON	

TS250 models

	Positive (+) Probe Pointer to Touch:			
Negative (−) pointer to touch:	Black Black/Yellow	Black/Red	Black/White (two)	White/Blue
Black Black/Yellow		CON	CON	CON
Black/Red	OFF		OFF	OFF
Black/White (two)	ON	ON		CON
White/Blue	ON	ON	ON	

4   If the CDI unit appears to be defective it should be removed from the machine and taken to a Suzuki dealer for checking, and if necessary, renewal. It is a sealed unit, and cannot be dismantled or repaired if defective.

5   To remove the unit, unscrew each of its two securing screws and lift the unit clear of its mounting points. Note that each screw has a plain and a spring washer fitted beneath its head. Renew the spring washer if it is seen to be flattened.

6   Fitting the new or original unit is a reversal of the removal procedure. Ensure that all electrical connections are clean and correctly made and the unit correctly positioned on the frame attachment points before fitting and tightening the retaining screws.

9   Flywheel generator: checking the output

1   The flywheel generator is instrumental in creating the power in the ignition system, and any failure or malfunction will affect the operation of the ignition system. If the machine will not start and there is no evidence of a spark at the spark plug, a check should first be made to ensure there is no fault in either the spark plug itself, the suppressor cap, the HT coil and lead assembly or, in the case of a machine equipped with contact breaker points, the contact breaker assembly or the condenser. TS185 and 250 models should have their CDI unit tested. TS100 and 125 models fitted with a PEI system should have their pulser coil tested for continuity.

2   If the above listed checks prove satisfactory, then the output from the generator itself should be suspected. Refer to Section 4 of Chapter 6 for the Section checking procedure by basic test methods, but before doing this, a thorough check of the circuit wiring should be made to ensure the wiring connections are not badly corroded or contaminated by dirt or moisture. Check the wiring itself for signs of chafing against the frame or engine compartment or any indication of a break in the wiring.

10   Ignition coil: location and testing

1   The ignition coil fitted to the machines covered in this Manual is a sealed unit which is designed to give long and trouble-free service without need for attention. The coil is located beneath the fuel tank and is mounted beneath the frame top tube. It follows therefore that the fuel tank must be removed in order to gain access to the coil. Refer to Section 2 of Chapter 2 for the tank removal and refitting procedures.

2   When working on a machine fitted with contact breaker points, bear in mind, before removing the fuel tank in order to gain access to the ignition coil, that a defective condenser in the contact breaker circuit can give the illusion of a defective coil and for this reason it is advisable to investigate the condition of the condenser before condemning the ignition coil. Refer to Section 6 of this Chapter for the appropriate details.

3   If, however, a weak spark and difficult starting causes the performance of the ignition coil to be suspect, then it should be tested as follows.

Contact breaker models

4   Trace the single thin low tension (LT) electrical lead from

10.1 The ignition coil is a sealed unit

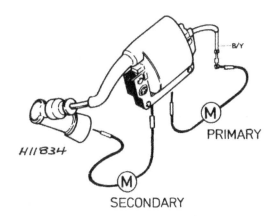

H11834

PRIMARY

SECONDARY

Fig. 3.7 Ignition coil resistance check – contact breaker models

the coil to its bullet connector and pull apart the connector. Disconnect the suppressor cap from the spark plug. Set a multimeter on its resistance function (K ohm scale). Connect one probe of the meter to the terminal of the LT lead and the other to the suppressor cap connection. If the secondary windings of the coil are in good order, then the resistance shown on the scale of the meter will coincide with the resistance given in the Specifications Section of this Chapter for the particular machine type.

5   To check the condition of the primary windings of the coil, reset the meter to its ohms scale and with one of its probes on earth, connect the other to the terminal of the LT lead. Again, the resistance shown on the scale of the meter should coincide with that given in the Specifications if the primary windings are in good order.

TS185 and 250 models

6   Trace the two thin low tension (LT) electrical leads from the coil to their block connector and pull apart the two halves of the connector. Disconnect the suppressor cap from the spark plug. Set a multimeter on its resistance function (ohms scale) and connect one of its probes to each of the LT terminals. If the

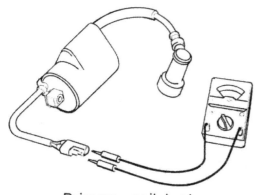

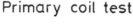

Primary coil test

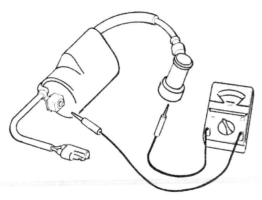

Secondary coil test

Fig. 3.8 Ignition coil resistance check – PEI models

primary windings of the coil are in good order, then the resistance measured across the two LT leads will coincide with the resistance given in the Specifications Section of this Chapter for the particular machine type.

7   To check the condition of the secondary windings of the coil, reset the meter to its K ohm scale and with one of its probes placed on a good earthing point, connect the other to the connection of the suppressor cap. Again, the resistance shown on the scale of the meter should coincide with that given in the Specifications if the secondary windings are in good order.

TS100 and 125 PEI models

8   The ignition coil should be checked for correct resistance of the windings in a manner similar to that for the TS185 and 250 models. In this case, however, the multimeter probes should be placed on the terminals shown in the following table.

All models

9   If the coil has failed, it is likely to have either an open or short circuit in its primary or secondary windings. This type of fault will be immediately obvious and will require the renewal of the coil. Where the fault is less clear cut, it is advisable to have the suspect coil tested on a spark gap tester by an official Suzuki service agent.

11 Spark plug: cleaning and resetting the gap

1   Suzuki fit NGK or ND spark plugs as standard equipment to all the model types covered in this Manual. Refer to the Specifications Section of this Chapter for the exact plug type fitted to each model. The recommended gap between the plug electrodes is 0.6 – 0.8 mm (0.024 – 0.031 in). The plug should be cleaned and the gap checked and reset at the service interval recommended in the Routine Maintenance Chapter at the beginning of this Manual. In addition, in the event of a roadside breakdown where the engine has mysteriously 'died' the spark plug should be the first item checked.

2   The plug should be cleaned thoroughly by using one of the following methods. The most efficient method of cleaning the electrodes is by using a bead blasting machine. It is quite possible that a local garage or motorcycle dealer has one of these machines installed on the premises and will be willing to clean any plugs for a nominal fee. Remember, before fitting a plug cleaned by this method, to ensure that there is none of the blasting medium left impacted between the porcelain insulator and the plug body. An alternative method of cleaning the plug electrodes is to use a small brass-wire brush. Most motorcycle dealers sell such brushes which are designed specifically for this purpose. Any stubborn deposits of hard carbon may be removed by judicious scraping with a pocket knife. Take great care not to chip the porcelain insulator round the centre electrode whilst doing this. Ensure that the electrode faces are clean by passing a small fine file between them; alternatively, use emery paper but make sure that all traces of the abrasive material are removed from the plug on completion of cleaning.

3   To reset the gap between the plug electrodes, bend the outer electrodes away from or closer to the central electrode and check that a feeler gauge of the correct size can be inserted between the electrodes. The gauge should be a tight sliding fit. Never bend the central electrode or the insulator will crack, causing engine damage if the particles fall in whilst the engine is running.

4   With some experience, the condition of the sparking plug electrodes and insulator can be used as a reliable guide to engine operating conditions. See the accompanying colour photographs.

5   Always carry a spare spark plug of the correct type. The plug in a two-stroke engine leads a particularly hard life and is liable to fail more readily than when fitted to a four-stroke.

6   Beware of overtightening the spark plug, otherwise there is risk of stripping the threads from the aluminium alloy cylinder head. The plug should be sufficiently tight to seat firmly on its sealing washer, and no more. Use a spanner which is a good fit to prevent the spanner from slipping and breaking the insulator.

7   If the threads in the cylinder head strip as a result of overtightening the spark plug, it is possible to reclaim the head by use of a Helicoil thread insert. This is a cheap and convenient method of replacing the threads, most motorcycle dealers operate a service of this nature at an economic price.

8   Before fitting the spark plug in the cylinder head, coat its threads sparingly with a graphited grease. This will prevent the plug from becoming seized in the head and therefore aid future removal.

9   When reconnecting the suppressor cap to the plug, make sure that the cap is a good, firm fit and is in good condition; renew its rubber seals if they are in any way damaged or perished. The cap contains the suppressor that eliminates both radio and TV interference.

		Positive (+) pointer to touch			
		Ground (core)	Black/Yellow	Red/White	High tension cord
Negative (−) pointer to touch	Ground (core)		OFF	OFF	4.4 - 6.6 kΩ
	Black/Yellow	2.7 - 4.1 kΩ		OFF	8.9 - 13.3 kΩ
	Red/White	2.5 - 3.7 kΩ	OFF		8.7 - 13.1 kΩ
	High tension cord	4.4 - 6.6 kΩ	OFF	OFF	

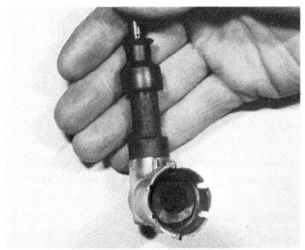

11.9 The rubber seals of the suppressor cap must be in good condition

12 High tension lead: examination

1   Erratic running faults and problems with the engine suddenly cutting out in wet weather can often be attributed to leakage from the high tension lead and spark plug. If this fault is present, it will often be possible to see tiny sparks around the lead and cap at night. One cause of this problem is the accumulation of mud and road grime around the lead, and the first thing to check is that the lead and cap are clean. It is often possible to cure the problem by cleaning the components and sealing them with an aerosol ignition sealer, which will leave an insulating coating on both components.

2   Water dispersant sprays are also highly recommended where the system has become swamped with water. Both these products are easily obtainable at most garages and accessory shops. Occasionally, the suppressor cap or the lead itself may break down internally. If this is suspected, the components should be renewed.

3   Where the HT lead is permanently attached to the ignition coil, it is recommended that the renewal of the HT lead is entrusted to an auto-electrician who will have the expertise to solder on a new lead without damaging the coil windings.

Chapter 4 Frame and forks

For modifications and information relating to later models, see Chapter 7

Contents

General description 1
Front fork legs: removal and refitting 2
Front fork legs: dismantling, examination, renovation
and reassembly 3
Steering head assembly: removal and refitting 4
Steering head bearings: examination and renovation 5
Fork yokes: examination 6
Steering lock: general description and renewal 7
Speedometer and tachometer heads: removal, inspection
and refitting 8
Speedometer and tachometer drive cables: examination
and renovation 9

Speedometer and tachometer drives: location and
examination 10
Swinging arm fork: removal, examination, renovation
and refitting 11
Rear suspension units: examination, adjustment, removal
and refitting 12
Frame: examination and renovation 13
Prop stand: examination and servicing 14
Footrests: examination and renovation 15
Rear brake pedal: examination and renovation 16
Kickstart lever: examination and renovation 17
Dualseat: removal and refitting 18

Specifications

Frame

Type	Cradle, welded tubular steel

Front forks

Type	Oil damped telescopic

Travel:
UK TS100 models	140 mm (5.51 in)
US TS100 T	145 mm (5.71 in)
TS125 and 185 models	180 mm (7.09 in)
TS250 models	195 mm (7.68 in)

Spring free length:
TS100 models	529.4 mm (20.8 in)
Service limit	517.4 mm (20.4 in)
US TS125 T	564.5 mm (22.22 in)
Service limit	539.5 mm (21.24 in)
UK TS125 N	177.8 mm (7.0 in) and 405.1 mm (15.9 in) overall
Service limit	539.5 mm (21.24 in)
UK TS125 T and X	554.5 mm (21.83 in)
Service limit	539.5 mm (21.24 in)
TS185 models	Not available
Service limit	473.0 mm (18.62 in)
TS250 models	Not available
Service limit	487.0 mm (19.2 in)

Oil capacity (per leg):
TS100 models	130 cc (4.39/4.58 US/Imp fl oz)
UK TS125 N	182 cc (6.15/6.40 US/Imp fl oz)
All other TS125	177 cc (5.98/6.23 US/Imp fl oz)
TS185 models	166 cc (5.61/5.84 US/Imp fl oz)
TS250 models	242 cc (8.17/8.52 US/Imp fl oz)

Oil type:
TS100 and 125 models	SAE 5W/20 mineral oil or ATF
US TS185 and 250 models	SAE 10 fork oil
UK TS185 and 250 models	50/50 mixture of SAE 10W/30 motor oil and ATF

Rear suspension

Type	Swinging arm fork, controlled by two hydraulic suspension units

Travel:
UK TS100 models	95 mm (3.74 in)
US TS100 models	115 mm (4.53 in)

UK TS125 N	105 mm (4.13 in)	
All other TS125	130 mm (5.12 in)	
TS185 models	130 mm (5.12 in)	
TS250 models	132 mm (5.20 in)	
Fork pivot runout	0.6 mm (0.02 in)	

Torque wrench settings

	kgf m	lbf ft
Handlebar clamp bolts	1.2 – 2.0	8.5 – 14.5
Steering stem top bolt	3.5 – 5.5	25.5 – 40.0
Upper yoke pinch bolts	2.0 – 3.0	14.5 – 21.5
Lower yoke pinch bolts	2.5 – 3.5	18.0 – 25.5
Steering stem pinch bolt:		
US TS250 models	1.5 – 2.5	11.0 – 18.0
Swinging arm fork pivot shaft nut:		
TS100 and 125	6.0 – 8.0	43.0 – 57.5
TS185	4.5 – 7.0	32.5 – 50.5
TS250	5.0 – 8.0	36.0 – 58.0
Rear suspension unit securing nuts	2.0 – 3.0	14.5 – 21.5

1 General description

Two types of frame design are utilised on the models covered in this Manual. Both are of conventional welded tubular steel construction; the only difference being that whereas the frame utilised in the TS185 and 250 models is of the full duplex cradle design, the frame utilised on the TS100 and 125 models has a single front downtube which divides into a duplex cradle at its base.

The front forks are of the conventional telescopic type, having internal oil-filled dampers. The fork springs are contained within the fork stanchions and each fork leg can be detached from the machine as a complete unit, without dismantling the steering head assembly.

Rear suspension is of the swinging arm type, using oil filled suspension units to provide the necessary damping action. The units are adjustable so that the spring ratings can be effectively changed within certain limits to match the load carried.

2 Front fork legs: removal and refitting

1   Before the fork legs can be removed, the machine must be supported in a stable position with the front wheel well clear of the ground. This can be accomplished with the aid of a stout wooden crate or blocks arranged beneath the crankcase. Make sure that there is no danger of the machine toppling whilst it is being worked on.

2   Remove the front wheel as instructed in Section 3 of the following Chapter.

3   Slacken the fork yoke pinch bolts to release each fork leg from the steering head assembly and remove each leg in turn by pulling it downwards to clear the lower yoke. On TS185 and 250 models, as a precursor to further dismantling, pull each fork leg downwards until it clears the upper yoke and so there is enough clearance between the top of the fork leg stanchion and the bottom of the upper yoke to allow removal of the rubber plug and the insertion of a removal tool into the fork spring retaining plug. The pinch bolt of the lower yoke can now be retightened to secure the fork stanchion in position. The spring retaining plug may now be loosened. This procedure will save the need to grip the fork stanchion between the jaws of a vice as described in the following Section. Once removed, each fork leg can be dismantled for examination and renovation by following the instructions given in the following Section.

4   Fitting of each fork leg is a straightforward reversal of the removal sequence, whilst noting the following points. With both forks and their related components in position, fully tighten each component securing nut or bolt, starting with the wheel spindle retaining nut and working upwards. Take note of the

torque settings given in the Specifications Section of this Chapter and of Chapter 5 and ensure that the top of each fork stanchion is flush with the top surface of the upper yoke.

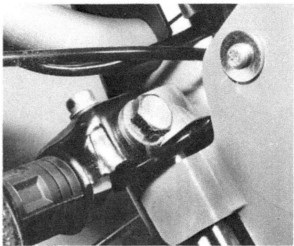

2.3a Slacken the pinch bolt of the fork upper yoke ...

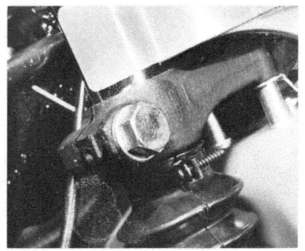

2.3b ... followed by the pinch bolt of the fork lower yoke ...

2.3c ... and pull the fork leg clear of the steering head assembly

3 Front fork legs: dismantling, examination, renovation and reassembly

1   It is strongly advised that each fork leg is dismantled separately whilst using an identical procedure. If this approach is adopted, it will mean that there is less chance of component parts being unwittingly exchanged between fork legs. Refer to the figure accompanying this text during the dismantling procedure and lay the component parts out on a clean work surface in their fitted order as they are removed.

2   Commence by loosening the fork gaiter retaining clamp(s) and sliding the gaiter clear of the fork stanchion. The fork leg should now be drained of damping oil. Where the fork leg has a drain plug fitted, this is a simple process of removing the plug with its sealing washer and allowing the oil to drain into a suitable receptacle. Moving the lower leg up and down the stanchion will assist in ejecting the oil from the fork leg.

3   Where the fork leg has no drain plug fitted, draining of the damping oil can be achieved by clamping the fork leg stanchion between the jaws of a vice, whilst keeping the leg roughly vertical, and removing the fork spring retaining plug (see following paragraphs). With this plug removed, the fork spring(s) may be withdrawn from the stanchion, the leg

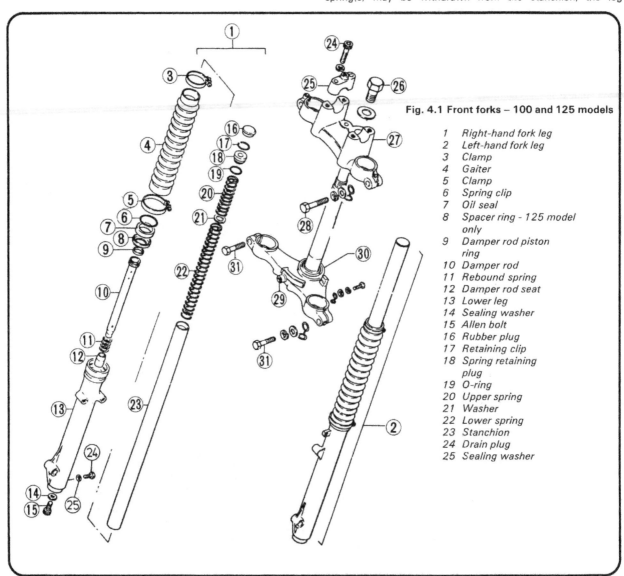

Fig. 4.1 Front forks – 100 and 125 models

1 Right-hand fork leg
2 Left-hand fork leg
3 Clamp
4 Gaiter
5 Clamp
6 Spring clip
7 Oil seal
8 Spacer ring - 125 model only
9 Damper rod piston ring
10 Damper rod
11 Rebound spring
12 Damper rod seat
13 Lower leg
14 Sealing washer
15 Allen bolt
16 Rubber plug
17 Retaining clip
18 Spring retaining plug
19 O-ring
20 Upper spring
21 Washer
22 Lower spring
23 Stanchion
24 Drain plug
25 Sealing washer

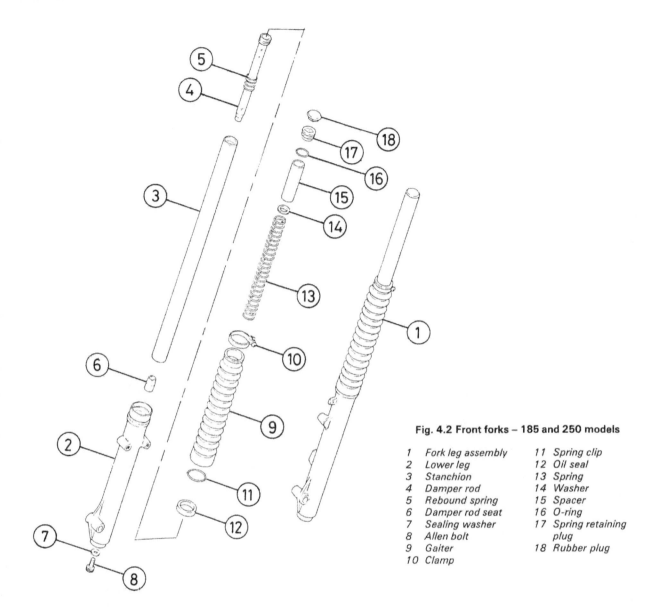

Fig. 4.2 Front forks – 185 and 250 models

1 Fork leg assembly	11 Spring clip
2 Lower leg	12 Oil seal
3 Stanchion	13 Spring
4 Damper rod	14 Washer
5 Rebound spring	15 Spacer
6 Damper rod seat	16 O-ring
7 Sealing washer	17 Spring retaining
8 Allen bolt	plug
9 Gaiter	18 Rubber plug
10 Clamp	

removed from the vice and inverted over a suitable receptacle, and the oil allowed to drain. Move the lower leg of the fork up and down the stanchion several times to eject any oil remaining within the leg on a clean work surface ready for further dismantling work to be carried out. Note that whenever any part of a fork leg is clamped between the jaws of a vice, the contact surfaces of the jaws must be protected to avoid their knurled surface causing damage to either the polished surface of the stanchion or the soft alloy of the lower leg. Thin wooden blocks or soft aluminium alloy pieces are ideal for providing the necessary protection, whereas the use of rag is not advised as components are more liable to slip from position.

4   On TS100 and 125 models, carefully lever the rubber plug from the top of the fork stanchion to expose the fork spring retaining plug located beneath it. With the fork stanchion clamped between the jaws of a vice (note the precautions given in the preceding paragraph) select a large tommy bar and place one of its ends in the centre of the plug. Push down on the plug and free its retaining clip by using the flat of a small screwdriver to lever it from position. With the clip removed, slowly reduce the downward pressure on the plug to allow it to be pushed

clear of the fork stanchion by the spring pressure behind it. Do not let go of the plug suddenly.

5   The fork spring retaining plug incorporated in the fork leg assembly fitted to TS185 and 250 models, is threaded into the top of the fork stanchion and requires the use of a $\frac{1}{2}$ inch socket drive to unscrew it. If no socket drive is available, a short length of $\frac{1}{2}$ inch square cross section steel bar can be used in conjunction with a close fitting spanner. Carefully lever the rubber plug from the top of the fork stanchion to expose the head of the fork spring retaining plug and with the fork stanchion clamped between the jaws of a vice (see the precautions given in paragraph 3), unscrew and remove the spring retaining plug. Note that the plug is under spring pressure, so care must be taken when unscrewing the plug the last few threads.

6   With the fork leg removed from the vice and with the fork spring(s) and spacer(s) removed from the stanchion, the next step in the dismantling procedure is to free the damper rod assembly by removing the Allen-headed bolt which fits in a recess in the base of the fork leg. In some cases the screw can be removed with ease, but it is likely that the damper rod, into

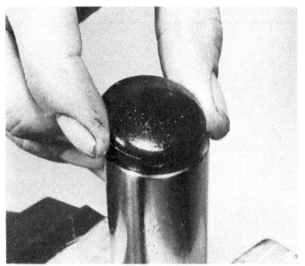

3.4a Remove the rubber plug from the fork stanchion

3.4b Remove the fork spring retaining plug and its retaining clip (100 and 125)

3.5 Remove the fork spring retaining plug with a half inch socket drive (185 and 250)

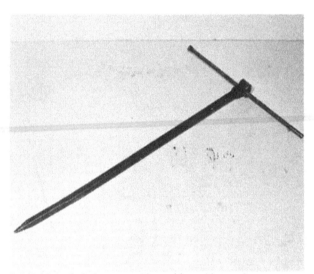

3.6 Hold the fork damper rod in position with a tool manufactured from a length of steel bar ...

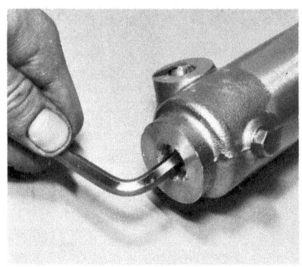

3.11 ... and then unscrew the damper rod retaining screw

3.12a With the circlip removed from its retaining groove ...

which the bolt screws, will rotate. In order to hold the rod in position whilst the bolt was removed, it was found necessary to manufacture a special tool from a length of square-section steel bar. Alternatively, Suzuki special tools Nos 09940-34512 and 09914-25811 (TS100 and 125 models) or 09940-34520 and 09940-34561 (TS185 and 250 models) may be obtained for this purpose.

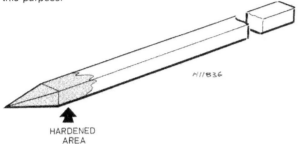

HARDENED
AREA
Fig. 4.3 Fabricated tool for holding the damper rod in position

7 To manufacture the tool, obtain a length of steel bar which is approximately 0.5 inch thick, square in section and approximately 20 inches long. Cut one end of this bar to a point, as shown in the figure accompanying this text. It is now necessary to harden this point so that when it is pushed into position against the circular recess in the end of the damper rod, its edges will bite into the softer metal of the recess walls thus providing enough grip to prevent the rod from rotating. The procedure which should be carried out in order to bring the point to the required degree of hardness is as follows.

8 Using the flame of a blowlamp or welding torch heat the bar to a cherry red for about half its length from the pointed end. Directly the bar turns red with the heat, quench its end in a large container full of water. Only 1-2 inches of the bar need be submerged beneath the surface of the water for this initial cooling procedure. Once the end of the bar has cooled, quickly remove it from the water and polish the edges of its pointed end with emery cloth. The heat remaining in the uncooled section of bar will soon travel by conduction, to the tip of the point. As this happens, the spectrum of tempering colours will be seen to progress up the edges of the point. Upon seeing the straw colour appear on the point edges, quench the complete tool in the water until it is completely cooled. The bar should now have a hardened point with the metal gradually decreasing in hardness towards the mid point of its length.

9 When carrying out the procedure described above, take great care to observe the following safety precautions. Always be aware of the dangers which come from naked flames and heated metal; have a means of extinguishing fire nearby and wear both eye and body protection. Thrusting red hot metal into cold water will produce a very violent reaction between the two; be prepared for this and guard against the possibility of scalding water being thrown from the container.

10 Once the bar has cooled, decide upon a means of preventing it from turning whilst the damper rod retaining bolt is being unscrewed. If a vice is available, then the problem is easily solved as all that needs to be done is to clamp the end of the bar between the jaws of the vice. Otherwise, clamp a self-grip wrench to the end of the bar or drill a hole through which to pass a tommy bar.

11 Locate the point of the special tool in the recess of the damper rod by passing it down inside the stanchion and giving its end a sharp tap. Unscrew the retaining bolt and withdraw the tool, followed by the damper rod assembly. In practice, it was found that several attempts were required in order to get the tool to grip the damper rod firmly, but with patience and a great deal of application this was eventually achieved. The fork stanchion can now be pulled out of the lower leg and the damper rod assembly allowed to slide out of the stanchion.

12 The oil seal fitted within the top of the lower leg should be removed only if it is to be renewed. This is because damage will almost certainly be inflicted upon the seal as it is prised from position. The spring clip which retains the seal in position may be displaced by inserting the flat of a small screwdriver into one of the clip indentations provided. With the clip thus removed (and the washer – TS185 and 250 models), the seal may be levered out of position by placing the flat of a screwdriver beneath its lower edge. Take great care when removing both of these items not to damage the alloy edge of the fork leg with the screwdriver. Any spacer ring fitted beneath the seal may be left in position unless the lower leg is to be renewed.

13 The type of fork legs fitted to the machines covered in this Manual do not contain bushes. The lower legs slide directly against the outer hard chrome surface of the fork stanchions. If wear occurs, indicated by slackness, the lower leg will have to be renewed, possibly also the fork stanchion. Wear of the fork stanchion is indicated by scuffing and penetration of the hard chrome surface.

14 Check the outer surface of the stanchion for scratches or roughness, it is only too easy to damage the oil seal during the re-assembly if these high spots are not eased down. The stanchions are unlikely to bend unless the machine is damaged in an accident. Any significant bend will be detected by eye, but if there is any doubt about straightness, roll down the stanchion tubes on a flat surface such as a sheet of plate glass. If the stanchions are bent, they must be renewed. Unless specialised repair equipment is available it is rarely practicable to effect a satisfactory repair to a damaged stanchion.

15 After an extended period of service, the fork springs may take a permanent set. If the spring lengths are suspect, then they should be measured and the readings obtained compared with the service limits given in the Specifications Section of this Chapter. It is always advisable to fit new fork springs where the length of the original items has decreased beyond the service limit given. Always renew the springs as a set, never separately. Where there are two springs fitted within each fork leg, take note of the condition of the seating ring which separates the two springs. If this ring shows signs of excessive wear or damage, then it must be renewed. The spacer tube fitted to the fork legs of TS185 and 250 models should not normally suffer wear or damage. It must, however, be renewed if thought to be defective.

16 The piston ring fitted to the damper rod may wear if oil changes at the specified intervals are neglected. If damping has become weakened and does not improve as a result of an oil change, the piston ring should be renewed. Check also that the oilways in the damper rod have not become obstructed. Suzuki provide no information as to the amount of set allowed on the damper rod spring before renewal is necessary. If in doubt as to the condition of this spring, ask the advice of a Suzuki service agent or compare the spring against a new item.

17 Closely examine each fork leg gaiter for splits or signs of deterioration. If found to be defective, it must be renewed as any ingress of dirt will rapidly accelerate wear of the oil seal and fork stanchion. It is advisable to renew any gasket washers fitted beneath the bolt heads as a matter of course. The same applies to the O-rings fitted to the cap bolts.

18 Reassembly of each fork leg is essentially a reversal of the dismantling procedure, whilst noting the following points. It is essential that all fork components are thoroughly washed in solvent and wiped clean with a lint-free cloth before assembly takes place. Any trace of dirt inside the fork leg assembly will quickly destroy the oil seal or score the stanchion to outer fork leg bearing surfaces.

19 Degrease the threads of the damper rod securing bolt and coat them with a thread locking compound. Suzuki recommend that on TS185 and 250 models, the shank of this bolt is coated with Suzuki Bond No 4 (No 1201 for US). A good quality RTV sealant will do if Suzuki's own compound cannot be obtained.

20 With the damper rod inserted into the fork stanchion and both components secured in the lower fork leg by means of the Allen-headed retaining bolt with its new sealing washer, commence fitting of the new oil seal, where required. Before fitting

3.12b ... the oil seal can be carefully levered from position

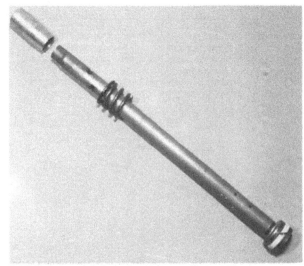

3.16 Examine and clean the damper rod assembly

3.17 Renew the O-ring fitted to the fork spring retaining plug (125 shown)

3.20a Commence fork leg assembly by fitting the damper rod into the fork stanchion

3.20b Fit the seat over the damper rod ...

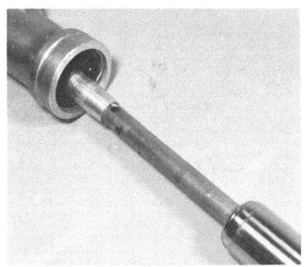

3.20c ... and insert the fork stanchion into the fork lower leg

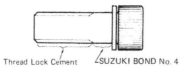

Thread Lock Cement   SUZUKI BOND No. 4

**Fig. 4.4 Damper rod securing bolt locking and sealing
compound applications**

the new seal, it should be coated with the recommended fork oil
on its inner and outer surfaces. This serves to make the fitting
of the seal into the lower fork leg easier and also reduces the
risk of damage to the seal when it is passed over the stanchion.
Take great care when fitting the seal over the stanchion.
21 Suzuki recommend the use of a service tool (No
09940-50110) with which to drive the seal into the fork leg
recess. With the seal partially located in the leg recess the tool,
which takes the form of a short length of metal tube approximately 3 in long, with an inner diameter just greater than the
outer diameter of the stanchion and an outer diameter just less
than that of the outer diameter of the oil seal, may be passed
over the stanchion and used to tap the seal home by using it as
a form of slide hammer. If this tool is not readily available it can,
of course, be fabricated from a piece of metal tubing of the
appropriate dimensions. Care should be taken however, to

3.20d Retain the damper rod in position by fitting and
tightening its securing screw

3.20e Fit the fork spring(s) into the stanchion ...

3.22a ... refit the drain plug with its serviceable sealing
washer ...

3.22b ... and replenish the fork leg with oil

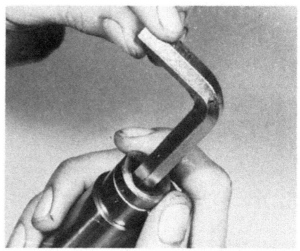

3.22c Use a socket to aid refitting of the spring retaining plug

ensure that the end of the tube that makes contact with the seal is properly chamfered, free of burrs and absolutely square to the fork stanchion. Alternatively, a suitable socket may be used to drive the seal into position before the stanchion is fitted. Carry out a final check to ensure that the seal has been driven home squarely before fitting the washer (TS185 and 250 models) and then locating the spring clip in its retaining slot.

22 Fit a serviceable gaiter to the fork leg and support the leg in a vertical position. Refer to the Specifications Section of this Chapter and replenish the fork leg with the correct quantity of the specified type of oil, pouring the oil into the top of the stanchion. On models where the fork leg has a drain plug fitted, check that the plug with its new sealing washer is properly tightened before replenishing the fork leg.

23 Fit the fork spring(s) into the stanchion along with the spacer and seating ring (where fitted). If in doubt as to the fitted position of any one component, then refer to the appropriate figure accompanying this text. Note that the single spring fitted in each leg of TS185 and 250 models must have its small diameter end facing the bottom of the fork leg. On TS250 models this end of the spring also has more widely spaced coils.

24 Refit the fork spring retaining plug with its serviceable O-ring. In practice, the fitting of this plug to the TS100 and 125 models proved to be a straightforward reversal of the removal procedure. On TS185 and 250 models however, it was difficult to locate the plug on the thread in the inside of the fork stanchion, owing to the spring pressure acting against it. This problem was eventually overcome by adapting a socket as shown in the photograph accompanying this text. A socket which is a close fit inside the bore of the stanchion will serve to keep the plug square to the thread whilst giving some means with which to bear down against the pressure of the spring. With the plug fully tightened, the fork leg is ready to be refitted to the machine.

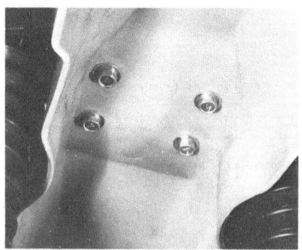
4.2 Remove the front mudguard retaining bolts ...

---

4 Steering head assembly: removal and refitting

1 The steering head will rarely require attention unless it becomes necessary to renew the bearings or if accident damage has been sustained. It is theoretically possible to remove the lower yoke together with the fork legs, but as this entails a considerable amount of unwieldy manoeuvring this approach is not recommended. A possible exception may arise if the fork stanchions have been damaged in an accident and are jammed in the lower yoke, and in this case a combination of this Section and Section 2 must be applied.

2 Refer to Section 3 of the following Chapter and remove the front wheel from the machine. Refer to Section 2 of this Chapter and remove the front fork legs. Where the machine has a front mudguard which is fitted to the lower yoke, detach the mudguard by removing its four retaining bolts. Note the fitted positions of the washers and spacers fitted beneath the head of each bolt for reference when refitting.

3 Before disconnecting any electrical leads from the multitude of electrical components mounted on the headstock assembly, it is advisable to isolate the battery from the machine's electrical system by removing one of the leads from its battery terminal. This will safeguard against any danger of components within the electrical system becoming damaged by short circuiting of the exposed connector ends.

4 Remove the three securing screws with plate washers that secure the headlamp nacelle in position. Lower the nacelle forward and detach the bulb connections from the headlamp shell so that the nacelle assembly can be removed from the machine and placed in safe storage until required for reassembly.

5 Unplug the bulb connection(s) from the base of each instrument head and release the electrical wires to these connections from their retaining clamp which is situated between the instrument heads. Manoeuvre all electrical wires clear of the steering head assembly.

4.4 ... and detach the headlamp nacelle

4.5 Unplug the bulb connections from the base of each instrument head ...

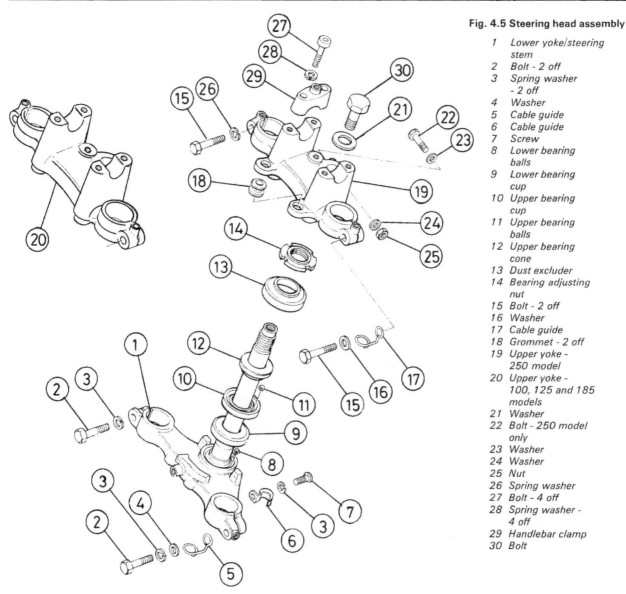

Fig. 4.5 Steering head assembly

1 Lower yoke/steering
   stem
2 Bolt - 2 off
3 Spring washer
   - 2 off
4 Washer
5 Cable guide
6 Cable guide
7 Screw
8 Lower bearing
   balls
9 Lower bearing
   cup
10 Upper bearing
   cup
11 Upper bearing
   balls
12 Upper bearing
   cone
13 Dust excluder
14 Bearing adjusting
   nut
15 Bolt - 2 off
16 Washer
17 Cable guide
18 Grommet - 2 off
19 Upper yoke -
   250 model
20 Upper yoke -
   100, 125 and 185
   models
21 Washer
22 Bolt - 250 model
   only
23 Washer
24 Washer
25 Nut
26 Spring washer
27 Bolt - 4 off
28 Spring washer -
   4 off
29 Handlebar clamp
30 Bolt

6   Disconnect the speedometer and tachometer (where fitted) drive cables by unscrewing each knurled retaining ring and then pulling the cable down to clear the instrument. Locate the two bolts which secure the instrument mounting bracket to the upper yoke. Remove these bolts and lift the complete instrument assembly clear of the machine.

7   Detach each direction indicator from its attachment on the upper yoke by removing its one securing bolt. The pinch bolt which passes through the yoke and the indicator mounting bracket will have to be withdrawn to allow each indicator to be manoeuvred clear of the steering head assembly.

8   Remove each of the two rubber plugs from the upper face of the warning light console. This will expose the heads of the two console securing screws which should now be removed to free the console from its upper yoke attachment points. Using a C-spanner or a small soft-metal drift and hammer, release the ignition switch retaining ring. With the ring removed, the console can be manoeuvred rearwards to clear the handlebar securing clamps.

9   Remove the two screws which retain the ignition switch mounting bracket to the upper yoke. Cover the forward part of the fuel tank with an old blanket, or similar item to protect its painted surface from damage. Slacken and remove the four handlebar clamp retaining bolts, lift off the top halves of the clamps and move the handlebars rearwards to rest on the protected area of tank. Carry out a final check around the steering head to ensure that no components remain connected to either of the fork yokes.

10 To remove the upper yoke, first unscrew the large chromium plated bolt with washer from its location through the centre of the yoke. On TS250 models, loosen the pinch bolt which passes through the rear of the yoke. Using a soft-faced hammer, give the yoke several gentle taps to free it from the steering stem and then lift it from position.

11 Support the weight of the lower yoke and, using a C-spanner of the correct size, remove the steering head bearing adjusting ring. If a C-spanner is not available, a soft metal drift may be used in conjunction with a hammer to slacken the ring.

12 Remove the dust excluder and the cone of the upper bearing. The lower yoke, complete with steering stem, can now be lowered from position. Ensure that any balls that fall from the bearings as the bearing races separate are caught and retained. It is quite likely that only the balls from the lower bearing will drop free, since those of the upper bearing will remain seated in the bearing cup. Full details of examining and renovating the steering head bearings are given in the following Section of this Chapter.

4.6a ... and disconnect each instrument drive cable

4.6b Remove the instrument mounting bracket retaining bolts

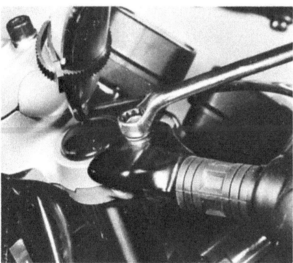

4.7 Detach each direction indicator mounting bracket by removing its two securing bolts

4.8 Expose the heads of the console securing screws

4.9a Detach the ignition switch mounting bracket from the fork upper yoke

4.9b Release the handlebar retaining clamps

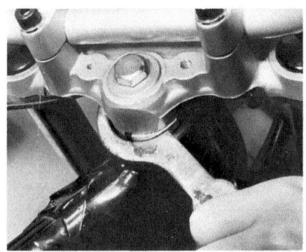

4.14 Use a C-spanner to tighten the steering head bearing adjuster ring

13 Fitting of the steering head assembly is a direct reversal of that procedure used for removal, whilst taking into account the following points. It is advisable to position all eighteen balls of the lower bearing around the bearing cone before inserting the steering stem fully into the steering head. Retain these balls in position with grease of the recommended type and fill both bearing cups with the same type of grease.

14 With the lower yoke pressed fully home into the steering head, place the twenty-two balls (eighteen for TS250 models) into the upper bearing cup and fit the bearing cone followed by the dust excluder. Refit the adjusting ring and tighten it, finger-tight. The ring should now be tightened until resistance to movement is felt and then loosened $\frac{1}{8}$ to $\frac{1}{4}$ of a turn. Move the lower yoke from side to side several times and then check tighten the ring by using the aforementioned procedure.

15 Adjustment of the steering head bearings is correct when all play in the bearings is taken up but the yoke will move freely from lock to lock. Note that it is possible to place several tons pressure on the steering head bearings if they are overtightened. The usual symptom of overtight bearings is a tendency for the machine to roll at low speeds even though the handlebars may appear to turn quite freely.

16 Finally, whilst refitting and reconnecting all disturbed components, take care to ensure that all control cables, drive cables, electrical leads, etc are correctly routed and that proper reference is made to the list of torque wrench settings given in the Specifications Section of this Chapter and of Chapter 5. When refitting the handlebars, note that they must be fitted so that the punch mark on the handlebar is directly in line with the rear mating faces of the handlebar clamps. Ensure that the clamp retaining bolts are fitted with serviceable spring washers beneath their heads and are tightened evenly so that the gap between the forward mating surfaces of the clamps is equal to that between the rear mating surfaces. Check that the headlamp beam height has not been disturbed and ensure that all controls and instruments function correctly before taking the machine on the public highway.

5 Steering head bearings: examination and renovation

1 Before commencing reassembly of the steering head component parts, take care to examine each of the steering head bearing races. The ball bearing tracks of their respective cup and cone bearings should be polished and free from any indentations or cracks. If wear or damage is evident, then the cups and cones must be renewed as complete sets.

2 Carefully clean and examine the balls contained in each

bearing assembly. These should also be polished and show no signs of surface cracks or blemishes. If any one ball is found to be defective, then the complete set should be renewed. Remember that a complete set of these balls is relatively cheap and it is not worth the risk of refitting items that are in doubtful condition.

3 TS250 models have eighteen balls fitted in both bearing races, whereas all other models have twenty-two balls fitted in the top bearing race and eighteen in the lower. This arrangement will leave a gap between any two balls but an extra ball must not be fitted, otherwise the balls will press against each other thereby accelerating wear and causing the steering action to be stiff.

4 The bearing outer races are a drive fit in the steering head and may be removed by passing a long drift through the inner bore of the steering head and drifting out the defective item from the opposite end. The drift must be moved progressively around the race to ensure that it leaves the steering head evenly and squarely.

5 The lower of the two inner races fits over the steering stem and may be removed by carefully drifting it up the length of the stem with a flat-ended chisel or a similar tool. Again, take care to ensure that the race is kept square to the stem.

6 Fitting of the new bearing races is a straightforward procedure whilst taking note of the following points. Ensure that the race locations within the steering head are clean and free of rust; the same applies to the steering stem. Lightly grease the stem and head locations to aid fitting of the races and drift each race into position whilst keeping it square to its location. Fitting of the outer races into the steering head will be made easier if the opposite end of the head to which the race is being fitted has a wooden block placed against it to absorb some of the shock as the drift strikes the race. It is a good idea to form a drift out of a short length of metal tube which has an outer diameter equal to that of the bearing outer races. The ends of this tube must be properly chamfered and absolutely square to its length. With the tube placed against the bearing race and a piece of wood or metal placed across it, strike the assembly firmly with a hammer to drive the bearing race into its location.

6 Fork yokes: examination

1 To check the top yoke for accident damage, push the fork stanchions through the bottom yoke and fit the top yoke. If it lines up, it can be assumed the yokes are not bent. Both yokes must also be checked for cracks, if they are damaged or cracked, fit serviceable replacements.

7 Steering lock: general description and renewal

1 A steering lock is attached to a lug on the steering head by means of two screws. When in the locked position, a bar extends from the body of the lock and abuts against a projection which forms part of the casting of the lower yoke. This effectively prevents the handlebars from being turned once they are set on full lock.

2 If the lock malfunctions, then it must be renewed. A repair is impracticable. Note that the steering stem must be lowered away from the steering head to allow removal of the lock securing screws. When the lock has been renewed, ensure a key which matches the lock is obtained and carried when the machine is in use.

8 Speedometer and tachometer heads: removal, inspection and refitting

1 The speedometer and tachometer heads are mounted side by side on a mounting bracket which itself is bolted to the upper

7.1 The steering lock is attached to a lug on the steering head

yoke of the steering head assembly. Note that certain models will have a warning light console mounted in place of a tachometer. Removal of this unit is dealt with in Section 14 of Chapter 6.

2   To gain full access to the base of each instrument, remove the three securing screws with plate washers that hold the headlamp nacelle in position and unclip the nacelle from the fork stanchions. Unplug the bulb connection from each instrument and then disconnect the drive cable by unscrewing its knurled retaining ring and pulling the cable down and clear of the instrument. Each instrument can now be detached from the mounting bracket by removing the nut from each of its base mounting studs and then pulling the instrument upwards to clear the bracket. Take care to retain the plate washer with each retaining nut.

3   If either instrument fails to record, check the drive cable first before suspecting the head. If the instrument gives a jerky response it is probably due to a dry cable, or one that is trapped or kinked.

4   The speedometer and tachometer heads cannot be repaired by the private owner, and if a defect occurs a new instrument has to be fitted. Remember that a speedometer in correct working order is required by law on a machine in the UK and also in many other countries.

5   Refer to the following Sections of this Chapter for details of servicing the instrument drive assemblies. On completion of servicing either the instruments or their drive assemblies, refit them by using the reverse procedure to that given for removal. Check that all disturbed electrical connections are properly remade and that the drive cables are correctly routed.

9   Speedometer and tachometer drive cables: examination and renovation

1   It is advisable to detach the speedometer and tachometer drive cables from time to time in order to check whether they are adequately lubricated and whether the outer cables are compressed or damaged at any point along their run. A jerky or sluggish movement at the instrument head can often be attributed to a cable fault.

2   Tp grease the cable, uncouple both ends and withdraw the inner cable. (On some model types this may not be possible in which case a badly seized cable will have to be renewed as a complete assembly). After removing any old grease, clean the inner cable with a petrol soaked rag and examine the cable for broken strands or other damage. Do not check the cable for

broken strands by passing it through the fingers or palm of the hand, this may well cause a painful injury if a broken strand snags the skin. It is best to wrap a piece of rag around the cable and pull the cable through it, any broken strands will snag on the rag.

3   Regrease the cable with high melting point grease, taking care not to grease the last six inches closest to the instrument head. If this precaution is not observed, grease will work into the instrument and immobilise the sensitive movement.

4   If the cable breaks, it is usually possible to renew the inner cable alone, provided the outer cable is not damaged or compressed at any point along its run. Before inserting the new inner cable, it should be greased in accordance with the instructions given in the preceding paragraph. Try to avoid tight bends in the run of the cable because this will accelerate wear and make the instrument movement sluggish.

10 Speedometer and tachometer drives: location and examination

1   The drive for the speedometer is taken from a gearbox which forms an integral part of the front brake backplate assembly. Drive is transmitted through the slotted end of the wheel bearing retaining boss which engages with a drive plate in the gearbox. Full details of servicing this particular unit are contained in Section 5 of the following Chapter.

2   The drive for the tachometer is taken from a point on the crankcase adjacent to the kickstart lever. The driven is taken from a gear which forms part of the oil pump drive assembly. Full details of removal and fitting of this gear are contained within the relevant Sections of Chapter 1. It is unlikely that this internal drive mechanism will give trouble during the normal service life of the machine, particularly since it is fully enclosed and effectively lubricated.

11 Swinging arm fork: removal, examination, renovation and refitting

1   The rear fork of the frame assembly is of the swinging arm type. This unit is of steel construction, its legs being formed from box section material and its pivot crossmember from tubular material. Heavy gussetting between each leg and the crossmember ensures that the assembly does not flex under the loads placed upon it. Fibre bushes provide an efficient bearing surface for the pivot, each of the two bushes being a press fit into the pivot crossmember. These bushes bear upon a steel shaft which passes through both the crossmember and lugs which are welded to the frame tubes.

2   Any wear in the pivot bearings of the swinging arm will cause imprecise handling of the machine, with a tendency for the rear end of the machine to twitch or hop. Worn bearings can be detected by placing the machine on its centre stand and then pulling and pushing vigorously on the rearmost end of one of the fork legs whilst holding on to a point on the main frame with the other hand. Any play in the bearings will be magnified by the leverage effect produced.

3   When wear develops in the bearings of the swinging arm, necessitating their renewal, the renovation procedure is quite straightforward. Commence by removing the rear wheel as described in Section 6 of Chapter 5.

4   The guard for the final drive chain is bolted to the left-hand leg of the swinging arm fork, as is the chain guide (where fitted). Removal of these components is not strictly necessary, although it will facilitate detachment and examination of the swinging arm.

5   Remove the cap nut and plain washer from each of the suspension unit to swinging arm fork attachment points. Grasp the lower section of each suspension unit and pull it outwards until it clears the fork leg. Swing the fork down and rest its leg on an area of padding, a wooden block is ideal. Leave the

Fig. 4.6 Swinging arm – 100 and 125 models

1   Swinging arm
2   Bush - 2 off
3   Centre spacer
4   Pivot shaft
5   Nut
6   Washer
7   Outer spacer -
     2 off
8   Dust cover -
     2 off
9   O-ring -
     2 off
10 Thrust washer -
     2 off
11 Chain guide
12 Chain guard
13 Reinforcement plate
14 Bolt - 3 off
15 Washer - 3 off
16 Bolt
17 Washer
18 Rear suspension
     unit - 2 off
19 Cap nut - 2 off
20 Nut - 2 off
21 Washer - 2 off
22 Washer - 6 off
23 Torque arm
24 Bolt
25 Bolt
26 Spring washer
     - 2 off
27 Nut - 2 off
28 Washer
29 Split pin -
     2 off

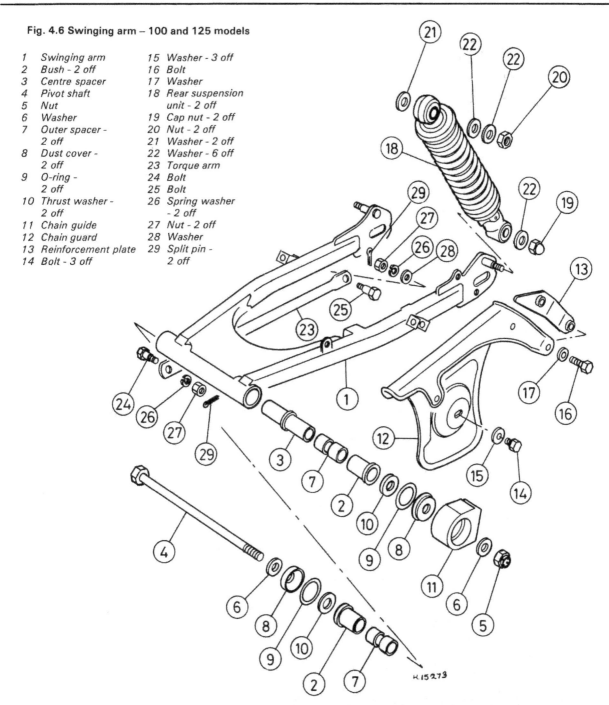

H.15273

suspension units hanging from the frame but slacken their frame mounting nuts so that each unit is free to move. This will make reattachment of each unit to the swinging arm fork a great deal easier.

6   Loosen and remove the pivot shaft retaining nut with washer. Where a guide for the spring of the stop lamp switch is located beneath this nut, detach the spring from the guide before removing the nut. Support the crossmember of the swinging arm and draw the pivot shaft out of position. If the shaft proves to be stubborn, use a hammer and soft metal drift to displace it. Manoeuvre the swinging arm rearwards so that it clears the frame mounting points and final drive chain and then lift it clear of the machine.

7   Dismantle the pivot assembly by first pulling the nylon chain protector from the end of the arm crossmember. Detach

the dust cover, O-ring and thrust washer from the protector and lay each component out in the order of removal on a clean work surface. Insert the tip of one finger in the centre of the outer spacer and withdraw it from the centre of the fibre bush. Remove the dust cover, O-ring, thrust washer and outer spacer from the end of the crossmember.

8   Each fibre bush can now be removed from its location in the arm crossmember by passing a long drift through the centre of the crossmember and drifting the bush out from the opposite end. Move the end of the drift progressively around the bush to ensure that it leaves its location evenly and squarely. The single centre spacer can now be removed from the crossmember. Give each component part a thorough clean before commencing the following examination and renovation procedures.

9   Check the pivot shaft for wear. If the shank of the shaft is

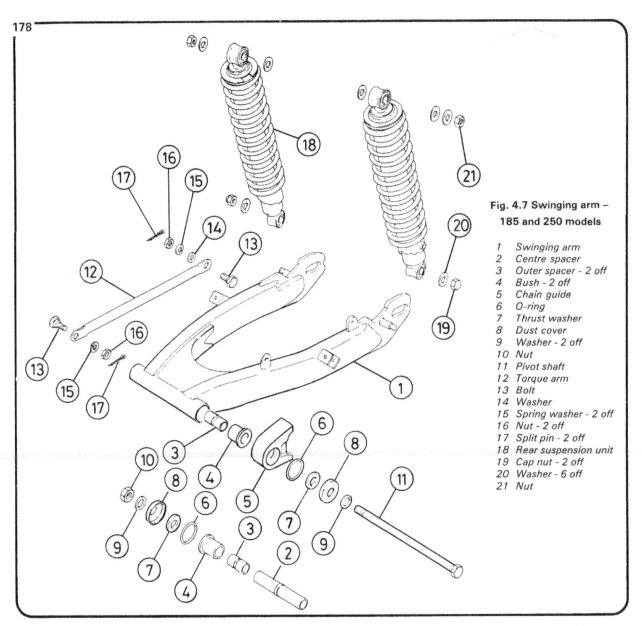

**Fig. 4.7 Swinging arm –
185 and 250 models**

1 Swinging arm
2 Centre spacer
3 Outer spacer - 2 off
4 Bush - 2 off
5 Chain guide
6 O-ring
7 Thrust washer
8 Dust cover
9 Washer - 2 off
10 Nut
11 Pivot shaft
12 Torque arm
13 Bolt
14 Washer
15 Spring washer - 2 off
16 Nut - 2 off
17 Split pin - 2 off
18 Rear suspension unit
19 Cap nut - 2 off
20 Washer - 6 off
21 Nut

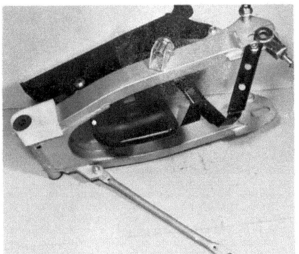

11.6 The swinging arm fork assembly

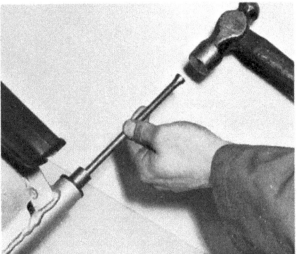

11.8 Each fibre bush can be driven from position

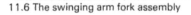

seen to be stepped or badly scored, then it must be renewed. Remove all traces of corrosion and hardened grease from the shaft before checking it for straightness by rolling it on a flat surface, such as a sheet of plate glass, whilst attempting to insert a feeler gauge of 0.6 mm (0.02 in) thickness beneath it. Alternatively, place the shaft on two V-blocks and measure the amount of runout on its shank with a dial gauge. If the amount of runout measured exceeds 0.6 mm (0.02 in), replace the shaft with a straight item. Note that a bent pivot shaft will prevent the swinging arm fork from moving smoothly about its axis.

10 Closely inspect each component part of the pivot assembly, renewing each defective component as necessary. Note that if one spacer is found to be worn them it is advisable to renew all three spacers as a set. Check for wear in each spacer by inserting the pivot shaft through it and feeling for wear between the two components. Both fibre bushes and both O-rings should be renewed as a matter of course.

11 Inspect the nylon protector block for signs of excessive wear. If it is considered that this item no longer serves as an adequate means of protecting the metal of the swinging arm crossmember against interference from the final drive chain, then it must be renewed.

12 Carefully inspect the structure of the swinging arm fork for signs of distortion, failure or any other damage which may lead to eventual failure of the component. It is worth taking steps to remove any corrosion from areas where the paint finish has been eroded away and then reprotecting the bared surface. Pay particular attention to the welds between component parts. If distortion is suspected, return the component to a Suzuki service agent who will be able to confirm whether or not replacement is necessary.

13 Upon completion of the examination procedure, reassemble the swinging arm pivot assembly whilst noting the following points. Check that the location of each fibre bush is clean of all dirt and grease before pushing the bush into position. If necessary, each bush can be tapped home by using a socket or length of tube of the same diameter as that of the bush in conjunction with a hammer. Each spacer should be well coated with a high-melting point lithium based grease before inserting it into position. The same applies to the dust cover with its O-ring.

14 Align the swinging arm in the frame and lightly grease the shank of the pivot shaft before inserting it into position; this will prevent it from becoming corroded and thus seizing to each of the spacers. Refit the pivot shaft retaining nut and tighten it to the specified torque loading. Do not omit to refit the washer or spring guide beneath this nut, as necessary. Finally, check that the fork pivots smoothly about its full axis before reconnecting the suspension units and tightening their retaining nuts to the specified torque loading.

12 Rear suspension units: examination, adjustment, removal and refitting

1   Rear suspension units of the hydraulically damped type are fitted to the machines used in this Manual. Each unit comprises a hydraulic damper, effective primarily on rebound, and a concentric spring. It is secured to the frame and swinging arm by means of rubber-bushed lugs at each end of the spring tension, giving five settings. The settings can be easily altered without moving the units from the machine by using a tommy bar in the hole provided in the adjuster ring. Turning this ring to increase the spring tension, that is to compress the spring, will stiffen the rear suspension. Turning the ring to lessen the spring tension will soften the ride. Suzuki recommend that for normal riding with no pillion passenger, the adjuster ring of each nut is set on its mid position. This is, however, a matter of personal preference but as a general guide, the heavier the load to be carried or the higher the speed to which the machine is taken, then the stiffer the suspension should be.

2   Note that in the interests of good roadholding, it is essential that both suspension units have the same load setting. If

11.13a Lubricate each outer spacer before insertion

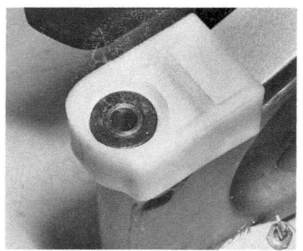
11.13b Fit a serviceable protector block

11.13c Pack each dust cover with grease and fit a serviceable O-ring

12.1 The setting of each rear suspension unit can be altered with a tommy bar

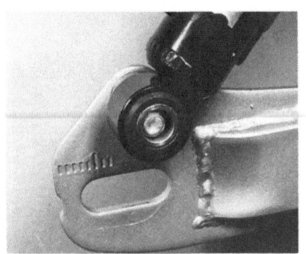

12.3 Check for deterioration of the suspension unit mounting bushes

renewal is necessary, the units must be replaced as a matched pair.

3   There is no means of draining the units or topping up, because the dampers are built as a sealed unit. If the damping fails or if the units start to leak, the complete damper assembly must be renewed. This applies equally if the damper rod has become bent, or its chromed surface badly corroded or damaged. Check also for deterioration of the rubber mounting bushes. The piston housing must be free of damage if the unit is to function correctly.

4   If necessary, each suspension unit can be removed from the frame and swinging arm attachment studs simply by removing the upper and lower retaining nuts and plain washers and pulling the unit out away from the machine. Remove and fit the units one at a time, otherwise the machine will have to be supported to prevent the back end from collapsing.

5   When fitting a suspension unit, employ a procedure which is a reversal of that used when removing the unit, whilst noting the following points. Before refitting the original or any replacement units, take the opportunity to give them a thorough clean. Do not, under any circumstances, grease the chromed surface of the damper rod; any break down of this chromed surface will quickly lead to failure of the unit seal, thereby rendering the unit

ineffective. Check that all the plain washers are correctly located before fitting and tightening the retaining nuts to the recommended torque loading.

13 Frame: examination and renovation

1   The frame is unlikely to require attention unless accident damage has occurred. In some cases, renewal of the frame is the only satisfactory remedy if the frame is badly out of alignment. Only a few frame specialists have the jigs and mandrels necessary for resetting the frame to the required standard of accuracy, and even then there is no easy means of assessing to what extent the frame may have been overstressed. Repair work of this nature can prove expensive and it is always worthwhile checking whether a good replacement frame of identical type can be obtained at a reasonable cost.

2   After the machine has covered a considerable mileage, it is advisable to examine the frame closely for signs of cracking or splitting at the welded joints. Minor damage can be repaired by welding or brazing, depending on the extent and nature of the damage. Check carefully areas where corrosion has occurred on the frame. Corrosion can cause a reduction in the material thickness and should be removed by use of a wire brush and derusting agents.

3   Remember that a frame which is out of alignment will cause handling problems and may even promote 'speed wobbles'. If misalignment is suspected, as a result of an accident, it will be necessary to strip the machine completely so that the frame can be checked, and if necessary, renewed.

14 Prop stand: examination and servicing

1   The prop stand bolts to a lug which is welded to the frame rear downtube on the left-hand side of the machine. An extension spring is fitted to ensure that the stand retracts when the weight of the machine is taken off the stand and the stand is kicked rearwards. When properly retracted, the stand should be tight against its stop and well out of the way.

2   Check to ensure that the nut which serves to retain the bolt about which the stand pivots is fully tightened and that the pivot surfaces are well lubricated with either grease or motor oil. The extension spring must be free from fatigue and in good condition. Smearing grease along the length of the spring will lessen the chances of its becoming corroded.

3   Finally, remember that a serious accident is almost inevitable if the stand extends whilst the machine is in motion.

14.1 Examine the prop stand return spring and pivot

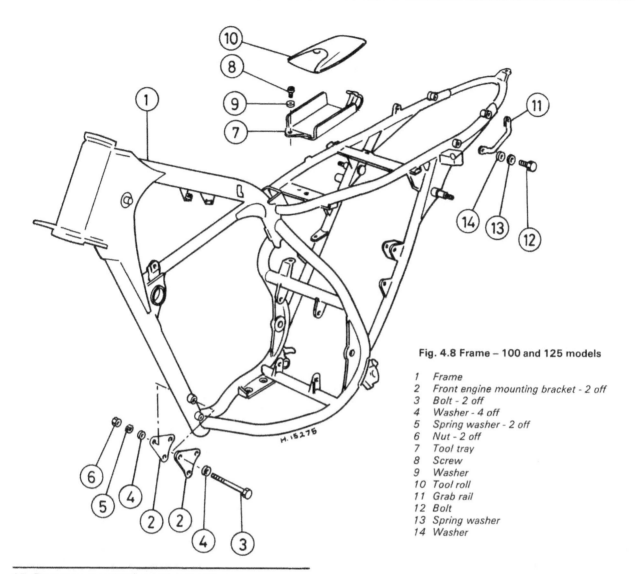

Fig. 4.8 Frame – 100 and 125 models

1   Frame
2   Front engine mounting bracket - 2 off
3   Bolt - 2 off
4   Washer - 4 off
5   Spring washer - 2 off
6   Nut - 2 off
7   Tool tray
8   Screw
9   Washer
10 Tool roll
11 Grab rail
12 Bolt
13 Spring washer
14 Washer

15 Footrests: examination and renovation

1   On TS250 models, each front footrest mounting comprises
a metal bracket which is bolted to bosses welded to each of the
frame lower tubes. Attached to each bracket is a pivoting metal
footrest. This footrest is attached to the mounting bracket by
means of a clevis pin which itself is secured in position by a
split-pin. A spring is fitted to each footrest to ensure that it is
returned to its horizontal position immediately after having been
knocked upwards about its pivot.
2   The front footrests fitted to all other models covered in this
Manual are similar in construction to those described for the
TS250 models; the one difference is in their method of
mounting. Whereas the TS250 has separate mounting brackets
for each footrest, other models have a mounting bar which
passes beneath the engine/gearbox unit and which is bolted to
lugs welded to the inside edge of each of the frame lower tubes.
3   In the case of either type of mounting bracket being bent in
a spill or through the machine falling over, remove the footrests
from the bracket, remove the bracket from the machine and
place it between the jaws of a vice. The bracket can now be
straightened by heating it to a dull red and carefully tapping it
straight with a hammer. Do not hammer it straight with the
metal cold as this will only cause stress fractures in the metal
which in turn will lead to failure of the component. A blow lamp

15.2 The front footrest assembly must be free of defects

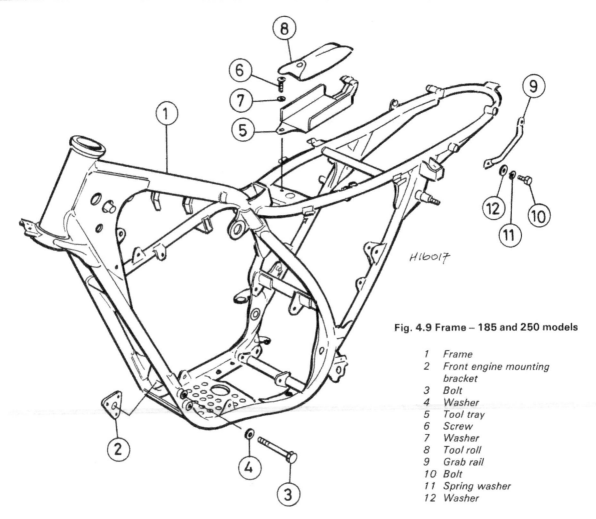

H16017

Fig. 4.9 Frame – 185 and 250 models

1   Frame
2   Front engine mounting
    bracket
3   Bolt
4   Washer
5   Tool tray
6   Screw
7   Washer
8   Tool roll
9   Grab rail
10 Bolt
11 Spring washer
12 Washer

or welding torch should be used to heat the bracket.
4   Regular maintenance on the front footrest brackets should
include lubrication of the pivot and inspection of the return
spring for fatigue or failure. A footrest which is allowed to stick
in the up position will constitute a serious hazard when riding
the machine. Check also that the split-pin is correctly fitted and
always use a new pin when fitting a footrest. A bent footrest
can be straightened by using a method similar to that described
for its mounting bracket.
5   The pillion footrests, like the front footrests, are pivoted on
clevis pins. If an accident occurs, it is probable that the footrest
peg involved will move and remain undamaged. To straighten a
bent peg, remove it from the swinging arm attachment and
carry out the procedure given for straightening of the front
footrest mounting bracket. The footrest rubber will, of course,
have to be removed as the heat will render it unfit for service.
Regular maintenance of the pillion footrests involves lubrication
of the pivot and checking that the split-pin is correctly fitted;
that is with both of the pin legs spread apart.
6   Note, with both types of footrest, that if there is evidence of
failure of the metal either before or after straightening, it is
advised that the damaged component is renewed. If a footrest
breaks, loss of machine control is almost inevitable.

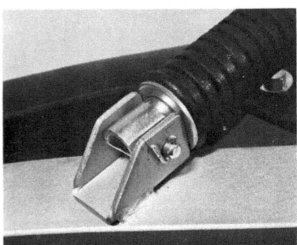

15.5 Check that each footrest pivot is functional and is correctly
assembled

16 Rear brake pedal: examination and renovation

1   Reference to the figures accompanying this text will clearly
indicate the method of attachment used for the rear brake pedal

fitted to each model type covered in this Manual. The procedure
for removing, renovating and fitting of the rear brake pedal
should be based on the following instructions.
2   In the event of damage occurring to the pedal, it should be
removed from the machine and treated in a similar manner to a

Fig. 4.10 Rear brake pedal assembly – 100, 125 and 185 N models

1 Rear brake pedal	6 Locknut	10 Adjusting nut
2 Pedal shaft	7 Operating rod	11 Washer
3 Bolt	8 Spring	12 Split pin
4 Return spring	9 Trunnion	13 Washer
5 Adjusting bolt		

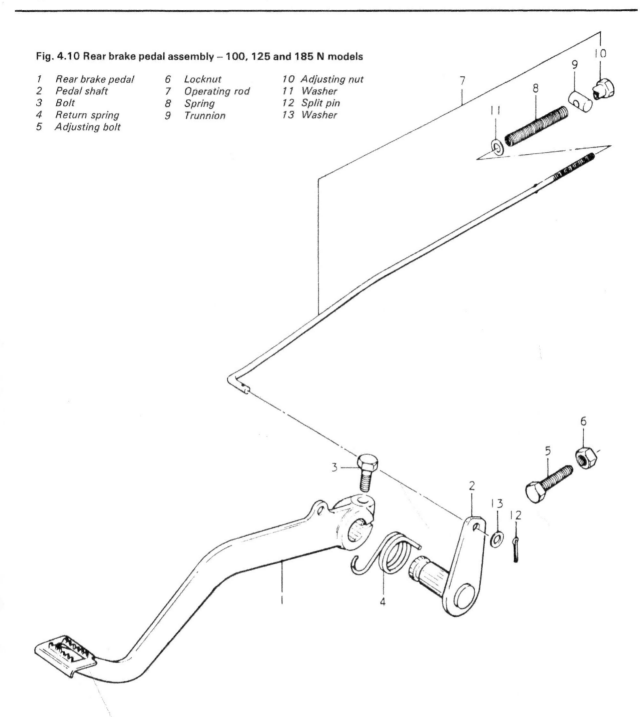

bent footrest mounting bracket as described in Section 15 of this Chapter. Note that the warning relating to footrest breakage applies equally to the brake pedal because it follows that failure is most likely to occur when the brake is applied firmly, which is when it is required most.

3   To dismantle the brake pedal assembly, refer to the aforementioned figures and proceed as follows. Where the brake operating rod is attached to the brake pedal, remove the split-pin with washer from the clevis pin, withdraw the clevis pin and move the forked end of the rod clear of the pedal. On all models, unhook the rear stop lamp switch operating spring from its pedal attachment. The pedal can now be detached from the machine after removal of its retaining bolt and, in the case of the TS250 model, detachment of the pedal return spring.

4   Where the brake operating rod is attached to the backplate of the pedal shaft, detach it by first removing its retaining pin with washer and then pulling it clear of the plate. The pedal shaft can now be removed for inspection.

5   With the complete brake pedal assembly laid out on a clean work surface, inspect the pedal return spring for signs of fatigue or failure and renew it if necessary. It is advisable, if the brake pedal has been in any way damaged, to check the condition of the pivot shaft. This is also the case if operation of the pedal is noticed to be stiff and the pedal fails to return immediately when released. Check the shaft for straightness by placing a straight edge alongside it for comparison. If bent, the shaft must be renewed.

6   If the pedal is seen to have seized through lack of

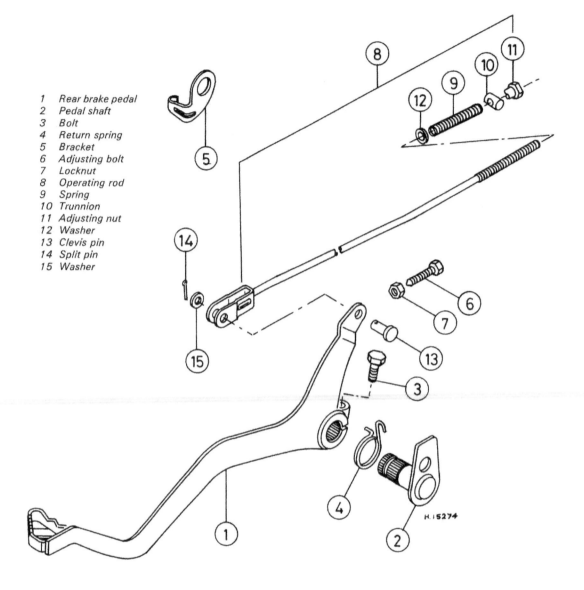

1   Rear brake pedal
2   Pedal shaft
3   Bolt
4   Return spring
5   Bracket
6   Adjusting bolt
7   Locknut
8   Operating rod
9   Spring
10 Trunnion
11 Adjusting nut
12 Washer
13 Clevis pin
14 Split pin
15 Washer

Fig. 4.11 Rear brake pedal assembly – 100, 125 and 185 T and X models

lubrication or if any of the bearing surfaces are found to be corroded, clean each surface with fine grade emery cloth before checking for excessive play between the components which make up the pivot assembly. If any one component is thought to be excessively worn then it must be renewed.

7   Where applicable, check for excessive wear in the splines of both the pedal and shaft. Unacceptable wear of these splines will mean renewal of both components. Note that TS250 models have a replaceable bush fitted between the pedal and its shaft.

8   Fitting of the brake pedal assembly is a direct reversal of the removal procedure, whilst noting the following points. Lubricate each bearing surface with a high quality lithium based grease. Grease should also be used to coat the return spring to lessen the risk of it being corroded by road salts. Always use a new split-pin when reconnecting the brake operating rod and spread the legs of the pin to retain it in position. On TS250 models, check the condition of the spring washer fitted to the pedal shaft and renew it if it is seen to be flattened. On all models, check that the pedal pivots smoothly over its full operating

range and returns to its correct position.

9   Reconnect the stop lamp switch operating spring and before riding the machine, carry out a final check to ensure that both the rear brake and the stop lamp switch are in correct adjustment and are functioning correctly.

17 Kickstart lever: examination and renovation

1   The kickstart lever is splined and is secured to its shaft by means of a pinch bolt. The kickstart crank swivels so that it can be tucked out of the way when the engine is started. It is held in position on the swivel by a washer and circlip. A spring-loaded ball bearing locates the kickstart arm in either the operating or folded position; if the action becomes sloppy it is probable that the spring behind the ball bearing needs renewing. It is advisable to remove the circlip and washer occasionally, so that the kickstart crank can be detached and the swivel greased.

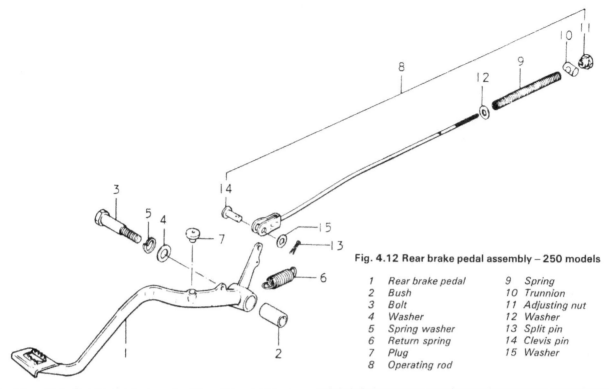

Fig. 4.12 Rear brake pedal assembly – 250 models

1	Rear brake pedal	9	Spring
2	Bush	10	Trunnion
3	Bolt	11	Adjusting nut
4	Washer	12	Washer
5	Spring washer	13	Split pin
6	Return spring	14	Clevis pin
7	Plug	15	Washer
8	Operating rod		

2   It is unlikely that the kickstart crank will bend in an accident unless the machine is ridden with the kickstart in the operating and not folded position. It should be removed and straightened, using the same technique as that recommended for the footrests.

3   Carry out a check at frequent intervals to ensure that the lever has not become loose on its shaft through slackening of the pinch bolt. Any movement between the lever and its shaft will lead to rapid wear of both sets of splines, resulting in expensive replacement items having to be purchased and fitted. If necessary, lock the pinch bolt in position by degreasing its threads and coating them with a thread locking compound.

18 Dualseat: removal and refitting

1   Removal of the dualseat is a simple and straightforward procedure. Commence by lifting the seat and removing the split-pin from each of the two seat pivot pins. Using a pair of pliers, draw each pivot pin from position and then lift the seat clear of the machine.

2   Refitting the seat is a direct reversal of the removal procedure, whilst noting the following points. Clean each pivot assembly and lubricate the pin with grease before inserting it and locking it in position with a new split-pin. Spread both legs of the split-pin to retain it in position. The seat must be properly fitted; should it become detached whilst the machine is in motion, the resulting loss in balance of the rider could well prove disastrous.

17.1 The kickstart lever must be correctly positioned on its shaft

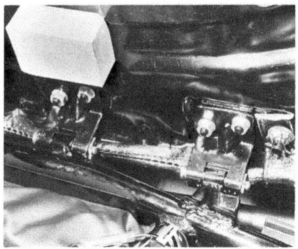

18.1 Detach the dualseat by withdrawing each of its two pivot pins

Chapter 5 Wheels, brakes and tyres

For modifications and information relating to later models, see Chapter 7

Contents

General description	1	Front brake: adjustment	11
Front wheel: examination and renovation	2	Rear brake: adjustment	12
Front wheel: removal and refitting	3	Front and rear brake assemblies: dismantling,	
Front wheel bearings: removal, examination and fitting	4	examination, renovation and reassembly	13
Speedometer drive gear: examination and renovation	5	Front and rear brakes: checking shoe wear	14
Rear wheel: examination, removal, renovation and		Tyres: removal, repair and refitting	15
refitting	6	Valve cores and caps	16
Rear wheel bearings: removal, examination and refitting	7	Security bolts	17
Rear wheel sprocket: examination and renewal	8	Wheel balancing	18
Cush drive: examination and renovation – TS250	9		
Final drive chain: examination, adjustment and			
lubrication	10		

Specifications

Wheels

Type	Conventional, steel spoked with chromed steel rim

	Front	Rear
Size:		
TS100 models	19 inch	18 inch
All other models	21 inch	18 inch

Rim runout:	
Radial and axial service limit	2.0 mm (0.08 in)
Spindle runout:	
Front and rear service limit	0.25 mm (0.01 in)

Brakes (front and rear)

Type	Internally expanding, single leading shoe, drum
Shoe lining thickness service limit	1.5 mm (0.06 in)
Drum internal diameter service limit:	
TS250 models	150.7 mm (5.93 in)
All other models	130.7 mm (5.15 in)

Tyres

	TS100 models	TS125 and 185 models	TS250 models
Size:			
Front	2.75-19 4PR	2.75-21 4PR	3.00-21 4PR
Rear	3.00-18 4PR	4.10-18 4PR	4.60-18 4PR

	TS100 models	All other models
Pressures:		
Solo:		
Front	21 psi (1.5 kg/cm²)	21 psi 1.5 kg/cm²
Rear	28 psi (2.0 kg/cm²)	25 psi (1.75 kg/cm²)
With pillion:		
Front	25 psi (1.75 kg/cm²)	21 psi (1.5 kg/cm²)
Rear	31 psi (2.25 kg/cm²)	28 psi (2.0 kg/cm²)

Torque wrench settings

	kgf m	lbf ft
Front wheel spindle nut:		
TS100 and 125	4.0 – 5.2	29.0 – 37.5
TS185	3.0 – 7.0	21.5 – 50.5
TS250	3.6 – 5.2	26.0 – 37.5
Front wheel spindle retaining clamp nuts:		
US TS125 only	1.5 – 2.5	11.0 – 18.0
Rear wheel spindle nut:		
TS100 and 125	6.0 – 8.0	43.0 – 57.5
TS185	3.0 – 7.0	21.5 – 50.5
TS250	5.0 – 8.0	36.0 – 57.5
Rear torque arm nuts:		
All models	1.0 – 1.5	7.5 – 10.5
Rear nut on TS250 only	2.0 – 3.0	14.5 – 21.5
Front and rear brake cam lever bolt or nut	0.5 – 0.8	3.5 – 6.0

1 General description

All of the models covered in this Manual utilise the traditional type of wheel, that is, a chromed steel rim laced to an aluminium alloy hub by steel spokes. Wheel sizes vary between model types and reference should be made to the Specifications Section of this Chapter for both wheel and tyre sizes. All models utilise conventional tubed tyres which have a block pattern tread suitable for both on and off road use.

The type of brake fitted to the front and rear wheels of all model types is a half-width hub design with single leading shoe (sls) operation.

2 Front wheel: examination and renovation

1 Wire spoked wheels are often viewed as being prone to problems when compared to the increasingly popular cast alloy and composite types. Whilst this is true to some extent, it is also true that wire spoked wheels are relatively easy and inexpensive to adjust or repair. Spoked wheels can go out of true over periods of prolonged use and like any wheel, as the result of an impact. The condition of the hub, spokes and rim should therefore be checked at regular intervals.

2 For ease of use an improvised wheel stand is invaluable but failing this, the wheel can be checked whilst in place on the machine after it has been raised clear of the ground. Make the machine as stable as possible, using blocks beneath the crankcase as a means of support. Spin the wheel and ensure that there is no brake drag. If necessary, slacken the brake adjuster until the wheel turns freely. In the case of rear wheels it is advisable, though not essential, to remove the final drive chain.

3 Slowly rotate the wheel and examine the rim for signs of serious corrosion or impact damage. Slight deformities, as might be caused by running the wheel along a curb, can often be corrected by adjusting spoke tension. More serious damage may require a new rim to be fitted, and this is best left to an expert. Whilst this is not an impossible undertaking at home, there is an art to wheel building, and a professional wheel builder will have the facilities and parts required to carry out the work quickly and economically. Badly rusted steel rims should be renewed in the interests of safety as well as appearance. Where light alloy rims are fitted, corrosion is less likely to be a serious problem, though neglect can lead to quite substantial pitting of the alloy.

4 If it has been decided that a new rim is required some thought should be given to the size and type of the replacement rim. In some instances the problem of obtaining replacement tyres for an oddly sized original rim can be resolved by having a more common rim size fitted. Do check that this will not lead to other problems, fitting a new rim whose tyre fouls some other part of the machine could prove a costly error. Remember that changing the size of the rear wheel rim will alter the overall gearing. In most cases it should be possible to have a light alloy rim fitted in place of an original plated steel item. This will have a marginal effect in terms of weight reduction, but will prove far more corrosion resistant.

5 Assuming the wheel to be undamaged, it will be necessary to check it for runout. This is best done by arranging a

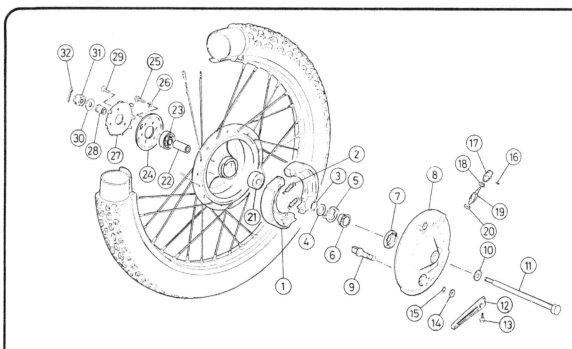

Fig. 5.1 Front wheel

1 Brake shoe - 2 off	10 Washer	19 Speedometer driven	25 Bolt - 3 off
2 Return spring -	11 Spindle	gear	26 Spring washer -
2 off	12 Brake operating arm	20 Washer	3 off
3 Circlip	13 Bolt	21 Left-hand wheel	27 Bearing retainer
4 Thrust washer	14 Washer	bearing	plate
5 Drive plate	15 O-ring	22 Spacer	28 Spacer
6 Speedometer drive gear	16 Grub screw	23 Right-hand wheel	29 Screw - 3 off
7 Oil seal	17 Speedometer cable	bearing	30 Washer
8 Brake backplate	terminal	24 Bearing retainer	31 Nut
9 Brake operating cam	18 Washer	plate hub	32 Split pin

temporary wire pointer so that it runs close to the rim. The wheel can now be turned and any distortion noted. Check for lateral distortion and for radial distortion, noting that the latter is less likely to be encountered if the wheel was set up correctly from new and has not been subject to impact damage. Note that it is possible that worn wheel bearings may cause rim runout.

6   The rim should be no more than 2.0 mm (0.1 in) out of true in either plane. If a significant amount of distortion is encountered, check that the spokes are of approximately equal tension. Tapping the spokes with a screwdriver is the best method of determining whether they are of equal tension. A loose spoke will produce a quite different sound to those around it. Adjustment is effected by turning the square-headed spoke nipples with the appropriate spoke key. This tool is obtainable from most motorcycle shops or tool retailers.

7   With the spokes evenly tensioned, any remaining distortion can be pulled out by tightening the spokes on one side of the hub and slackening the corresponding spokes from the opposite hub flange. This will allow the rim to be pulled across whilst maintaining spoke tension.

8   If more than slight adjustment is required, it should be noted that the tyre and inner tube should be removed first to give access to the spoke ends. Those which protrude through the nipple after adjustment should be filed flat to avoid the risk of puncturing the tube. It is essential that the rim band is in good condition as an added precaution against chafing. In an emergency, use a strip of duct tape as an alternative; unprotected tubes will soon chafe on the nipples.

9   Should a spoke break, a replacement item can be fitted and retensioned in the normal way. Wheel removal is usually necessary for this operation, although complete removal of the tyre can be avoided if care is taken. A broken spoke should be attended to promptly because the load normally taken by that spoke is transferred to adjacent spokes which may fail in turn.

10 Remember to check wheel condition regularly. Normal maintenance is confined to keeping the spokes correctly tensioned and will avoid the costly and complicated wheel rebuilds that will inevitably result from neglect. When cleaning the machine do not neglect the wheels. If the rims are kept clean and well polished many of the corrosion related maladies will be prevented.

---

3   Front wheel: removal and refitting
---

1   Before attempting removal of the front wheel, support the machine in a stable position with the front wheel well clear of the ground. This can be accomplished by positioning a stout wooden crate or blocks beneath the engine crankcase guard. Make sure that there is no danger of the machine toppling whilst it is being worked on.

2   Displace the split-pin from the wheel spindle retaining nut and remove the nut with its plain washer. Where a wheel spindle retaining clamp is fitted to the base of the right-hand fork leg, release the clamp by loosening its two retaining nuts. Disconnect the speedometer cable from the speedometer gearbox by unscrewing its knurled retaining ring and pulling it out of its location. Allow the cable to hang clear of the wheel.

3   Disconnect the brake operating cable from the brake cam operating arm. Before doing this, it is necessary to release the cable from its retaining clamp at the top of the left-hand lower fork leg. Pull the small rubber boot off the end of the threaded adjuster at the brake backplate end of the cable and release both adjuster locknuts. Wind the lower of these locknuts off the end of the adjuster. Displace the split-pin from the clevis pin which secures the cable end bracket to the cam operating arm and remove the clevis pin. Pull the adjuster up and clear of the backplate and allow the cable to hang clear of the wheel.

4   Fit a tommy bar through the hole provided in the end of the wheel spindle, support the wheel and pull the spindle clear of the fork legs. Carefully lower the wheel and then manoeuvre it

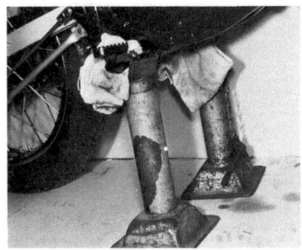

3.1 Support the machine securely with the front wheel clear of the ground

3.5a Check that the wheel spacer is correctly fitted ...

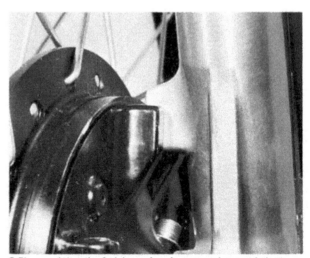

3.5b ... and that the fork leg spigot is engaged correctly in the brake backplate ...

Tyre changing sequence - tubed tyres

 A Deflate tyre. After pushing tyre beads away from rim flanges push tyre bead into well of rim at point opposite valve. Insert tyre lever adjacent to valve and work bead over edge of rim.

 **B** Use two levers to work bead over edge of rim. Note use of rim protectors

 C Remove inner tube from tyre

 D When first bead is clear, remove tyre as shown

 E When fitting, partially inflate inner tube and insert in tyre

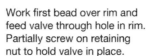 **F** Work first bead over rim and feed valve through hole in rim. Partially screw on retaining nut to hold valve in place.

 G Check that inner tube is positioned correctly and work second bead over rim using tyre levers. Start at a point opposite valve.

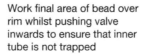 **H** Work final area of bead over rim whilst pushing valve inwards to ensure that inner tube is not trapped

clear of the machine. Note the fitted position of the spacer located in the right-hand side of the wheel hub.

5   Fitting of the wheel is a direct reversal of the removal procedure, whilst noting the following points. When lifting the wheel into position between the fork legs, ensure that the spacer is correctly located in the right-hand side of the wheel hub and that the spigot which forms part of the fork lower leg engages correctly in the slot cast in the brake backplate. Note that if the brake backplate is allowed to rotate, due to its not being engaged correctly with the fork leg, the wheel will lock on the first application of the brake, with disastrous consequences.

6   Before inserting the wheel spindle, apply a light film of grease along its length. Align the wheel between the fork legs and push the spindle into position, giving it a light tap with a soft-faced hammer to seat it. Fit the spindle retaining nut, with its plain washer, and tighten it to the specified torque loading. Where a spindle retaining clamp is fitted, tighten the nuts of the clamp to the specified torque setting. Lock the spindle retaining nut in position with a new split-pin.

7   Reconnect the speedometer drive cable and the brake operating cable. Use a new split-pin to retain the brake cable clevis pin in position and check the brake for correct operation and adjustment. Finally, before taking the machine on the road, recheck all disturbed connections for security.

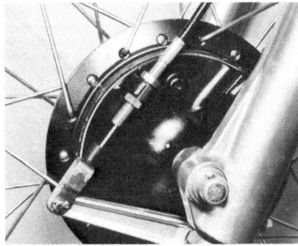

3.7b Reconnect and adjust the brake operating cable ...

3.6 ... before locking the tightened spindle retaining nut in position with a new split pin

3.7c ... before clamping the cable to the fork leg

3.7a Reconnect the speedometer drive cable

4   Front wheel bearings: removal, examination and fitting

1   Access to the bearings of the front wheel may be gained after removal of the wheel from the machine and withdrawal of the brake assembly from the wheel hub. It will also be necessary to detach the bearing retainer plate from the right-hand side of the wheel hub by removing its three retaining screws.

2   Position the wheel on a work surface with its hub well supported by wooden blocks so that enough clearance is left beneath the wheel to drive one bearing out. Ensure the blocks are placed as close to the bearing as possible, to lessen the risk of distortion occurring to the hub casting whilst the bearings are being removed or fitted.

3   Place the end of a long-handled drift against the upper face of the lower bearing and tap the bearing downwards out of the wheel hub. The spacer located beneath the two bearings may be moved sideways slightly in order to allow the drift to be positioned against the face of the bearing. Move the drift around the face of the bearing whilst drifting it out of position, so that the bearing leaves the hub squarely.

4   With the one bearing removed, the wheel may be lifted and

the spacer withdrawn from the hub, invert the wheel and remove the second bearing, using a similar procedure to that used for the first.

5   Remove all the old grease from the wheel hub and bearings, giving the latter a final wash in fuel whilst observing the necessary fire precautions. Check the bearings for signs of play or roughness when they are turned. If there is any doubt about the condition of a bearing, it should be renewed.

6   If the original bearings are to be refitted, then they should be repacked with the recommended grease before being fitted into the hub. New bearings must also be packed with the recommended grease. Ensure that all the bearing recesses in the hub are clean and both bearings and recess mating surfaces lightly greased to aid fitting. Check the condition of the hub recesses for evidence of abnormal wear which may have been caused by the outer race of a bearing spinning. If evidence of this happening is found, and the bearing is a loose fit in the hub, then it is best to seek advice from a Suzuki service agent or a competent motorcycle engineer. Do not proceed with fitting a bearing that is a loose fit in the hub.

7   With the wheel hub and bearing thus prepared, proceed to fit the bearings and central spacer as follows. With the hub again well supported by the wooden blocks, drift the first of the two bearings into position. Note that each bearing must be fitted with its sealed side facing outboard; the bearing with the rubberised side being fitted into the right-hand side of the hub.

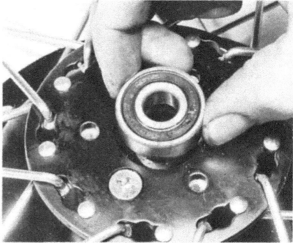
4.7a Each bearing must be fitted with its sealed side facing outboard

4.1 Remove the wheel bearing retainer plate ...

4.7b Do not omit to insert the spacer into the wheel hub ...

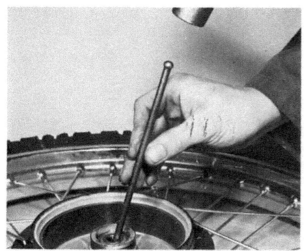
4.3 ... and drift each wheel bearing from position

4.7c ... before fitting the second of the wheel bearings

To fit the bearing, use a soft-faced hammer in conjunction with a socket or length of metal tube which has an overall diameter which is slightly less than that of the outer race of the bearing. Invert the wheel, insert the spacer and fit the second bearing, using the same procedure as given for the first. Take great care to ensure that each of the bearings enters its housing correctly, that is, square to the housing, otherwise the housing surface may be broached.

8   With both bearings fitted, position the bearing retainer plate on the right-hand side of the wheel hub so that the holes in the plate are aligned with the threaded holes in the wheel hub. Degrease the threads of each plate retaining screw and coat them with a thread locking compound before fitting and tightening each screw. Note that this is now the ideal time for examining the components which make up the brake and speedometer drive assemblies.

5   Speedometer drive gear: examination and renovation

1   The speedometer drive gear forms part of the front wheel brake backplate assembly. To gain access to the component parts which make up the drive assembly, remove the front wheel from the machine and detach the brake backplate from the wheel hub.

2   To dismantle the drive gear, remove the circlip from the centre of the backplate and lift the thrust washer, drive plate and drive gear from position. Invert the backplate and remove the small grub screw from the side of the worm gear housing. The complete worm gear assembly can now be pulled from its housing. Take care to retain the two small thrust washers. Clean the component parts and lay them out on a clean work surface, ready for inspection.

3   Damage to any one component part should be immediately obvious and that part should be placed to one side, ready for renewal. Pay particular attention to the condition of the tangs on the drive plate; broken or badly worn tangs are the usual cause of the gearbox failing to drive the speedometer cable.

4   Note the condition of the large oil seal which surrounds the drive plate. If this seal shows signs of damage or deterioration, then it must be renewed otherwise grease will work through from the drive assembly to contaminate the brake linings. Use the flat of a screwdriver to lever the seal from position. Normally, the seal will slide out of its location quite easily but care must be taken not to damage the metal of the backplate with the screwdriver. To fit the new seal, push it as far as it will go into its location with finger pressure, place a strip of wood across it and then tap down on the wood with a hammer to drive the seal fully home. Ensure that the seal enters its location squarely.

5   Reassemble the component parts of the drive gear in the

5.2 Remove the grub screw which retains the speedometer worm gear assembly in position

5.4 Carefully lever the damaged oil seal from position

5.5a Lubricate the speedometer drive gears ...

5.5b ... before fitting the drive plate over the drive gear ...

5.5c ... followed by the thrust washer ...

5.5d ... and the drive assembly retaining circlip

reverse order of dismantling. Lightly lubricate the drive assembly with a good quality high melting point grease and refer to the figure accompanying this text if in doubt as to the fitted position of any one part.

6 Rear wheel: examination, removal, renovation and refitting

1   Before examining and, if necessary, attempting removal of the rear wheel, support the machine in a stable position with the rear wheel clear of the ground. This can be accomplished by positioning a stout wooden crate or blocks beneath the rear section of the frame lower tubes. Ensure that there is no danger of the machine's toppling during wheel removal. Examine the wheel as described in Section 2 of this Chapter for the front wheel.
2   Commence removal of the wheel by inserting a tommy bar through the hole provided in the end of the wheel spindle; displace the split-pin from the spindle retaining nut and remove the nut together with any plain washers located beneath it. Detach the torque arm from the brake backplate by removing the split-pin or spring clip which passes through the threaded section of its retaining bolt and then removing the retaining nut with its spring washer and plain washer to allow the bolt to be pushed through the torque arm.
3   Detach the brake operating rod from the cam shaft operating arm by removing the nut from the threaded end of the rod and then depressing the brake pedal so that the rod leaves the trunnion in the arm. Refit the nut, the trunnion from the operating arm, the return spring and the washer to the rod to prevent loss and then rest the end of the rod on the ground beneath the machine.
4   Lay a length of clean rag or paper beneath the lower run of the final drive chain. Rotate the rear wheel until the split link in the chain appears at the wheel sprocket. Using a pair of flat-nose pliers, remove the spring clip from the link and withdraw the link from the chain. Lift both ends of the chain clear of the sprocket teeth and place them on the length of rag or paper. Doing this will prevent any dirt attaching itself to the chain. Reassemble the component parts of the split link to prevent their loss.
5   Support the wheel and, using the tommy bar in the end of the wheel spindle as a handle, pull the spindle clear of the machine. Allow the wheel to drop clear of the swinging arm fork and note the fitted position of the wheel spacer(s) before detaching each one from the wheel hub. The wheel can now be manoeuvred clear of the machine.

6.7a Check that the left-hand spacer ...

6.7b ... and the right-hand spacer are correctly fitted before inserting the rear wheel spindle (125 shown)

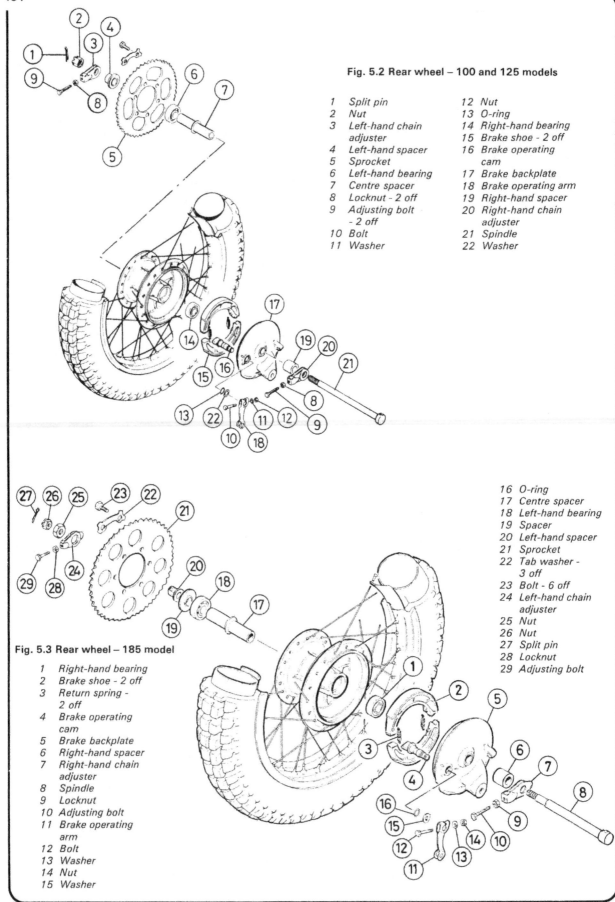

Fig. 5.2 Rear wheel – 100 and 125 models

1　Split pin
2　Nut
3　Left-hand chain adjuster
4　Left-hand spacer
5　Sprocket
6　Left-hand bearing
7　Centre spacer
8　Locknut - 2 off
9　Adjusting bolt - 2 off
10　Bolt
11　Washer
12　Nut
13　O-ring
14　Right-hand bearing
15　Brake shoe - 2 off
16　Brake operating cam
17　Brake backplate
18　Brake operating arm
19　Right-hand spacer
20　Right-hand chain adjuster
21　Spindle
22　Washer

16　O-ring
17　Centre spacer
18　Left-hand bearing
19　Spacer
20　Left-hand spacer
21　Sprocket
22　Tab washer - 3 off
23　Bolt - 6 off
24　Left-hand chain adjuster
25　Nut
26　Nut
27　Split pin
28　Locknut
29　Adjusting bolt

Fig. 5.3 Rear wheel – 185 model

1　Right-hand bearing
2　Brake shoe - 2 off
3　Return spring - 2 off
4　Brake operating cam
5　Brake backplate
6　Right-hand spacer
7　Right-hand chain adjuster
8　Spindle
9　Locknut
10　Adjusting bolt
11　Brake operating arm
12　Bolt
13　Washer
14　Nut
15　Washer

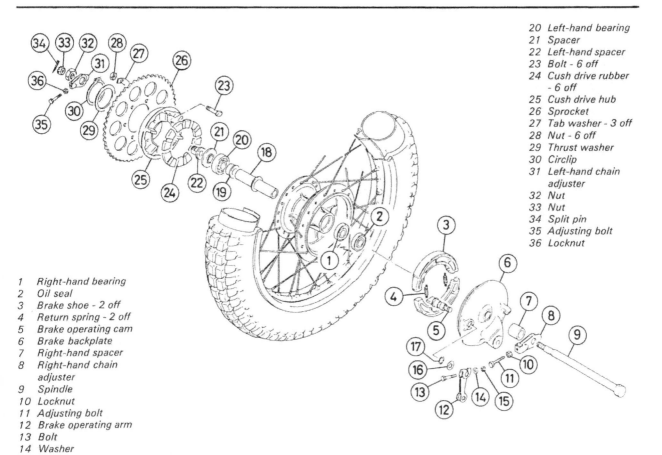

20 Left-hand bearing
21 Spacer
22 Left-hand spacer
23 Bolt - 6 off
24 Cush drive rubber
   - 6 off
25 Cush drive hub
26 Sprocket
27 Tab washer - 3 off
28 Nut - 6 off
29 Thrust washer
30 Circlip
31 Left-hand chain
   adjuster
32 Nut
33 Nut
34 Split pin
35 Adjusting bolt
36 Locknut

1 Right-hand bearing
2 Oil seal
3 Brake shoe - 2 off
4 Return spring - 2 off
5 Brake operating cam
6 Brake backplate
7 Right-hand spacer
8 Right-hand chain
   adjuster
9 Spindle
10 Locknut
11 Adjusting bolt
12 Brake operating arm
13 Bolt
14 Washer
15 Nut
16 Washer
17 O-ring
18 Centre spacer
19 Bearing holder

Fig. 5.4 Rear wheel – 250 model

6   Once any faults found in the wheel have been rectified, the wheel should be fitted by reversing the procedure used for its removal. If working on a TS250 model, take note of the information given in Section 9 of this Chapter concerning the cush drive assembly. If in doubt as to the fitted position of any one component part, then refer to the figure accompanying this text. Note the following information.

7   Before inserting the wheel spindle, apply a light film of grease along its length. With the wheel and hub spacer(s) aligned between the legs of the swinging arm, push the spindle into position, giving its end a tap with a soft-faced hammer to seat it. Fit the spindle retaining nut together with any plain washers and tighten the nut finger-tight.

8   When reconnecting the brake operating rod, feed the rod through the brake lever trunnion and fit the adjusting nut onto the end of the rod to retain it in position. Ensure that the spring is correctly positioned between the washer on the operating rod and the forward facing face of the lever trunnion.

9   Fit and tighten the torque arm retaining nut to the specified torque loading. Note that the spring washer fitted beneath this nut must be in a serviceable condition. Fit a new split-pin through the arm retaining bolt and spread both of its legs. Where a spring clip is fitted, the original clip may be reused. Note that the torque arm must be properly attached to the brake backplate. Failure to ensure this will mean that on the first application of the rear brake, the wheel will lock, with disastrous consequences.

10 Reconnect the ends of the final drive chain and check the tension of the chain in accordance with the instructions given in Section 10 of this Chapter. Take particular note of the instruc-

6.8 Reconnect the rear brake operating rod

tions which refer to fitting of the split link. With this done, tighten the wheel spindle retaining nut to the specified torque loading and then lock it in position with a new split-pin.

11 Refer to Section 12 of this Chapter and check that the rear brake is correctly adjusted. Check that the wheel spins freely and, before taking the machine on the road, recheck all disturbed connections for security.

6.9 Fit a new split-pin through the torque arm retaining bolt

6.10 Torque load the spindle retaining nut and lock it in position with a new split-pin

7   Rear wheel bearings: removal, examination and refitting

1   Access to the bearings of the rear wheel may be gained after removal of the wheel from the swinging arm fork and withdrawal of the brake assembly from the wheel hub. The two bearings contained within the wheel hub may be drifted out of position by using a procedure similar to that given for removal of the front wheel bearings.

2   Reference to the figure accompanying this text will show that TS250 models have an oil seal fitted over the right-hand bearing. This seal will be drifted out of position along with the bearing and should be renewed as a matter of course. To fit the new seal, use a method similar to that employed for fitting each bearing.

3   Note when fitting the bearings, that the bearing with the rubberised sealed side must be fitted in the left-hand side of the wheel hub. Both bearings must have their sealed side facing outboard. On TS250 models, lightly smear the lip of the bearing oil seal with grease. The procedure for fitting the bearings is similar to that given for fitting of the front wheel bearings.

4   Before refitting the wheel to the machine, it is advisable to carry out a careful examination of the brake assembly, the rear wheel sprocket and, on TS250 models, the cush drive assembly.

8   Rear wheel sprocket: examination and renewal

1   The rear wheel sprocket need only be renewed if the teeth are hooked or badly worn. It is considered bad practice to renew one sprocket on its own; both drive sprockets should be renewed as a pair, preferably with a new final drive chain. If this recommendation is not observed, rapid wear resulting from the running of old and new parts together will necessitate even earlier replacement on the next occasion.

2   On TS250 models, the sprocket is secured to the cush drive hub by six bolts whose nuts are locked in pairs by tab washers. All other models have the sprocket bolted direct to the wheel hub; six bolts are used and these are secured in position in pairs by tab washers.

3   To release the sprocket from the wheel, bend back the locking tab at each end of the washers and remove the securing bolts (or nuts). Fitting of the replacement sprocket is simply a reversal of the procedure given for removal of the worn item, whilst noting the following points. Tighten the securing bolts evenly and in a diagonal sequence. Do not rebend the locking

8.3 Lock each wheel sprocket retaining bolt in position with its tab washer (125 shown)

tab previously used; if necessary, renew the tab washers as a set.

9   Cush drive: examination and renovation – TS250 models

1   The cush drive assembly is contained within the left-hand side of the rear wheel hub. It takes the form of six rubber pads which are a push fit in the wheel hub. The cush drive hub, which is bolted to the rear wheel sprocket, incorporates six substantial vanes each one of which engages with the central slot in one of the rubber pads. This arrangement permits the sprocket to move within certain limits, therefore absorbing any surge or roughness in the transmission. The rubbers should be renewed when movement of the sprocket in relation to the wheel indicates bad compaction of the rubbers. The rubbers should also be renewed if they are seen to be breaking up.

2   To gain access to the cush drive rollers, remove the circlip which retains the sprocket and hub assembly to the rear wheel. With this circlip removed, lift the sprocket clear of the wheel

and detach the large thrust washer from its location beneath the circlip retaining groove.

3 Before fitting the new rubbers, carry out a careful inspection of the cush drive hub for any signs of fatigue or failure, especially around the base of the vanes. Fatigue will usually be indicated by the presence of hairline cracks. If in doubt as to the condition of this component, then return it to an official Suzuki service agent for further examination.

4 Before reassembling the cush drive, it should be noted that it may be difficult to insert the vanes of the cush drive hub into new rubbers unless the rubbers are first lubricated with a solution of soapy water around the area of the slots. With the hub and sprocket pushed into position, refit the thrust washer and retain the assembly in position with the circlip.

10 Final drive chain: examination, adjustment and lubrication

1 As the final drive chain is fully exposed on all models, it requires lubrication and adjustment at regular intervals. To adjust the chain, place the machine on its stand, take out the split-pin from the rear wheel spindle and slacken the spindle nut. Slacken also the nuts securing the brake torque arm. Undo the locknut on the chain adjusters and turn the adjuster bolts inwards to tighten the chain. Marks on the adjusters must be in line with identical marks on the frame fork to align the rear wheel correctly. A final check can be made by laying a straight wooden plank alongside the wheels, each side in turn. Chain tension is correct when there is the specified amount of slack in the centre of the lower chain run.

Model type	Chain movement
TS100	30–40 mm (1.2–1.6 in)
TS125	40–50 mm (1.6–2.0 in)
TS185	35–45 mm (1.4–1.8 in)
US TS250	35–45 mm (1.4–1.8 in)
US TS250	50–60 mm (2.0–2.4 in)

Note that each measurement listed in the above table must be made with the rider off the machine and with the machine supported on its prop stand. The measurement of 15 – 20 mm (0.6 – 0.8 in) given by Suzuki in some of their service literature applies only when the rider is seated and with the rear suspension compressed. Remember that the chain may not have worn evenly. Rotate the rear wheel to check along the chain length for its tight spot and make the necessary measurement with this tight spot in the centre of the lower chain run.

2 Under no circumstances run the chain overtight to compensate for uneven wear. A tight chain will place excessive stresses on the gearbox and rear wheel bearings leading to their early failure. It will also absorb a surprising amount of power.

3 The chain may be checked for wear with it fitted to the machine. Commence by washing the chain thoroughly with a petrol-soaked brush. Wipe the chain dry and stretch it to its full length by turning the adjuster bolts inwards. Select a length of chain in the middle of the lower run and count of 21 pins, that is, a 20 pitch length. Measure the distance between the 1st and 21st pins. If this distance is found to be in excess of the service limit specified in the following table, then the chain must be renewed.

Model type	Wear limit
TS100 and 125	259 mm (10.2 in)
TS185	323 mm (12.72 in)
TS250	324 mm (12.8 in)

4 Should the chain require renewal, move the rear wheel fully forward and then rotate it until the split link appears at the rear sprocket. Using a pair of flat-nosed pliers, remove the spring clip from the link and withdraw the link from the chain. The new chain should now be attached to the end of the upper run of old chain and the old chain used to pull the new chain into position

10.1a Check that wheel alignment is correct by referring to the marks provided

10.1b On completion of adjustment, lock each adjuster bolt in position with its locknut

10.4 The closed end of the final drive chain spring clip must face the normal direction of chain travel

over the gearbox sprocket. Once in position, connect the ends of the new chain with the new split link provided. Note that the spring clip which retains the link in position must have the side plate fitted beneath it, be seated correctly and have its closed end facing the direction of chain travel.

5   Note that replacement chains are now available in standard metric sizes from Renold Limited, the British chain manufacturer. When ordering a new chain, always quote the size, the number of chain links and the type of machine to which the chain is to be fitted.

6   Remember, on completion of examination and, if necessary, adjustment or replacement of the chain, to ensure that both adjuster bolts are locked in position by tightening their locknuts. The wheel spindle retaining nut should be tightened to the specified torque setting and locked in position by fitting a new split-pin. Finally, spin the rear wheel to ensure that it rotates freely, adjust the rear brake operating mechanism, when necessary, and retighten the torque arm retaining nuts to their specified torque setting.

7   After a period of running, the chain will require lubrication. Lack of oil will accelerate the rate of wear of both chain and sprockets and will lead to harsh transmission. The application of engine oil will act as a temporary expedient, but it is preferable to use one of the proprietary graphited greases contained within an aerosol can. This type of lubricant is thrown off the chain less easily than engine oil. Ideally the chain should be removed at regular intervals, and immersed in a molten lubricant such as Linklyfe or Chainguard after it has been cleaned in a paraffin bath. These latter lubricants achieve better penetration of the chain links and rollers and are less likely to be thrown off when the chain is in motion.

10.7 An aerosol chain lubricant should be used at frequent intervals

11 Front brake: adjustment

1   Adjustment of the front brake is correct when there is 20 – 30 mm (0.8 – 1.2 in) of clearance between the end of the handlebar lever and the throttle twistgrip with the lever fully applied.

2   To adjust the clearance between the lever and twistgrip, simply loosen the locknut(s) at the cam shaft operating arm end of the cable and turn the cable adjuster the required amount in the appropriate direction before retightening the locknut(s). Any minor adjustments necessary may then be made with the cable adjuster at the handlebar lever bracket. To use this adjuster, simply loosen the lock ring, turn the knurled adjuster the required amount and then retighten the lock ring.

3   On completion of adjustment, check the brake for correct

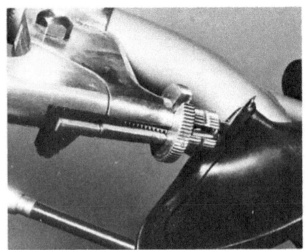

11.2 Fine adjustment of the front brake may be made with the adjuster at the handlebar lever bracket

operation by spinning the wheel and applying the brake lever. There should be no indication of the brake binding as the wheel is spun. If the brake shoes are heard to be brushing against the surface of the wheel drum back off on the cable adjuster slightly until all indication of binding disappears. The brake may be readjusted after a period of bedding-in has been allowed for the brake shoes.

12 Rear brake: adjustment

1   Adjustment of the rear brake is correct when there is 20 – 30 mm (0.8 – 1.2 in) of movement, measured at the forward point of the brake pedal, between the point at which the brake pedal is fully depressed and the point where it abuts against its return stop.

2   The range of pedal movement may be adjusted simply by turning the nut on the wheel end of the brake operating rod in the required direction.

3   On completion of brake adjustment, check that the stop lamp switch operates the stop lamp as soon as the brake pedal is depressed. If necessary, adjust the height setting of the switch in accordance with the instructions given in Chapter 6.

13 Front and rear brake assemblies: dismantling, examination, renovation and reassembly

1   The brake assembly, complete with the brake backplate, can be withdrawn from its wheel hub after removal of the wheel from the machine. With the wheel laid on a work surface, brake backplate uppermost, the brake backplate may be lifted away from the hub. It will come away quite easily, with the brake shoe assembly attached to it.

2   Examine the condition of the brake linings. If they are worn beyond the specified limit then the brake shoes should be renewed. The linings are bonded on and cannot be supplied separately.

3   If oil or grease from the wheel bearings has badly contaminated the linings, the brake shoes should be renewed. There is no satisfactory way of degreasing the lining material. Any surface dirt on the linings can be removed with a stiff-bristled brush. High spots on the linings should be carefully eased down with emery cloth.

4   Examine the drum surface for signs of scoring, wear beyond the service limit or oil contamination. All of these conditions will impair braking efficiency. Remove all traces of dust, preferably

13.4 The brake drum should not be worn beyond the stated service limit (125 shown)

13.10a Relocate the cleaned and lubricated brake cam shaft in the brake backplate ...

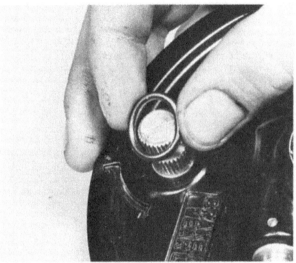

13.10b ... fit the new O-ring over the shaft end ...

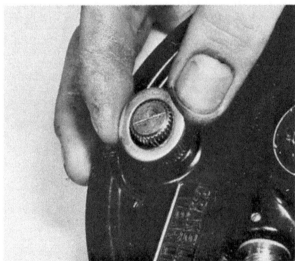

13.10c ... followed by the plain washer ...

13.10d ... and then position the brake operating arm correctly on the shaft before fitting and tightening its securing bolt

13.12 Carefully push the brake shoe assembly into position

using a brass wire brush, taking care not to inhale any of it, as it is of an asbestos nature, and consequently harmful. Remove oil or grease deposits with a rag soaked in fuel whilst observing the necessary fire precautions.

5 If deep scoring is evident, due to the linings having worn through to the shoe at some time, the drum must be skimmed on a lathe, or renewed. Whilst there are firms who will undertake to skim a drum whilst it is fitted to the wheel, it should be borne in mind that excessive skimming will change the radius of the drum in relation to the brake shoes, therefore reducing the friction area until extensive bedding in has taken place. Also full adjustment of the shoes may not be possible. If in doubt about this point, the advice of one of the specialist engineering firms who undertake this work should be sought.

6 Note that it is a false economy to try to cut corners with brake components; the whole safety of both machine and rider being dependent on their good condition.

7 Removal of the brake shoes is accomplished by folding the shoes together so that they form a 'V'. With the spring tension relaxed, both shoes and springs may be removed from the brake backplates as an assembly. Detach the springs from the shoes and carefully inspect them for any signs of fatigue or failure. If in doubt, compare them with a new set of springs.

8 Before fitting the brake shoes, check that the brake operating cam is working smoothly and is not binding in its pivot. The cam can be removed by withdrawing the retaining bolt of the operating arm and pulling the arm off the shaft. Before removing the arm, it is advisable to mark its position in relation to the shaft, so that it can be relocated correctly.

9 Remove any deposits of hardened grease or corrosion from the bearing surface of the brake cam shaft by rubbing it lightly with a strip of fine emery paper or by applying solvent with a piece of rag. Lightly grease the length of the shaft and the face of the operating cam prior to reassembly. Clean and grease the pivot stud which is set in the backplate.

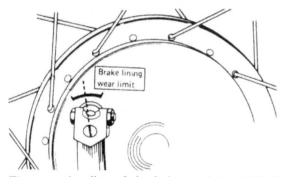

The extension line of the index mark is within the range.

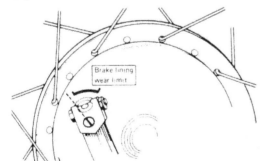

The extension line of the index mark is beyond the range.

Fig. 5.5 Brake shoe lining wear indicator mark

10 Check the condition of the O-ring which prevents the escape of grease from the end of the cam shaft. If it is any way damaged or perished, then it must be renewed before the shaft is relocated in the backplate. Relocate the cam shaft and align and fit the operating arm with the O-ring and plain washer. The bolt which retains the arm in position on the shaft should be tightened to the specified torque loading.

11 Before refitting existing shoes, roughen the lining surface sufficiently to break the glaze which will have formed in use. Glasspaper or emery cloth is ideal for this purpose but take care not to inhale any of the asbestos dust that may come from the lining surface.

12 Fitting the brake shoes and springs to the brake backplate is a reversal of the removal procedure. Some patience will be needed to align the assembly with the pivot and operating cam whilst still retaining the springs in position; once they are correctly aligned though, they can be pushed back into position by pressing downwards in order to snap them into position. Do not use excessive force, or there is risk of distorting the brake shoes permanently.

14 Front and rear brakes: checking shoe wear

1 An indication of brake shoe lining wear is provided by an indicator line which is cast into the brake backplate. If, when the brake is correctly adjusted and applied fully, the line on the end of the brake cam spindle is seen to align with a point outside the arc shown by the indicator line, then the lining on the brake shoes can be assumed to have worn beyond limits and should be renewed at the earliest possible opportunity.

15 Tyres: removal, repair and refitting

1 At some time or other the need will arise to remove and replace the tyres, either as a result of a puncture or because replacements are necessary to offset wear. To the inexperienced, tyre changing represents a formidable task, yet if a few simple rules are observed and the technique learned, the whole operation is surprisingly simple.

2 To remove the tyre from either wheel, first detach the wheel from the machine. Deflate the tyre by removing the valve core, and when the tyre is fully deflated, push the bead away from the wheel rim on both sides so that the bead enters the centre well of the rim. On TS250 models the nuts securing the security bolts must be slackened off so that the bolts can be pushed in towards the tyre. Remove the locking ring and push the tyre valve into the tyre itself.

3 Insert a tyre lever close to the valve and lever the edge of the tyre over the outside of the rim. Very little force should be necessary; if resistance is encountered it is probably due to the fact that the tyre beads have not entered the well of the rim all the way round. If aluminium rims are fitted, damage to the soft alloy by tyre levers can be prevented by the use of plastic rim protectors. Suitable rim protectors may be fabricated very easily from short lengths (4 – 6 inches) of thick-walled nylon petrol pipe which have been split down one side using a sharp knife.

4 Once the tyre has been edged over the wheel rim, it is easy to work round the wheel rim, so that the tyre is completely free from one side. At this stage the inner tube can be removed.

5 Now working from the other side of the wheel, ease the other edge of the tyre over the outside of the wheel rim that is furthest away. Continue to work around the rim until the tyre is completely free from the rim.

6 If a puncture has necessitated the removal of the tyre, reinflate the inner tube and immerse it in a bowl of water to trace the source of the leak. Mark the position of the leak, and deflate the tube. Dry the tube, and clean the area around the puncture with a petrol soaked rag. When the surface has dried, apply rubber solution and allow this to dry before removing the backing from the patch, and applying the patch to the surface.

7 It is best to use a patch of self vulcanizing type, which will form a permanent repair. Note that it may be necessary to remove a protective covering from the top surface of the patch after it has sealed into position. Inner tubes made from a special synthetic rubber may require a special type of patch and adhesive, if a satisfactory bond is to be achieved.

8 Before replacing the tyre, check the inside to make sure that the article that caused the puncture is not still trapped inside the tyre. Check the outside of the tyre, particularly the tread area to make sure nothing is trapped that may cause a further puncture.

9 If the inner tube has been patched on a number of past occasions, or if there is a tear or large hole, it is preferable to discard it and fit a replacement. Sudden deflation may cause an accident, particularly if it occurs with the rear wheel.

10 To fit the tyre, inflate the inner tube just sufficiently for it to assume a circular shape but only to that amount, and then push the tube into the tyre so that it is enclosed completely. Lay the tyre on the wheel at an angle, and insert the valve through the rim tape and the hole in the wheel rim. Attach the locking ring on the first few threads, sufficient to hold the valve captive in its correct location.

11 Starting at the point furthest from the valve, push the tyre bead over the edge of the wheel rim until it is located in the central well. Continue to work around the tyre in this fashion until the whole of one side of the tyre is on the rim. It may be necessary to use a tyre lever during the final stages.

12 Make sure there is no pull on the tyre valve and again commencing with the area furthest from the valve, ease the other bead of the tyre over the edge of the rim. Finish with the area close to the valve, pushing the valve up into the tyre until the locking ring touches the rim. This will ensure that the inner tube is not trapped when the last section of bead is edged over the rim with a tyre lever.

13 Check that the inner tube is not trapped at any point. Reinflate the inner tube, and check that the tyre is seating correctly around the wheel rim. There should be a thin rib moulded around the wall of the tyre on both sides, which should be an equal distance from the wheel rim at all points. If the tyre is unevenly located on the rim, try bouncing the wheel when the tyre is at the recommended pressure. it is probable that one of the beads has not pulled clear of the centre well.

14 Always run the tyres at the recommended pressures and never under or over inflate. The correct pressures are given in the Specfications Section of this Chapter.

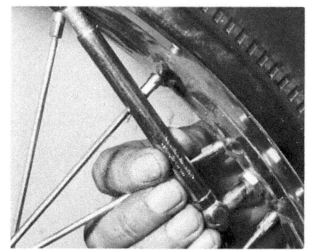

15.14 Each tyre must be inflated to the recommended pressure whilst cold

15 Tyre replacement is aided by dusting the side walls, particularly in the vicinity of the beads, with a liberal coating of french chalk. Washing up liquid can also be used to good effect, but this has the disadvantage, where steel rims are used, of

causing the inner surface of the wheel rim to rust. Do not be overgenerous in the application of lubricant of tyre creep may occur.

16 Never replace the inner tube and tyre without the rim tape in position. If this precaution is overlooked there is a good chance of the ends of the spoke nipples chafing the inner tube and causing a crop of punctures.

17 Never fit a tyre that has a damaged tread or sidewalls. Apart from legal aspects, there is a very great risk of blowout, which can have very serious consequences on a two wheeled vehicle.

18 Tyre valves rarely give trouble, but is is always advisable to check whether the valve itself is leaking before removing the tyre. Do not forget to fit the dust cap, which forms an effective extra seal.

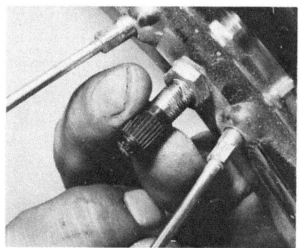

15.18 The dust cap forms an effective extra seal

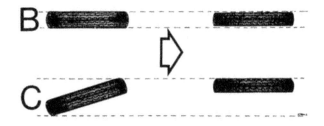

Fig. 5.6 Method of checking wheel alignment

A & C - Incorrect   B - Correct

16 Valve cores and caps

1 Valve cores seldom give trouble, but do not last indefinitely. Dirt under the seating will cause a puzzling 'slow-puncture'. Check that they are not leaking by applying spittle to the end of the valve and watching for air bubbles.

2 A valve cap is a safety device, and should always be fitted. Apart from keeping dirt out of the valve, it provides a second seal in case of valve failure, and may prevent an accident resulting from sudden deflation.

17 Security bolts

1   It is often necessary to fit a security bolt to the rear wheel
of a competition model because the initial take up of drive may
cause the tyre to creep around the wheel rim and tear the valve
from the inner tube. The security bolt retains the bead of the
tyre to the wheel rim and prevents this occurrence. Similarly,
the same device may be employed on the front wheel to prevent
tyre creep when the front brake is applied. Security bolts are
fitted as standard on both wheel of the TS250 models.

18 Wheel balancing

1   It is customary on all high performance machines to
balance the wheels complete with tyre and tube. The out of
balance forces which exist are elimiated and the handling of the
machine is improved in consequence. A wheel which is badly
out of balance produces through the steering a most unpleasant
hammering effect at high speeds.

2   Some tyres have a balance mark on the sidewalls, usually
in the form of a coloured spot. This mark must be in line with
the tyre valve, when the tyre is fitted to the inner tube. Even
then the wheel may require the addition of balance weights, to
offset the weight of the tyre valve itself.

3   If the wheel is raised clear of the ground and is spun, it will
probably come to rest with the tyre valve or the heaviest part
downward and will always come to rest in the same position.
Balance weights must be added to a point diametrically
opposite this heavy spot unti the wheel will come to rest in ANY
position after it is spun.

4   Balance weights which clip around the wheel spokes are
normally available from official Suzuki service agents. These
weights are supplied in 20 or 30 gram sizes. If they are not
available, wire solder, wrapped around the spokes close to the
spoke nipples, forms a good substitute.

5   Although the rear wheel is more tolerant to out-of-balance
forces than is the front wheel, ideally this too should be
balanced if a new tyre is fitted. Because of the drag of the final
drive components the chain must be removed from the rear
sprocket. Balancing can then be carried out as for the front
wheel.

Chapter 6 Electrical system

For modifications and information relating to later models, see Chapter 7

Contents

General description 1
Testing the electrical system 2
Wiring: layout and examination 3
Flywheel generator: checking the output 4
Rectifier unit: location and testing 5
Voltage regulator unit: function and testing –
US TS100 6
Ballast resistor: function and location –
UK TS100 and 125 models 7
Battery: examination and maintenance 8
Battery: charging procedure 9
Fuse: location, function and renewal 10
Headlamp: bulb renewal and beam alignment 11
Stop and tail lamp: bulb renewal 12
Flashing indicator lamps: bulb renewal 13
Warning light console: bulb renewal 14
Speedometer and tachometer heads: bulb renewal 15
Handlebar switches: general information and testing 16
Ignition switch: location, testing and renewal 17
Front brake stop lamp switch: testing and renewal 18
Rear brake stop lamp switch: adjustment and testing 19
Neutral indicator switch: location and testing 20
Horn: location, adjustment and replacement 21
Flasher unit: location and renewal 22

Specifications

Battery

Make	Yuasa or Furukawa
Type	6N4B-2A
Voltage	6 volt
Capacity	4 Ah
Electrolyte specific gravity	1.60 at 20°C (68°F)
Earth	Negative

Fuse

Rating:

TS100 and 125 models	10 amp
TS185 and 250 models	15 amp

Resistor

Rating:

UK TS100 and TS125 only	4 ohm, 22.5 W

Regulated voltage

TS100 and TS125 models	7.7 – 8.9 volt
TS185 models	6.7 – 7.3 volt
TS250 models	7.7 – 8.0 volt

Flywheel generator

Charging coil resistance(s):

TS100 and 125 PEI models	Below 1.0 ohm
All points models:	
Generator 038000 –3840 and –3470	0.1 ohm
Generator 038000 –3810 and –3480	0.3 ohm
TS185 models:	
Y/W to earth	1.0 – 2.0 ohm
G to earth	Below 1.0 ohm
TS250 models:	
Y/W to earth	Below 1.0 ohm
G to earth	Below 1.0 ohm
Lighting coil resistance	
All points models:	
Generator 038000 –3840 and –3470	0.3 ohm
Generator 038000 –3810 and –3480	0.25 ohm
TS100 and 125 PEI models,	
TS185 and 250 models	Below 1.0 ohm

Charging rate	Lights on	Lights off
TS100 T and X models	See tables included in text	See tables included in text
UK TS125		
UK TS100 N:		
At 4000 rpm	Above 0.6 amp	Above 0.7 amp
At 8000 rpm	Below 2.8 amp	Below 2.8 amp
US TS125		
At 4000 rpm	Above 0.7 amp	Above 0.8 amp
At 8000 rpm	Below 2.8 amp	Below 2.8 amp
TS185 and 250 models:		
At 2000 rpm	Above 1.1 amp	–
At 8000 rpm	Below 3.2 amp	–

Bulbs

	UK models	US models
Headlamp	6V, 25/25W	6V, 30/30W
Tail/stop lamp	6V, 5/21W	6V, 5.3/25W
Direction indicators	6V, 18W	6V, 17W
Speedometer light:		
UK TS125	6V. 1.7W	–
All other models	6V, 3W	6V, 3W
Tachometer light:		
UK TS125	6V, 1.7W	–
All other models (where fitted)	6W, 3W	6V, 3W
Main beam indicator	6V, 1.7W	6V, 1.7W
Neutral indicator	6V, 3W	6V, 3W
Direction indicator warning light	6V, 3W	6V, 3W
Pilot lamp	6V, 3W	–

1　General description

All Suzuki TS models covered in this Manual are fitted with a flywheel generator, the stator assembly of which incorporates a coil to provide ignition source power and coils which power the battery charging system and the lighting. Although, on TS100 and 125 models, the charging and the lighting coils are both wound on a common core, they may still be considered as separate component systems for the purposes of testing and fault isolating.

The charging coil of the flywheel generator produces alternating current (ac) which must be converted to direct current (dc) to make it compatible with the battery and components of the electrical system. This conversion is achieved by means of a silicon diode rectifier, which effectively blocks half of the output wave by acting as a one way electronic switch. For obvious reasons, this system is known as half-wave rectification. The resulting dc current is used to charge the 6 volt 4 amp hour battery which in turn provides power to such components as the stop lamp, direction indicators and the horn. This circuit is protected by a 10 amp (TS100 and 125) or 15 amp (TS185 and 250) fuse which is incorporated in the positive lead to the battery.

The lighting coil of the flywheel generator feeds ac current direct to the main lighting circuit. On all but UK TS100 and 125 models, this current is monitored by a voltage regulator unit. On UK TS100 and 125 models, a ballast resistor is incorporated in the lighting circuit to prevent overcharging of the battery when the main lighting circuit is not in use.

2　Testing the electrical system

1　Simple continuity checks, for instance when testing switch units, wiring and connections, can be carried out using a battery and bulb arrangement to provide a test circuit. For most tests described in this Chapter, however, a pocket multimeter should be considered essential. A basic multimeter capable of measuring volts and ohms can be bought for a very reasonable sum and will prove an invaluable tool. Note that separate volt and ohm meters may be used in place of the multimeter, provided those with the correct operating ranges are available. In addition, if the generator output is to be checked, an ammeter of 0-4 amperes range will be required.

2　Care must be taken when performing any electrical test, because some of the electrical components can be damaged if they are incorrectly connected or inadvertently shorted to earth. This is particularly so in the case of electronic components. Instructions regarding meter probe connections are given for each test, and these should be read carefully to preclude accidental damage occurring.

3　Where test equipment is not available, or the owner feels unsure of the procedure described, it is strongly recommended that professional assistance is sought. Errors made through carelessness or lack of experience can so easily lead to damage and need for expensive replacement parts.

4　A certain amount of preliminary dismantling will be necessary to gain access to the components to be tested. Normally, removal of the seat and side panels will be required, with the possible addition of the fuel tank and headlamp unit to expose the remaining components.

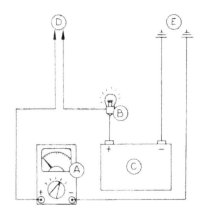

Fig. 6.1 Simple testing arrangement for checking the wiring

A　Multimeter　　　　　D　Positive probe
B　Bulb　　　　　　　　E　Negative probe
C　Battery

3 Wiring: layout and examination

1 The wiring harness is colour-coded and will correspond with the accompanying wiring diagram. When socket connectors are used, they are designed so that reconnection can be made in the correct position only.

2 Visual inspection will usually show whether there are any brakes or frayed outer coverings which will give rise to short circuits. Occasionally a wire may become trapped between two components, breaking the inner core but leaving the more resilient outer cover intact. This can give rise to mysterious intermittent or total circuit failure. Another source of trouble may be the snap connectors and sockets, where the connector has not been pushed fully home in the outer housing, or where corrosion has occurred.

3 Intermittent short circuits can often be traced to a chafed wire that passes through, or is close to, a metal component such as a frame member. Avoid tight bends in the lead or situations where a lead can become trapped between casings.

4 Flywheel generator: checking the output

Charging coil performance

1 The purpose of this test is to check the amount of electrical current being fed to the battery from the charging coil of the flywheel generator. This test should be carried out at the positive (+) connection of the battery so that the amount of direct current (dc) flowing from the voltage rectifier unit is measured.

2 Commence the test by unplugging the positive battery connection. Ensure that the two halves of the connection are clean and connect a multimeter between them. Set the meter on its dc ampere range (0-20 amps), check that it is properly supported so that its connections will not become detached from the battery leads and then start the engine. The readings obtained in the meter scale should correspond with those given below.

UK TS100 T and X and 125 N, T and X

	Engine r/min. (×1000)	1	2	3	4	5	6	7	8
Day-time	Current (A)	0.1	0.55	1.05	1.35	1.6	1.8	1.9	2.0
Night-time	Current (A)	0	0.35	0.7	0.95	1.1	1.2	1.3	1.35

US TS 100
Figures for night-time (lights on)

Engine r/min. (x 1000)	1	2	3	4	5	6	7	8
Current (A)	0	0.15	0.85	1.35	1.7	2.0	2.2	2.4

Model type	Engine speed	Lighting state	Charging current
UK TS100 N	4000 rpm	On	Above 0.6 amp
	8000 rpm	On	Below 2.8 amp
	4000 rpm	Off	Above 0.7 amp
	8000 rpm	Off	Below 2,8 amp
US TS125	4000 rpm	On	Above 0.7 amp
	8000 rpm	On	Below 2.8 amp
	4000 rpm	Off	Above 0.8 amp
	8000 rpm	Off	Below 2.8 amp
All TS185 and 250 models	2000 rpm	On	Above 1.1 amp
	8000 rpm	On	Below 3.2 amp

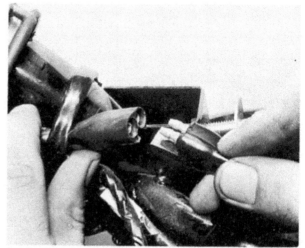
3.2 Carefully examine all wiring connections for signs of contamination

4.1 Both the charging coil and lighting coil are mounted on the flywheel generator stator (UK 125 shown)

Lighting coil performance
TS100 and 125 models

3 The purpose of this test is to check the amount of voltage being fed to the headlamp bulb from the lighting coil of the flywheel generator. Prepare for the test by removing the headlamp reflector unit. Trace the yellow lead which runs from the headlamp dip switch to the headlamp bulb holder and disconnect it by pulling apart the two halves of the connector through which it passes.

4 Set a multimeter to its 0 – 10 ac volt range and connect its positive probe to the yellow wire terminal. Connect the negative probe of the meter to a good earth point, set the switch button to the On position and start the engine. The readings obtained on the meter scale should correspond with those given below.

Model type	Engine speed	Voltage reading
UK TS100 N	2500 rpm	Above 5.7 volt
	8000 rpm	Below 8.7 volt
US TS100	3000 rpm	Above 5.5 volt
	8000 rpm	Below 7.2 volt
UK TS100 T and X, UK TS125 models	2500 rpm	Above 5.7 volt
	8000 rpm	Below 8.4 volt

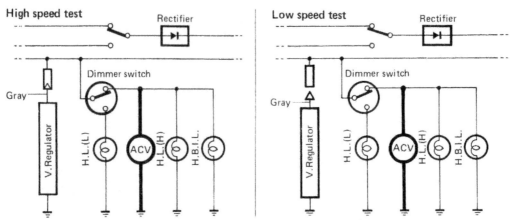

Fig. 6.2 Lighting coil test – 185 and 250 models

US TS125 models   2500 rpm   Above 6.0 volt
                         8000 rpm   Below 8.7 volt

TS185 and 250 models

5 Reference to the figures accompanying this text will show that two separate tests are necessary to determine the performance of the lighting coil. Set a multimeter to its 0-10 ac volt range and connect it as shown in each of the two figures. Disconnect the grey lead which runs from the voltage regulator at its nearest push connector. With the headlamp dip switch set to its 'Hi' position, start the engine and gradually increase its speed to 2500 rpm. Do not exceed this engine speed. The reading observed on the meter scale should be above 5.7 volts if the coil is serviceable.

6 With the low speed test successfully completed, stop the engine and reconnect the voltage regulator. Check that the multimeter is properly connected and that the dip switch is still set to its 'Hi' position before restarting the engine and gradually increasing its speed to 8000 rpm. Observe the reading on the meter scale directly the engine reaches this speed and then shut off the engine. If the coil is serviceable, then the meter reading should be below 7.8 volts for TS185 models, or below 8.0 volts for TS250 models.

Elimination checks – all models

7 If the results obtained in any of the above tests are unsatisfactory, then it is possible that the coil in question has failed and will need renewing. Before assuming that this is the case, carry out the following checks.

8 Refer to the wiring diagram at the end of this Chapter for the machine in question and carry out a check for continuity on the wires running between the coil and the point in the circuit at which the test was carried out. Separate each block or bullet connector in the circuit and check that it is free of all traces of dirt, moisture and corrosion.

9 Visually check each wire for signs of its having chafed against an engine or frame component and check any component contained within the circuit for serviceability by following the procedure given in the relevant Section of this Chapter. Finally, the condition of the coil in question can be checked by determining whether continuity exists. On those models fitted with a voltage regulator, this component should be checked for performance by a process of elimination.

Continuity checks

10 To check any one coil for continuity, refer to the appropriate wiring diagram at the end of this Chapter and identify the lead(s) running from the coil in question by the colour coding. Trace the lead(s) from the point where they leave the stator housing to their nearest push connector and pull apart the two halves of the connector. Set a multimeter to its resistance function and carry out a measurement for resistance across the coil windings. Note that where it is necessary to connect one

probe of the meter to earth, then select a point on the engine crankcase which is both clean and free from protective finish.

11 Note the reading on the meter scale and compare it with the readings given in the Specifications of this Chapter (charging and lighting coils) or the readings given in the Specifications Section of Chapter 3 (ignition source coil). If the reading shown on the meter scale does not correspond with that given in the appropriate Specifications, then the coil in question is unserviceable and must be renewed.

5 Rectifier unit: location and testing

1 The rectifier unit is mounted on the forward facing wall of the battery tray and is retained in position by a single screw. Access to the unit may be gained by removal of the right-hand side panel from the machine.

5.1 The rectifier is mounted on the wall of the battery tray

2 The function of this unit is to convert the alternating current (ac) from the flywheel generator to direct current (dc) which can then be used to charge the battery. This it does by offering a very low resistance to current flow in one direction and an extremely high resistance in the reverse direction. As the ratio of these resistances is very high, one can generalise by saying that an effective current will flow in one direction only.

3 The rectifier unit is mounted in such a way that it is not exposed to direct contamination by road salts, water or oil, yet

has free circulation of air to permit cooling. It should be kept clean and dry. The unit should not give trouble during normal service. It can however be damaged by inadvertently reversing the battery connections.

4   To test the unit for continuity, set the multimeter to its resistance function (X1 ohm range) and connect its positive (+) probe to the negative (male) connection of the unit. Connect the negative (–) probe to the positive (female) connection of the unit and note the reading on the meter scale.

5   Reverse the position of the meter probes and once again note the meter reading. If an indication of continuity is given by the first meter reading and an indication of non-continuity shown by the second, then the unit is serviceable.

6   The test can, of course be carried out by connecting the rectifier unit to a simple battery and bulb test circuit. If the bulb lights with the circuit leads connected in one position and fails to light when the leads are reversed, then this indicates that the unit is doing its intended task, namely, allowing the current to flow in one direction only.

6   Voltage regulator unit: function and testing – US TS100 and 125 models, all TS185 and 250 models

1   A voltage regulator unit is incorporated in the lighting system of these models for the purpose of controlling the lighting coil output to the lights. This unit is normally very reliable, there being no possibility of mechanical failure. It can, however, become damaged in the event of a short circuit in the electrical system or by poor or intermittent earth connections.

2   If trouble with blowing bulbs is experienced, or if the lighting should suddenly dim, then there is evidence that the regulator has failed. The regulator performance can best be deduced by checking the lighting coil performance, as detailed in Section 4, and eliminating other possible causes of voltage output control faults. To renew the unit, simply remove its two mounting screws and ease it clear of its frame mounting. Disconnect the earth and the power lead to the unit and remove the unit from the machine. Fitting of the replacement unit is a direct reversal of the removal procedure.

7   Ballast resistor: function and location – UK TS100 and 125 models

1   A ballast resistor is incorporated in the lighting system of these models to help prevent overcharging of the battery when the lights of the machine are not in use. It does this by absorbing the same amount of power as used by the lights directly the lighting switch is moved from its 'On' to its 'Off' position. The drain on the lighting coil of the flywheel generator is therefore kept constant, obviating any risk of the current generated in the lighting coil combining with that of the charging coil to overcharge the battery.

2   The resistor is clamped within a pressed steel case which is attached to the base of the battery tray by two crosshead screws. If removal of the unit is necessary, simply remove these two screws and disconnect the two wires leading to the unit at their nearest push connectors.

3   If overcharging is experienced, indicated by a rapid level drop of the battery electrolyte, or if poor charging occurs when the lights are off, there is evidence that the resistor has failed. To test the resistor, set a multimeter to its resistance function, disconnect the two wires leading to the resistor and place the probes of the meter on the wire terminals to measure the resistance across the resistor. If the reading obtained differs markedly from the specified resistance of 4 ohm, then the resistor is unserviceable and must be renewed.

8   Battery: examination and maintenance

1   A Yuasa or Furukawa 6N4B-2A battery is fitted as standard

6.1 The voltage regulator unit (UK 185 shown)

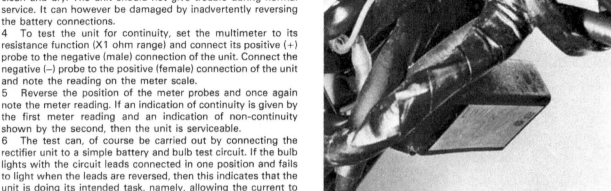

7.2 The ballast resistor is attached to the base of the battery tray (UK100 and 125)

8.2 Observe the level of electrolyte through the battery case

to all of the models covered in this Manual and is of the conventional lead-acid type. The battery is retained in a tray which is located behind the right-hand side panel of the machine.

2   The level of electrolyte may be observed through the translucent plastic case of the battery. Maintenance is normally limited to keeping the electrolyte level between the prescribed upper and lower limits and making sure that the vent tube is not blocked. The lead plates and their separators are also visible through the case, a further guide to the general condition of the battery. If electrolyte level drops rapidly, suspect over-charging and check the system.

3   Unless acid is spilt, as may occur if the machine falls over, the electrolyte should always be topped up with distilled water to restore the correct level. If acid is split onto any part of the machine, it should be neutralised with an alkali such as washing soda or baking powder and washed away with plenty of water, otherwise serious corrosion will occur. Top up with sulphuric acid of the correct specific gravity (1.260) only when spillage has occurred. Check that the vent pipe is well clear of the frame or any of the other cycle parts.

4   It is seldom practicable to repair a cracked battery case because the acid present in the joint will prevent the formation of an effective seal. It is always best to renew a cracked battery, especially in view of the corrosion which will be caused if the acid continues to leak.

5   If the machine is not used for a period of time, it is advisable to remove the battery and give it a 'refresher' charge every six weeks or so from a battery charger. The battery will require recharging when the specific gravity falls below 1.260 (at 20°C – 68°F). The hydrometer reading should be taken at the top of the meniscus with the hydrometer vertical. If the battery is left discharged for too long, the plates will sulphate. This is a grey deposit which will appear on the surface of the plates, and will inhibit recharging. If there is sediment on the bottom of the battery case, which touches the plates, the battery needs to be renewed. Warping of the plates or separators is also indicative of an expiring battery, and will often be evident in only one or two of the cells. It can often be caused by old age, but a new battery which is overcharged will show the same failure. There is no cure for the problem, and the need to avoid overcharging cannot be overstressed. Prior to charging the batery, refer to the following Section for correct charging rate and procedure. If charging from an external source with the battery on the machine, disconnect the leads, or the rectifier will be damaged.

6   Note that when moving or charging the battery, it is essential that the following basic safety precautions are taken:

(a) Before charging check that the battery vent is clear or, where no vent is fitted remove the combined vent/filler caps. If this precaution is not taken the gas pressure generated during charging may be sufficient to burst the battery case, with disastrous consequences.
(b) Never expose a battery on charge to naked flames or sparks. The gas given off by the battery is highly explosive.
(c) If charging the battery in an enclosed area, ensure that the area is well ventilated
(d) Always take great care to protect yourself against accidental spillage of the sulphuric acid contained within the battery. Eyeshields should be worn at all times. If the eyes become contaminated with acid they must be flushed with fresh water immediately and examined by a doctor as soon as possible. Similar attention should be given to a spillage of acid on the skin. Note also that although, should an emergency arise, it is possible to charge the battery at a more rapid rate than that stated in the following Section, this will shorten the life of the battery and should therefore be avoided if at all possible.

7   Occasionally, check the condition of the battery terminals to ensure that corrosion is not taking place, and that the electrical connections are tight. If corrosion has occurred, it should be cleaned away by scraping with a knife and then using emery cloth to remove the final traces. Remake the electrical connections whilst the joint is still clean, then smear the assembly with petroleum jelly (NOT grease) to prevent recurrence of the corrosion. Badly corroded connections can have a high electrical resistance and may give the impression of complete battery failure.

9   Battery: charging procedure

1   The normal charging rate for the battery fitted to the machines covered in this Manual is 0.4 amp. It is permissible to charge at a more rapid rate in an emergency but this effectively shortens the life of the battery and should therefore be avoided. Because of this, go for the smallest charging rate available. Avoid quick charge services offered by garages. This will indeed charge the battery rapidly but it will also overheat it and may halve its life expectancy.

2   Never omit to remove the battery cell caps or neglect to check that the side vent is clear before recharging a battery, otherwise the gas created within the battery when charging takes place might burst the case with disastrous consequences. Do not attempt to charge the battery with it in situ and with the leads still connected. This can lead to failure of the rectifier unit.

3   Make sure that the battery charger connections are correct; red to positive and black/white to negative. When the battery is reconnected to the machine, the black/white lead must be connected to the negative terminal and the red lead to positive. This is most important, as the machine has a negative earth system. If the terminals are inadvertently reversed, the electrical system will be damaged permanently. The rectifier unit can be destroyed by a reversal of the current flow.

4   A word of caution concerning batteries. Sulphuric acid is extremely corrosive and must be handled with great respect. Do not forget that the outside of the battery is likely to retain traces of acid from previous spills, and the hands should always be washed promptly after checking the battery. Remember too that battery acid will quickly destroy clothing.

5   Note the following rules concerning battery maintenance:
Do not allow smoking or naked flames near batteries.
Do avoid acid contact with skin, eyes and clothing.
Do keep battery electrolyte level maintained.
Do avoid over-high charge rates.
Do avoid leaving the battery discharged.
Do avoid freezing.
Do use only distilled or demineralised water for topping up.

10 Fuse: location, function and renewal

1   The fuse is contained within a plastic holder which is clipped to the wall of the battery tray. A spare fuse of the same specified rating is supplied with the machine when new and is contained within a clear plastic holder which is attached to the wire leading from the fuse in circuit.

2   The fuse is fitted to protect the electrical system in the event of a short circuit or sudden surge. It is, in effect, an intentional 'weak link' which will blow in preference to the circuit burning out.

3   Before replacing a fuse that has blown, check that no obvious short circuit has occurred, otherwise the replacement fuse will blow immediately it is inserted. It is always wise to check the electrical circuit thoroughly, to trace the fault and eliminate it.

4   When a fuse blows while the machine is running and no spare is available, a 'get you home' remedy is to remove the blown fuse and wrap it in silver paper before replacing it in the fuseholder. The silver paper will restore the electrical continuity by bridging the broken fuse wire. This expedient should **never**

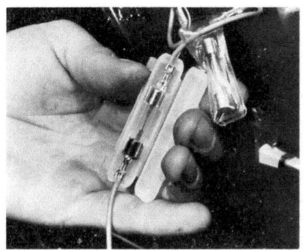

10.1 The fuse is contained within a plastic holder

be used if there is evidence of a short circuit or other major electrical fault, otherwise more serious damage will be caused. Replace the 'doctored' fuse at the earliest possible opportunity to restore full circuit protection. It follows that spare fuses that are used should be replaced as soon as possible to prevent the above situation from arising.

11 Headlamp: bulb renewal and beam alignment

UK models

1   In order to gain access to the headlamp bulbs of these models, it is necessary to detach the headlamp rim, complete with reflector and headlamp glass, from the headlamp nacelle. The rim is retained in position by two crosshead screws, each with a spring washer and a plain washer fitted beneath its head. These screws are sited at the 4 o'clock and the 8 o'clock positions around the headlamp and pass through the nacelle into a threaded lug projecting from the rim edge. With each screw removed, the rim can be drawn out of the nacelle and pulled forward to expose the rear face of the reflector.

2   The main headlamp bulb is a push fit in the central bulb holder of the reflector. This holder is covered by a protective rubber cover which should be carefully eased back along the electrical wires which pass through it in order to allow the holder to be removed by turning it anti-clockwise and then pulling it from its location. Note that this holder can be fitted in one position only to ensure that the bulb is always correctly focussed. The bulb has a fitting of the bayonet type and can be released from its holder by pressing inwards, turning anti-clockwise and then pulling it outwards. A twin filament bulb of a 6 volt, 25/25 watt rating is used.

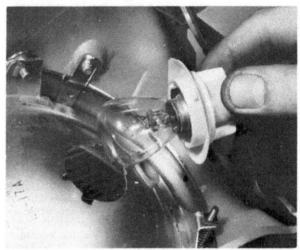

11.2a Remove the headlamp bulb holder from its location in the reflector ...

3   The holder for the pilot lamp bulb is a direct push fit into the reflector. The bulb is of a 6 volt, 3 watt rating and has a bayonet

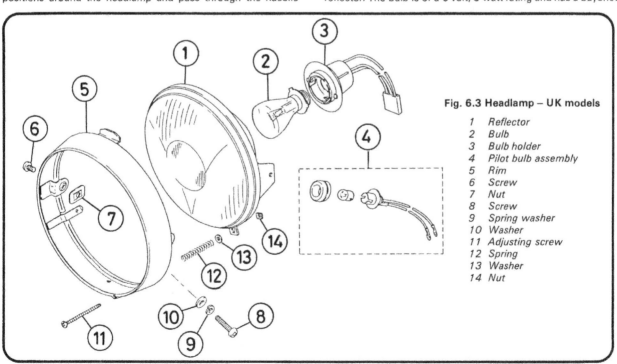

Fig. 6.3 Headlamp – UK models

1   Reflector
2   Bulb
3   Bulb holder
4   Pilot bulb assembly
5   Rim
6   Screw
7   Nut
8   Screw
9   Spring washer
10   Washer
11   Adjusting screw
12   Spring
13   Washer
14   Nut

11.2b ... and then remove the headlamp bulb from its holder

11.3 Pull the pilot bulb holder from its location in the headlamp reflector

fitting in the holder for the pilot lamp bulb is a direct push fit into the reflector. The bulb is of a 6 volt, 3 watt rating and has a bayonet fitting in the holder.

US models

4   These models have a sealed beam unit of a 6 volt, 30/30 watt rating fitted within the headlamp assembly. In order to remove this unit, it is first necessary to follow the procedure stated in paragraph 1 of this Section and detach the headlamp rim from the headlamp nacelle. With the rim pulled forward to expose the rear face of the sealed beam unit, unplug the wiring connection from the centre of the unit and place the unit on a protected work surface.

5   Remove the two headlamp alignment screws from the headlamp rim and detach the rim from the sealed beam unit retaining ring. The sealed beam unit can now be freed from its retaining ring by careful removal of its securing clip. Use the flat of a screwdriver or a pair of long-nose pliers to ease one end of the clip free of the retaining ring whilst taking great care not to slip and thereby cause damage to the sealed beam unit. Take note of the fitted position of the sealed beam unit in relation to its retaining ring and fit the replacement unit in exactly the same position.

All models

6   Upon completion of relocating the headlamp rims within

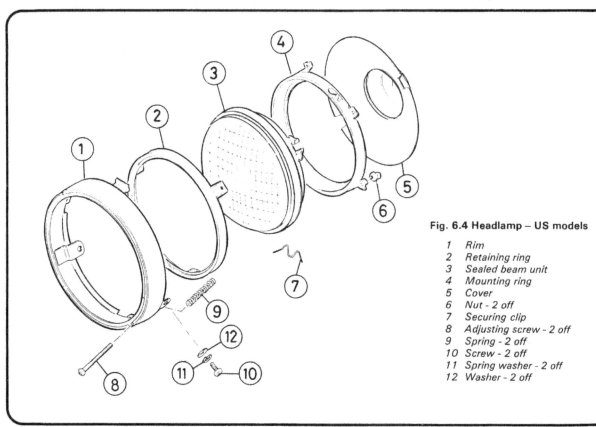

Fig. 6.4 Headlamp – US models

1   Rim
2   Retaining ring
3   Sealed beam unit
4   Mounting ring
5   Cover
6   Nut - 2 off
7   Securing clip
8   Adjusting screw - 2 off
9   Spring - 2 off
10   Screw - 2 off
11   Spring washer - 2 off
12   Washer - 2 off

11.6 Adjust the headlamp beam alignment

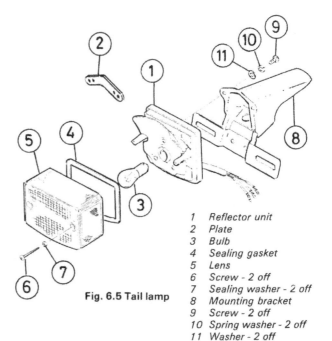

Fig. 6.5 Tail lamp

1   Reflector unit
2   Plate
3   Bulb
4   Sealing gasket
5   Lens
6   Screw - 2 off
7   Sealing washer - 2 off
8   Mounting bracket
9   Screw - 2 off
10 Spring washer - 2 off
11 Washer - 2 off

the nacelle, check the headlamp beam for correct alignment. Provision is made for adjusting the alignment of the beam on both the horizontal and vertical planes. To effect the horizontal alignment, turn the screw which is located at a 10 o'clock position through the headlamp rim. To effect vertical alignment of the beam, turn the screw which is located at the exact opposite point through the headlamp rim.

7   In the UK, regulations stipulate that the headlamp must be aligned so that the light will not dazzle a person standing at a distance greater than 25 feet from the lamp, whose eye level is not less than 3 feet 6 inches above that place. It is easy to approximate this setting by placing the machine 25 feet away from a wall, on a level road, and setting the dip beam height so that it is concentrated at the same height as the distance of the centre of the headlamp from the ground. The rider must be seated normally during this operation and also the pillion passenger, if one is carried regularly.

8   Most other areas have similar regulations controlling head-lamp beam alignment, and these should be checked before any adjustment is made.

12 Stop and tail lamp: bulb renewal

1   The combined tail and stop lamp unit fitted to these models is fitted with a double filament bulb having offset pins to prevent its unintentional reversal in the bulb holder. The lamp unit serves a two-fold purpose – to illuminate the rear of the machine and the rear number plate, and to give visual warning when the rear brake is applied.

2   To gain access to the bulb, remove the two screws with sealing washers which hold the red plastic lens in position. The bulb is released by pressing inwards and with an anti-clockwise turning action; the bulb will now come out. When lifting the lens from the backplate, take care not to tear the sealing gasket located between the two. This gasket is essential to keep moisture and dirt from entering the electrical contacts within the unit and to stop corrosion or dulling of the backplate. The sealing washers on the lens securing screws serve a similar function and should be renewed if seen to be damaged.

3   Refitting of the bulb and lens is a reversal of the removal procedure. Check the inside of the bulb holder for any signs of corrosion or moisture, ensuring that the contacts are free to move when depressed. When refitting the lens securing screws, take care not to overtighten them as it is possible that the lens may crack.

12.2a Remove the tail lamp lens securing screws

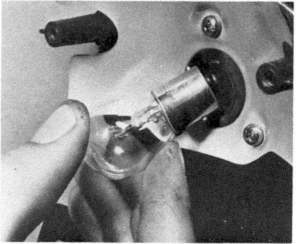

12.2b The stop/tail lamp bulb is a bayonet fit in its holder

13 Flashing indicator lamps: bulb renewal

1   Flashing indicator lamps are fitted to the front and rear of
the machine. They are mounted on short stalks through which
the wires pass. Access to each bulb is gained by removing the
two screws with sealing washers which hold the plastic lens
cover in position. When removing the lens, take care not to tear
the sealing gasket located beneath it. Both the gasket and
washers are essential in keeping moisture and dirt from entering
the electrical connections within the unit.
2   Each bolt is of the bayonet type and may be released by
pushing in, turning anti-clockwise and pulling out of the holder.
3   Refitting of the bulb and lens is a reversal of the removal
procedure. Check the inside of the bulb holder for any signs of
corrosion or moisture, ensuring that the contact is free to move
when depressed. Do not omit to fit a lens seal which is in good
condition and take great care not to overtighten the lens
securing screws as it is possible to crack the lens by doing so.

13.2 Each indicator bulb has a bayonet fitting

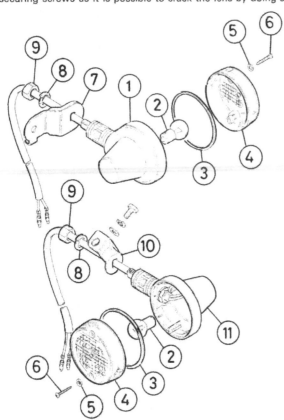

Fig. 6.6 Front and rear indicator assemblies

1	Rear indicator unit		bracket
2	Bulb	8	Spring washer
3	O-ring	9	Nut
4	Lens	10	Front indicator mounting
5	Sealing washer - 2 off		bracket
6	Screw - 2 off	11	Front indicator unit
7	Rear indicator mounting		

14 Warning light console: bulb renewal

All TS100, US TS125 and 185

1   The warning light console fitted to these models is incorporated in what is the casing for the tachometer instrument
head on all other models covered in this Manual. To remove any
one unserviceable bulb from this unit, it is first necessary to gain

13.3 Do not overtighten the indicator lens securing screws

full access to the base of the unit. Instructions for achieving this
and for subsequent bulb renewal are given in the following
Section of this Chapter, as for removal of the tachometer
illumination bulb.
2   If it is found necessary to remove the complete console
assembly from the machine, then simply expose the base of the
console by detaching the headlamp nacelle and then remove
the nut from each of the console base mounting studs. With the
bulb holders released from the console, it can be lifted clear of
its mounting bracket.

All TS250, US TS125 and 185

3   To renew the bulbs incorporated in the type of warning light
console fitted to these models first detach the console from its
mountings on the handlebar securing clamps by removing each
of the two rubber bungs from its upper surface and then
removing each of its two securing screws. Using a C-spanner or
a small soft-metal drift and hammer, release the ignition switch
retaining ring. The console can now be manoeuvred upwards
just enough to allow the bulb holder to be unplugged from its
base. Place the console to one side and remove the defective
bulb from its location in the holder by pushing it inwards,
turning anti-clockwise and then pulling it from position.

4   Before fitting the replacement bulb, check that the contact within its holder is free to move when depressed and remove any signs of moisture or corrosion from within the holder. Fitting the bulb and holder and refitting the console is a direct reversal of the above listed procedure.

15 Speedometer and tachometer heads: bulb renewal

1   Before attempting renewal of the one bulb located in the base of each instrument head, it is first necessary to expose the base of the instrument head in question by detaching the headlamp nacelle from its three mounting points on the steering head assembly and then pulling it forwards to unclip it from the fork stanchions.
2   Each bulb holder is a straight push-fit in the base of its respective instrument head. With the holder pulled from position, its bulb may be removed by pushing it inwards, turning it anti-clockwise and then pulling it out of the holder.
3   Before fitting the replacement bulb, check that the contact within the holder is free to move when depressed and remove any signs of moisture or corrosion from within the holder. Fitting the bulb and holder and then fitting the headlamp nacelle is a direct reversal of the above listed procedure.

16 Handlebar switches: general information and testing

1   It will be seen that each individual switch incorporated in the handlebar switch assemblies is clearly marked. Generally speaking these switch assemblies give little trouble, but if necessary, they can be dismantled by separating their halves which form a split clamp around the handlebars.
2   Always disconnect the battery before removing any of the switches, to prevent the possibility of a short circuit. Most troubles are caused by dirty contacts, but in the event of the breakage of some internal part, it will be necessary to renew the complete switch.
3   Because the internal components of each switch are very small, and therefore difficult to dismantle and reassemble, it is suggested a special electrical contact cleaner be used to clean corroded contacts. This can be sprayed into each switch, without the need for dismantling.
4   Each individual switch may be tested by carrying out the following procedure. Obtain a multimeter and set it to its resistance function. Refer to the back of this Chapter and select the correct wiring diagram for the machine being worked on. It will be seen that each switch is symbolised by a block which is divided into switch positions and test connections for each one of these positions. The method of testing is to trace the wires from the switch assembly concerned to their nearest block connector. With the two halves of this connector separated, place the probes of the multimeter on the test connections shown in the wiring diagram for each switch position and then carry out a check for continuity. If continuity is found to exist in all of the tests for any one switch then that switch is serviceable. If either one of the tests shows non-continuity, then the switch must be renewed.
5   Bear in mind the danger which will occur if a machine is ridden with defective lighting switches during the hours of darkness whilst on an unlit road. Should any failure occur within a switch, then the rider will be plunged into darkness with disastrous consequences.

17 Ignition switch: location, testing and renewal

1   The ignition switch is mounted on a support bracket which itself is attached to the upper yoke of the steering head assembly by two screws with spring washers.
2   The correct procedure for testing the switch is similar to

14.3a Remove the console securing screws and the ignition switch retaining ring ...

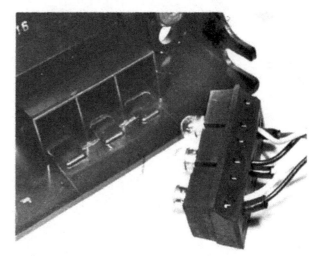

14.3b ... and with the console manoeuvred clear of the handlebars, unplug the bulb holder

15.2 Each bulb holder is a push fit in the instrument head

16.1 Each handlebar switch forms a split clamp around the handlebars

18.2 The front brake stop lamp switch is fitted to the brake lever mounting bracket

19.1 The rear brake stop lamp switch is frame-mounted

20.1 The neutral indicator switch is adjacent to the gearbox sprocket

21.1 The horn is mounted on the lower of the frame top tubes

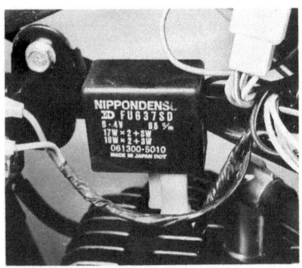

22.1 The flasher unit is easily identified ...

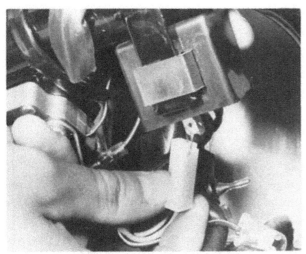

22.3 ... and can easily be detached from the machine

that given for testing of the handlebar switches. If the switch is found to be unserviceable, then it must be renewed as repair is impracticable, but before doing this, carry out a physical check of the wiring from the switch to its block connector to ensure that there is no indication of the wire having been cut or frayed by any of the nearby cycle parts.

3   To remove an unserviceable switch, commence by detaching the plastic console from both the switch and the handlebar securing clamps. Remove each of the two rubber plugs from the upper face of this console to expose the heads of its securing screws and remove each one of the screws. Using a C-spanner or a small soft-metal drift and hammer, release the switch retaining ring and then manoeuvre the console clear of the switch.

4   With the ignition switch thus exposed, release the second of its retaining rings and detach it from its mounting bracket. With its wiring connector separated, the switch can now be removed from the machine. Fitting a replacement switch is a direct reversal of the removal procedure.

18   Front brake stop lamp switch: testing and renewal

1   Testing the front brake stop lamp switch is a simple matter of checking for continuity between the two switch terminals with the front brake lever fully applied, that is, with the switch in the 'On' position. If no continuity is found, then the switch must be renewed as it is not possible to effect a satisfactory repair.

2   The switch is retained in position on the underside of the front brake lever mounting bracket by two crosshead screws with washers. With these screws removed, the switch can be eased from its location and its electrical leads disconnected at their nearest connections. Fitting of a replacement switch is a direct reversal of this removal procedure.

19   Rear brake stop lamp switch: adjustment and testing

1   The rear brake stop lamp switch is located on a frame-mounted bracket which is situated directly above the pivot for the swinging arm on the right-hand side of the machine. It is operated by movement of the rear brake pedal, the two components being adjoined by an extension spring. The body of

the switch is threaded to permit adjustment and is secured to the bracket by a single special nut.

2   If the stop lamp is seen to be late in operating, raise the switch body by rotating the adjuster nut in a clockwise direction whilst holding the switch steady. If the stop lamp is seen to be permanently on, then the switch body should be lowered in relation to its mounting bracket. As a guide to operation, the stop lamp should illuminate immediately after the brake pedal is depressed.

3   Testing the switch is a simple matter of checking for continuity between the switch terminals with the switch fully extended, that is in the 'On' position. Before testing, check the switch adjustment by referring to the above text. Disconnect the two electrical leads from the switch terminals. Set a multimeter to its resistance function and with the probes of the meter connected one to each terminal, operate the switch. If no continuity is found, then the switch must be renewed as it is not possible to effect a satisfactory repair.

20   Neutral indicator switch: location and testing

1   The neutral indicator switch is sited over the left-hand end of the gearchange selector drum. Access to the switch may be gained by removal of the gearbox sprocket cover.

2   To test the switch, set a multimeter to its resistance function and carry out a check for continuity between the switch terminal and earth with the gearchange lever set in the neutral position. If continuity is found, then the switch is serviceable. If this is not the case, then the switch should be removed by following the instructions given in Section 9 of Chapter 1. Full instructions for fitting a switch are given in Section 42 of Chapter 1.

21   Horn: location, adjustment and replacement

1   The horn is mounted on the left-hand side of the machine at a point where the lower of the two frame top tubes adjoins the base of the steering head gusset plates.

2   The horn should not normally require adjustment. If, however, this becomes necessary, then adjustment is provided by means of a screw at the rear of the horn case. This screw should be turned fractionally clockwise to increase the horn volume.

3   If necessary, the horn may be removed from the machine by disconnecting the two electrical wires from the horn terminals, after having first noted their fitted positions, and then detaching either the horn unit from its mounting bracket or the bracket from its frame attachment. Fitting a horn is a direct reversal of the removal procedure.

22   Flasher unit: location and renewal

1   The flasher relay unit is supported on an anti-vibration mounting which itself is attached to a lug welded to one of the frame top tubes which runs beneath the fuel tank.

2   If the flasher unit is functioning correctly, a series of audible clicks will be heard when the indicator lamps are in operation. If the unit malfunctions and all the bulbs are in working order, the usual symptom is one initial flash before the unit goes dead; it will be necessary to replace the unit complete if the fault cannot be attributed to any other cause.

3   If renewal is necessary, then removal of the existing unit is simple. Unplug the block connector from the base of the unit and move the unit upwards to unhook it from its mounting. When fitting the new unit, take great care not to subject it to any sudden shock; it is easily damaged if dropped.

The TS100 ERZ model

Chapter 7 The UK TS100 and 125 ERZ models

Contents

General description 1
Crankcase and cover: modifications 2
Oil pump and drive: removal and refitting 3
Clutch release mechanism: general 4
Gearbox components: general 5
Carburettor: modifications 6
Fuel tap: modifications 7
Oil pump: adjustment 8
Air filter: general 9
Ignition system: general description and testing 10
Front forks: modifications 11

'Full Floater rear' suspension: general description 12
Rear suspension: dismantling and reassembly 13
Rear suspension linkage: examination and renovation 14
Footrests: examination and renovation 15
Final drive chain: adjustment 16
Electrical system: general 17
Electrical system: testing the charging output 18
Flywheel generator: testing the no-load performance 19
Regulator/rectifier unit: checking the resistances 20
Instrument panel: removal and bulb renewal 21
Headlamp, stop/tail lamp and turn signal lamps: general .. 22

Specifications

Note: *except where entered below, specifications for the models covered in this Chapter are the same for the earlier TS100 and 125 UK models at the beginning of each Chapter.*

Specifications relating to Chapter 1

Engine
Compression ratio:
   TS100 .... 6.8:1
   TS125 .... 6.7:1
Cylinder barrel standard bore service limit – TS100 .... 50.085 mm (1.971 in)
Cylinder bore to piston clearance – TS100 .... 0.045 – 0.055 mm (0.0018 – 0.0021 in)
Piston outside diameter – TS100 .... 49.950 – 49.965 mm (1.966 – 1.967 in)
Gudgeon pin hole bore diameter .... 14.002 – 14.010 mm (0.5512 – 0.5515 in)
Piston ring to groove clearance .... 0.03 – 0.07 mm (0.0011 – 0.0027 in)
Piston ring end gap – TS125:
   Free – Teikoku (T) .... 7.5 mm (0.29 in)
   Service limit – Teikoku (T) .... 6.0 mm (0.24 in)
Crankshaft total web width .... 52.0 ± 0.1 mm (2.047 ± 0.0039 in)

Clutch
Spring free length service limit .... 29.8 mm (1.173 in)
Friction plate tang width service limit .... 11.3 mm (0.44 in)
Cable free play .... 4.0 mm (0.157 in)

Gearbox
2nd gear ratio (no. of teeth) 1.857:1 (26/14)
Final reduction ratio (no. of teeth):
   TS100 .... 3.142:1 (44/14)
   TS125 .... 3.000:1 (45/15)

Torque wrench settings

	kgf m	lbf ft
Cylinder head nuts	2.3 – 2.7	16.5 – 19.5
Cylinder barrel nuts:		
Type A (M8)	2.3 – 2.7	16.5 – 19.5
Type B (M6)	0.9 – 1.1	6.5 – 8.0
Oil pump drive pinion nut	3.0 – 4.0	21.5 – 29.0
Exhaust pipe cylinder mounting	1.0 – 1.6	7.5 – 11.5
Engine mounting bolts:		
Front 65 mm	2.8 – 3.4	20.0 – 24.5
Lower 200 mm	3.7 – 4.5	26.5 – 32.5

Specifications relating to Chapter 2

Fuel tank
Total capacity	9.0 litres (1.97 Imp gal)
Reserve capacity	1.5 litres (2.64 Imp pt)

Carburettor
	TS100	TS125
Type	VM22SS	VM24SS
ID no.	48750	48700
Bore size	22 mm	24 mm
Main jet	130	130
Needle jet	0-6	0-4
Jet needle	5F55-3	5F55-3
Air jet	0.5	0.6
Pilot jet	22.5	22.5
Pilot air screw	1½ turns out	1¼ turns out
Starter jet	70	70
Float height	25.8 ± 1.0 mm (1.016 ± 0.020 in)	25.8 ± 1.0 mm (1.016 ± 0.020 in)
Idle speed	1350 ± 100 rpm	1350 ± 100 rpm

Lubrication – engine
Oil pump discharge rate (fully open) over 2 mins @ 2000 rpm	1.31 – 1.60 cc (0.0461 – 0.0563 fl oz)

Lubrication – transmission
Capacity at oil change	900 cc (1.58 Imp pt)
Capacity at engine overhaul	950 cc (1.67 Imp pt)

Specifications relating to Chapter 3

Ignition system
Type	Suzuki PEI Electronic
Ignition timing	22° ± 3° BTDC @ 4000 rpm
Ignition coil resistance:	
Primary	150 – 220 ohm
Secondary	14 – 22 K ohm
Pulser coil resistance	120 – 200 ohm

Spark plug
Type	NGK B8ES or ND W24ES

Specifications relating to Chapter 4

Front forks
	TS100	TS125
Travel	155 mm (6.102 in)	180 mm (7.086 in)
Spring free length	Not available	Not available
Spring free length service limit	529 mm (20.82 in)	525 mm (20.66 in)
Oil capacity (per leg)	162 cc (5.70 fl oz)	179 cc (6.30 fl oz)
Oil level	180 mm (7.08 in)	180 mm (7.08 in)
Oil type	SAE 10 fork oil	SAE 10 fork oil

Rear suspension
	TS100	TS125
Travel	142 mm (5.59 in)	163 mm (6.41 in)

Torque wrench settings
	kgf m	lbf ft
Lower yoke pinch bolts	4.0 – 5.0	29.0 – 36.0
Front fork damper rod Allen bolt	1.5 – 2.5	11.0 – 18.0
Swinging arm pivot shaft nut	5.0 – 8.0	36.0 – 58.0
Rear suspension unit mounting nuts	4.0 – 6.0	29.0 – 43.5
Suspension link arm upper nut	4.0 – 6.0	29.0 – 43.5
Suspension link arm lower nuts	1.8 – 2.8	13.0 – 20.0
Suspension rocker arm nut	4.5 – 7.0	32.5 – 50.5

Specifications realting to Chapter 5

Wheels
Size	21 inch

Brakes

Drum internal diameter service limit:
Front .... 130.7 mm (5.145 in)
Rear .... 110.7 mm (4.358 in)

Tyres

Size – TS100 front .... 2.50 – 21 4PR
Tread minimum depth .... 4.0 mm (0.157 in)

Pressures:	Front	Rear
Solo	21 psi (1.5 kg/cm^2)	24 psi (1.75 kg/cm^2)
Pillion	21 psi (1.5 kg/cm^2)	28 psi (2.0 kg/cm^2)

Torque wrench settings

	kgf m	lbf ft
Front wheel spindle nut	3.6 – 5.2	26.0 – 37.5
Rear wheel spindle nut	5.0 – 8.0	36.0 – 57.5
Rear sprocket nuts	2.5 – 4.0	18.0 – 29.0

Specifications relating to Chapter 6

Battery

Type YB4L-B
Voltage .... 12 volt
Electrolyte specific gravity .... 1.28 at 20°C (68°F)

Flywheel generator

Regulated voltage .... 13.5 – 15.5 volts @ 5000 rpm
No load voltage .... 40 volts or more @ 5000 rpm
Charging coil resistance .... 0.3 – 1.0 ohm

Bulbs

Headlamp 12V, 35/35W
Tail/stop lamp 12V, 5/21W
Direction indicators 12V, 21W
Instrument lights 12V, 3.4W
Main beam indicator light 12V, 3.4W
Neutral indicator light 12V, 3.4W
Direction indicator warning light 12V, 3.4W
Oil level warning light 12V, 3.4W
Parking lamp 12V, 3.4W

1 General description

Of the models covered in the preceding Chapters of this manual, the US model range was discontinued in favour of the four-stroke models. Of the UK machines, the TS250 ERX was discontinued in June 1984 and the TS185 ERX in June 1984. It is likely, however, that these models were sold for some time after being officially discontinued, while remaining stocks were sold.

The UK TS125 ERX was replaced by the TS125 ERZ in September 1982 and the TS100 ERX was replaced by the TS100 ERZ in November of that year. This update Chapter covers the two ERZ machines. The TS125 ERZ remained in production until 1984, when superseded by the liquid-cooled TS125 X, and the TS100 ERZ has continued without significant change.

With the exception of the rising-rate Full Floater rear suspension, the ERZ machines were broadly similar to their predecessors, changes being confined to detail modifications and cosmetic alterations. These are described in the following Sections.

2 Crankcase and cover modifications

1   The crankcase castings were redesigned to accommodate the relocated oil pump and the revised rear suspension. The pump was moved from its recess behind the gearbox sprocket to a new position at the front of the right-hand outer cover. The swinging arm pivot bolt doubles as an engine mounting bolt, passing through lugs at the rear of the unit.

2   When removing the engine unit from the frame, note that the oil pump control cable can be released from the pump arm after the pump cover has been removed. The cover is retained by three screws. When removing the engine mounting bolts, take care not to withdraw the swinging arm pivot bolt completely; place a dummy shaft in from the left-hand end to support the swinging arm while the engine is out of the frame.

3   Refitting the engine unit should pose no new problems, but note the engine mounting bolt torque wrench settings shown below. Note also that Suzuki recommend that the self-locking nuts are renewed each time they are removed.

Engine mounting bolt torque wrench settings:
Front mounting
bolts .... 2.8 – 3.4 kgf m (20 – 25 lbf ft)
Lower mounting
bolt .... 3.7 – 4.5 kgf m (27 – 33 lbf ft)
Rear mounting/
swinging arm
pivot bolt .... 5.0 – 8.0 kgf m (36 – 58 lbf ft)

4   As may be expected, the crankcase modifications have led to associated changes to the outer covers. The left-hand cover is shortened, and a separate sprocket cover is fitted, retained by two screws. The right-hand cover now houses the oil pump and thus incorporates an inspection cover near the front edge. This is held in place by three screws.

3 Oil pump and drive: removal and refitting

1   As has been mentioned, the oil pump is housed in a chamber at the front of the right-hand outer cover. It can be removed with the cover in place, if necessary. Remove the inspection cover and disconnect the pump cable. Disconnect the pipe from the oil tank, plugging the outlet to contain the oil. Prise off the oil delivery pipe at the cylinder barrel stub. The pump is retained by two screws and may be lifted away when these have been removed.

2   The pump is driven from an extra gear on the end of the crankshaft, via an idler pinion. To gain access to these, drain the transmission oil and remove the outer cover. The cover is held in place by a total of nine screws, two of which are inside the pump recess. Note that it is not necessary to remove the pump to allow the cover to be removed.

3   The idler pinion locates over a short shaft forward of the crankshaft end. In the case of the TS100 ERZ model, a plain washer is fitted outboard of the pinion and a smaller washer and a tubular spacer inboard of it. The assembly is normally retained by the cover and can be slid off the shaft. On TS125 ERZ models, the idler gear assembly also incorporates the tachometer drive gear in place of the tubular spacer. The two arrangements can be compared in the accompanying line drawings, the removal procedure being similar.

4   To remove the drive pinion it will be necessary to lock the crankshaft as described in Chapter 1, Section 11, paragraphs 10 and 11. Slacken and remove the nut, followed by the washer and pinion. The crankshaft primary drive pinion retaining nut will now be revealed and may be removed in the same way. When refitting the pump drive pinion, tighten the retaining nut to 3.0 – 4.0 kgf m (21.7 – 28.9 lbf ft).

4 Clutch release mechanism: general

1   The rack and pinion release mechanism of the earlier models was replaced by a conventional pushrod arrangement on the ERZ versions. The actuating lever is relocated on the left-hand outer cover and controls a cam-type operating spindle running vertically in the cover bore. This in turn acts on a pushrod inside the gearbox input shaft. The pushrod moves a mushroom-headed thrust piece via a single steel ball. The thrust piece incorporates a screw and locknut adjuster.

2   When removing the clutch assembly, take care not to lose the small steel ball; this isolates the revolving thrust piece from the static pushrod and prevents wear. When refitting the clutch, especially if the plates have been renewed, the clutch adjustment must be checked. Start by screwing the clutch cable adjusters fully home. Set the adjuster at the centre of the clutch to give about 4.0 mm (0.16 in) free play between the lever stock and blade. Subsequent adjustment can then be made using the cable adjusters.

5 Gearbox components: general

1   The gearbox input shaft on the ERZ models is redesigned to allow the clutch pushrod to run through the hollow centre of the shaft. In addition, the 5th gear pinion is now bushed, rather than running directly on the shaft. In most other respects the shaft is similar to the earlier type, though not interchangeable with it.

2   Other minor changes include the repositioning of the gear selector drum neutral detent plunger assembly; this is now fitted at a different angle. The kickstart lever on the ERZ models is retained by a single bolt which screws into the end of the kickstart shaft. The pinch bolt arrangement used on earlier models is not used.

6 Carburettor: modifications

The carburettor specifications differ somewhat from those given in Chapter 2, the new settings and jet sizes being shown in the specifications at the beginning of this Chapter. Other than this, the only significant alteration is to the method of mounting the instrument to the inlet stub. On the ERZ models, the old flange mounting is replaced by a circular stub. This pushes into an inlet adaptor and is secured by a hose clip.

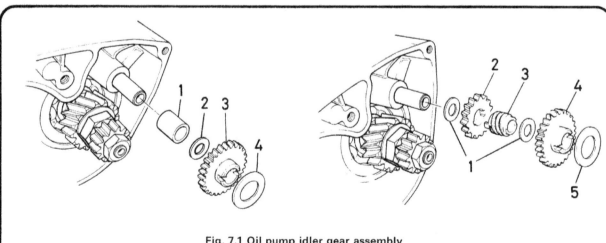

Fig. 7.1 Oil pump idler gear assembly

TS100

1   *Spacer*
2   *Washer*
3   *Idler pinion*
4   *Washer*

TS125

1   *Washers*
2   *Tachometer drive gear*
3   *Tachometer worm drive*
4   *Idler gear*
5   *Washer*

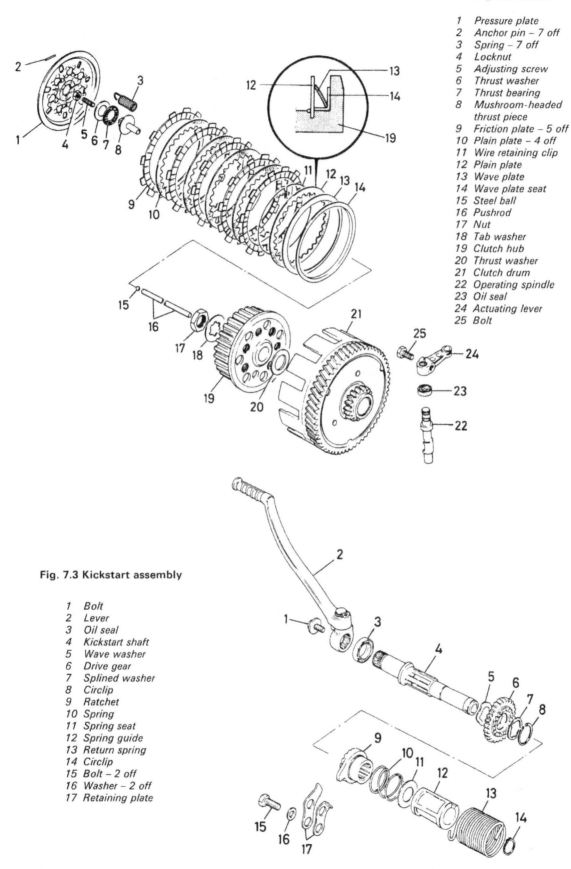

Fig. 7.2 Clutch

1 Pressure plate
2 Anchor pin – 7 off
3 Spring – 7 off
4 Locknut
5 Adjusting screw
6 Thrust washer
7 Thrust bearing
8 Mushroom-headed thrust piece
9 Friction plate – 5 off
10 Plain plate – 4 off
11 Wire retaining clip
12 Plain plate
13 Wave plate
14 Wave plate seat
15 Steel ball
16 Pushrod
17 Nut
18 Tab washer
19 Clutch hub
20 Thrust washer
21 Clutch drum
22 Operating spindle
23 Oil seal
24 Actuating lever
25 Bolt

Fig. 7.3 Kickstart assembly

1 Bolt
2 Lever
3 Oil seal
4 Kickstart shaft
5 Wave washer
6 Drive gear
7 Splined washer
8 Circlip
9 Ratchet
10 Spring
11 Spring seat
12 Spring guide
13 Return spring
14 Circlip
15 Bolt – 2 off
16 Washer – 2 off
17 Retaining plate

7 Fuel tap: modifications

The fuel tap fitted to the ERZ models is similar to that fitted to the earlier versions, with the exception of the sediment bowl; this is omitted on the later machines.

8 Oil pump: adjustment

As has been mentioned, the oil pump is relocated in a chamber at the front of the right-hand outer cover. To gain access to the pump to check the adjustment, remove the three screws which retain the cover and lift it away. The adjustment procedure is as described in Chapter 2 Section 15.

9 Air filter: general

The ERZ models are fitted with a slightly modified air filter arrangement. The polyurethane foam element is supported on a plastic frame, rather than a metal type as used on previous machines. The filter assembly can be removed after the cover has been released, this being held by three screws. The element can be examined, cleaned and re-oiled in the same way as described for the earlier machines. When refitting the foam to the frame, make sure that the edge of the foam overlaps the side of the frame to ensure a good seal.

10 Ignition system: general description and testing

1   Both of the ERZ models use a capacitor discharge ignition system, known as pointless electronic ignition (PEI) by Suzuki. The system is similar to that used on earlier US models and is described in Chapter 3, but in this case is a 12 volt arrangement rather than 6 volt. The main system checks are outlined below, but note that all wiring connections and the condition of the spark plug must be checked before turning attention to the ignition components proper.

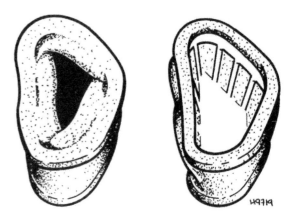

Fig. 7.4 Correct fitted position of air filter element

CDI unit/ignition coil

2   The CDI unit is built into the ignition coil, and thus the unit must be renewed if either circuit proves faulty. The best method of testing the assembly is to allow it to operate for at least five minutes on a Suzuki electro tester, and many dealers will be able to perform this test at a reasonable cost. Failing this, and in the absence of a new unit which would allow checking by substitution, a simple check can be made using a multimeter set on the K ohms x1 scale. The readings shown on the accompanying table should be obtained, but note that there may be some variation if meters other than Suzuki's own are used.

Pulser coil

3   Trace the wiring back from the alternator stator and separate the six-pin connector. Measure the resistance between the black/red and black/white leads. If this is outside the range 120 – 200 ohms, the pulser coil is faulty.

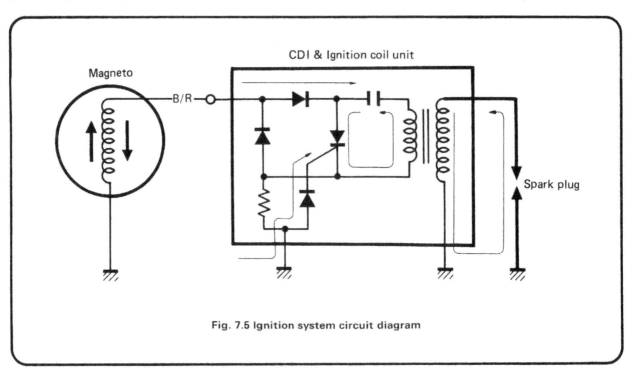

Fig. 7.5 Ignition system circuit diagram

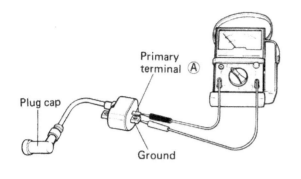

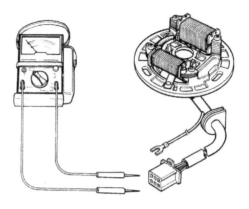

Fig. 7.7 Pulser coil test

Unit: kΩ

Negative ⊖ probe of tester to:	Positive ⊕ probe of tester to:		
	Ⓐ	Plug cap	Ground
Ⓐ		OFF	OFF
Plug cap	20 – 30		13 – 20
Ground	2.5 – 3.5	13 – 20	

Fig. 7.6 CDI unit/ignition coil test

Note: *The results shown in the table are the expected values when using a Suzuki pocket tester. If another tester is used the readings may differ slightly*

11 Front forks: modifications

The front forks are virtually unchanged from the earlier models, and the description of the overhaul of the fork legs given in Chapter 4, Section 3 can be applied. Note that the ERZ models have only a single compression spring, and that both 100 and 125 models have a backing ring fitted below the oil seal in the top of the lower leg. Apart from these modifications the forks remain the same as shown in Chapter 4, Fig. 4.1

12 Full Floater rear suspension: general description

1   The ERZ models employ a rising-rate rear suspension system comprising a box section swinging arm connected to the frame via a linkage arrangement. The single rear suspension unit is attached at its lower end to the swinging arm and at its upper end to a rocker arm. The system derives its name from the fact that at no point is the suspension unit attached directly to the frame, and is thus 'floating' within the suspension linkage.
2   The purpose of rising-rate suspension is to allow soft springing during small movements of the rear wheel, this in turn allowing the wheel to track the surface accurately and compliantly. Over rougher terrain the wheel is deflected further, and to offset this the effective spring rate is increased progressively as the wheel travels further. This is achieved by way of the suspension linkage between the frame, swinging arm and suspension unit. The way in which the linkage operates is much easier to grasp by observation, and it is recommended that you ask an assistant to bounce the machine up and down so that the linkage movement and its effect on the suspension unit can be viewed and understood.

13 Rear suspension: dismantling and reassembly

1   Open the seat and remove both side panels. Remove the tool tray from the frame below the seat. Remove the rear wheel, then release the three screws which hold the suspension unit mudflap at the front of the rear mudguard. Remove the two nuts and bolts which retain the A-shaped link arm to the swinging arm lugs, then remove the suspension unit lower mounting bolt.
2   Disconnect the oil tank outlet pipe and plug the tank opening or drain the tank. Disconnect the oil level switch wiring, then remove the tank mounting screws and those retaining the mudflap to the lower edge of the tank and lift the tank away.
3   Remove the rocker arm pivot nut and displace the pivot bolt to free the linkage from the frame. The linkage and the suspension unit can now be removed as an assembly, lifting it upwards and clear of the frame. The swinging arm can now be removed. Start by releasing the chain guide at the rear of the swinging arm, then remove the pivot bolt nut. Pull out the pivot bolt, using a soft metal drift if it proves stubborn, and place the swinging arm to one side.
4   If the linkage is to be dismantled for examination, remove the pivot bolt between the link arm and the rocker arm and separate them. Each pivot point on the link arm has two headed spacers, two dust seals and a single bush. These can be displaced for cleaning and examination, but should not be mixed up; place each set of components in a marked container to avoid this.
5   Remove the pivot bolt between the suspension unit and the rocker arm. Two headed spacers and a bush are fitted at each end of the suspension unit, and these can be displaced as described above. In the case of the rocker arm, pull off the dust seals and thrust washers and displace the inner sleeve. If the rocker arm bushes are worn they should be removed using a slide hammer, available as Suzuki tools, Part Numbers 09923-73210 and 09930-30102, or tapped out using a drift. If any of the bushes prove difficult to remove, it may be worth making up a drawbolt arrangement such as that shown in the accompanying illustration. A drawbolt must be used to fit the new needle roller bearings in the rocker arm.
6   The swinging arm pivot bushes can be reached after removing the dust seals and thrust washers and lifting away the rubber buffer on the left-hand side. The bushes can be removed in the same way as described above for the rocker arm bearings.
7   Reassembly is carried out by reversing the dismantling sequence, but before doing so clean all parts thoroughly. Any

dirt trapped in the bearings or bushes will cause rapid wear. Renew any worn or damaged bearings or bushes, and examine all dust seals carefully. It may be considered worth renewing these as a matter of course to prevent the ingress of dirt at a later date. Grease all pivots during assembly, preferably using molybdenum disulphide grease. Tighten the various mounting bolts as follows.

Component	kgf m	lbf ft
Suspension unit mounting bolts	4.0 – 6.0	29 – 43
Link arm to rocker arm bolt	4.0 – 6.0	29 – 43
Rocker arm pivot bolt	4.5 – 7.0	33 – 51
Link arm to swinging arm bolts	1.8 – 2.8	13 – 20
Swinging arm pivot bolts	5.0 – 8.0	36 – 58
Rear brake torque arm bolts	1.0 – 1.5	7 – 11

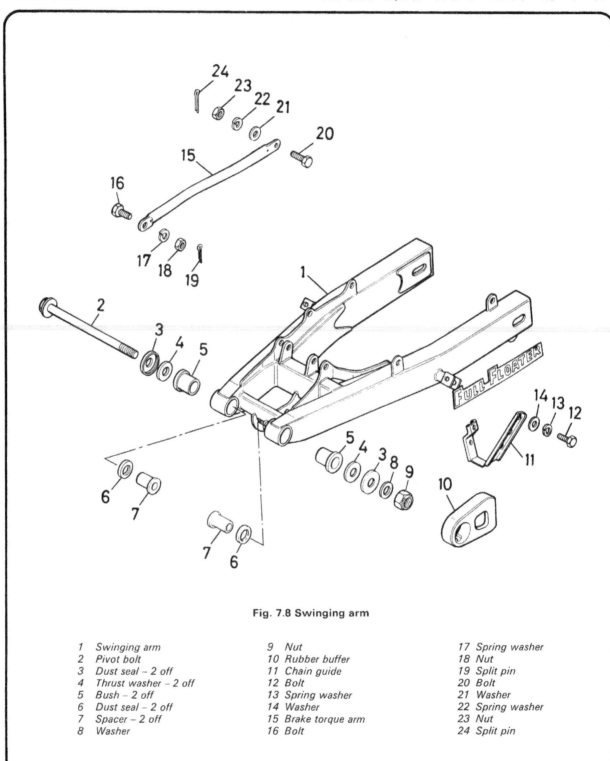

Fig. 7.8 Swinging arm

1	Swinging arm	9	Nut	17	Spring washer
2	Pivot bolt	10	Rubber buffer	18	Nut
3	Dust seal – 2 off	11	Chain guide	19	Split pin
4	Thrust washer – 2 off	12	Bolt	20	Bolt
5	Bush – 2 off	13	Spring washer	21	Washer
6	Dust seal – 2 off	14	Washer	22	Spring washer
7	Spacer – 2 off	15	Brake torque arm	23	Nut
8	Washer	16	Bolt	24	Split pin

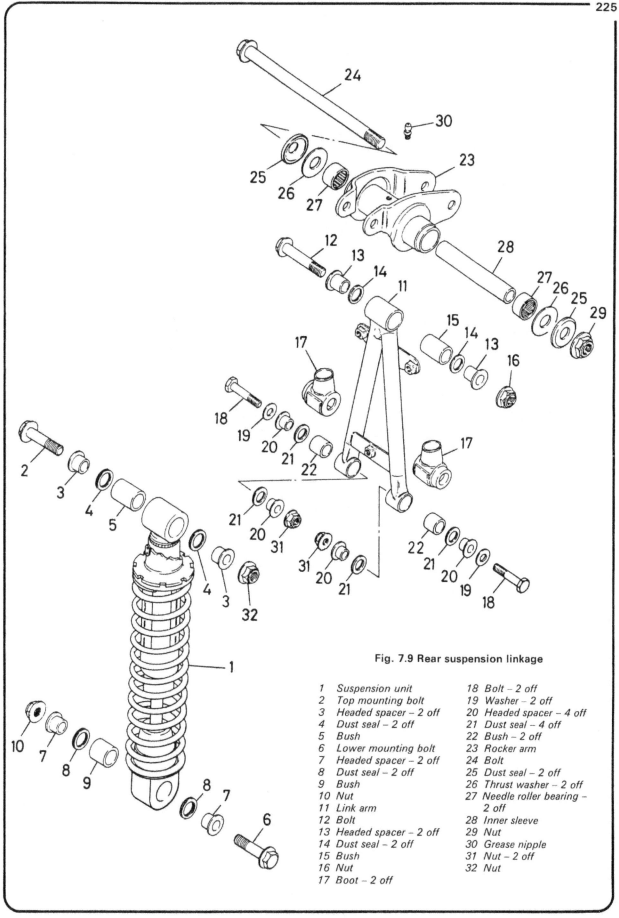

Fig. 7.9 Rear suspension linkage

1	Suspension unit	18	Bolt – 2 off
2	Top mounting bolt	19	Washer – 2 off
3	Headed spacer – 2 off	20	Headed spacer – 4 off
4	Dust seal – 2 off	21	Dust seal – 4 off
5	Bush	22	Bush – 2 off
6	Lower mounting bolt	23	Rocker arm
7	Headed spacer – 2 off	24	Bolt
8	Dust seal – 2 off	25	Dust seal – 2 off
9	Bush	26	Thrust washer – 2 off
10	Nut	27	Needle roller bearing – 2 off
11	Link arm	28	Inner sleeve
12	Bolt	29	Nut
13	Headed spacer – 2 off	30	Grease nipple
14	Dust seal – 2 off	31	Nut – 2 off
15	Bush	32	Nut
16	Nut		
17	Boot – 2 off		

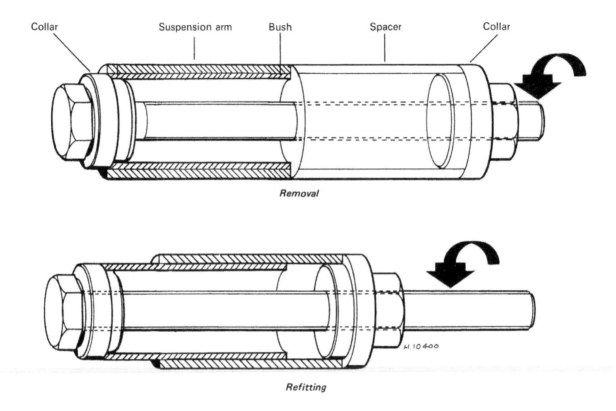

Removal

Refitting

Fig. 7.10 Drawbolt tool for removing and refitting suspension bushes

14 Rear suspension linkage: examination and renovation

1 The rear suspension linkage and pivots should be checked for wear or damage as described in Section 11 of Chapter 4, bearing in mind the removal and refitting procedures described in the previous Section of this Chapter. The various pivots are exposed to a great deal of dirt, particularly where the machine is used off road. To extend the service life of the various bushes and bearings, keep the pivot areas clean at all times.

2 During winter it is a good idea to coat the linkage and pivots with grease to provide a barrier to the elements. It goes without saying that if a seal fails in service, the ingress of dirt will cause rapid wear of the bush or bearing. Check these regularly, and if necessary renew them before the pivots are damaged. When overhauling the suspension, check each seal carefully, and renew it if in anything other than perfect condition.

15 Footrests: examination and renovation

The footrests on the ERZ models are similar in construction to those fitted to previous models, but note that they bolt directly to frame lugs instead of the bolt-on crossbar mounting of their predecessors. In other respects they can be dealt with as described in Chapter 4 Section 15.

16 Final drive chain: adjustment

This can be carried out as described in Section 10 of Chapter 5. Note that the adoption of the box section swinging arm on the later machines has resulted in a revised chain adjuster, but the general procedure is largely unchanged. The recommended free play is 25 – 30 mm (1.0 – 1.2 in) for the TS100 ERZ and 35 – 40 mm (1.4 – 1.6 in) in the case of the TS125 ERZ.

17 Electrical system: general

1 The main alteration to the electrical system on the ERZ models is the change to 12 volts in place of the 6 volt system used in earlier machines. The alternating output from the flywheel generator is fed to a combined regulator/rectifier unit. This converts the output to direct current and controls the charge to the battery.

2 Less significant alterations are the use of revised instruments and lamps. These are mostly of a cosmetic nature and are described later in this Chapter. The new electrical system requires a slightly different approach when tracing faults and checking the various components. This is discussed in the following Sections.

18 Electrical system: testing the charging output

1 Remove the right-hand side panel to gain access to the battery. Note that this test presupposes that the battery is fully charged and in good condition; false readings will result if this is not the case. Set a multimeter to the 0 – 20 volts dc range and connect the positive (+) probe to the battery positive terminal and the negative (–) probe to the battery negative terminal.
2 Start the engine, and note the reading at 5000 rpm. If the reading shown is below 13.5 volts or above 15.5 volts, check the generator no-load performance and the condition of the regulator/rectifier unit as described below.

19 Flywheel generator: testing the no-load performance

1 Trace the output leads from the flywheel generator and disconnect them at the block connector. Using an ac voltmeter or a multimeter set on an appropriate range and connected to the two yellow leads, measure the generator output at 5000 rpm. If a reading of 40V or less is shown, the generator windings should be considered faulty.
2 The windings can be checked for continuity with the engine off and a continuity tester, or multimeter set on its resistance scale, connected to the two yellow output leads. The specified resistance is 0.3 – 1.0 ohm, but more importantly, if infinite resistance is shown, it can be assumed that the windings are broken internally.
3 Next, check for insulation between the stator core and each of the two leads. If anything other than infinite resistance is shown, the windings have shorted to the stator core.

20 Regulator/rectifier unit: checking the resistances

1 Locate the regulator/rectifier unit and unplug the wiring connector. Note that it can be identified by its finned alloy case. The unit is mounted below the frame top tubes, to the rear of the CDI/ignition coil unit.
2 Using a multimeter set on the K ohms x1 scale, check the resistances as shown in the accompanying illustration. If the readings obtained differ significantly from those specified, the unit must be renewed. Note that it is of sealed construction and thus cannot be repaired.

21 Instrument panel: removal and bulb renewal

1 The TS125 ERZ employs a one-piece instrument panel assembly housing the speedometer and tachometer heads, lamps for internal illumination and the various warning lamps. The TS100 ERZ uses a similar panel arrangement, but without the tachometer.
2 The various bulbs and connections can be checked for continuity using a multimeter, thus avoiding unnecessary dismantling work. Start by removing the headlamp unit from its shell. Trace the instrument panel wiring back to the connector inside the shell and separate it. Using the accompanying diagrams, check for continuity on each bulb circuit.
3 If a bulb has failed, remove the instrument panel to gain access to the bulb holders. Push the wiring connector out of the headlamp shell and disconnect the speedometer cable, and in the case of the TS125 ERZ, the tachometer cable. Remove the bolts which retain the panel to the top yoke and lift it away.
4 Remove the nuts or bolts which retain the bottom cover and lift it clear. The various bulb holders are a push fit in the underside of the panel. Note that the bulbs are of the capless type and are removed by pulling them out of the bulbholder.

22 Headlamp, stop/tail lamp and turn signal lamps: general

The design of the above-mentioned lamps has been altered slightly from the types used on previous models. Apart from cosmetic differences, there are no significant changes and the text in Chapter 6 can be applied. Note that all bulbs are rated at 12 volts on the ERZ models.

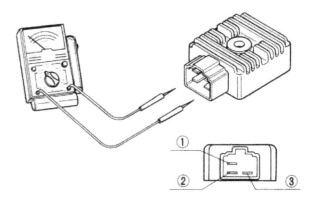

Unit: kΩ

		⊕ Probe of tester			
		①	②	③	Ground
Probe of tester	①		OFF	OFF	OFF
	②	2 – 4		OFF	OFF
	③	2 – 4	OFF		OFF
	Ground	7 – 10	2 – 4	2 – 4	

Fig. 7.12 Regulator/rectifier test

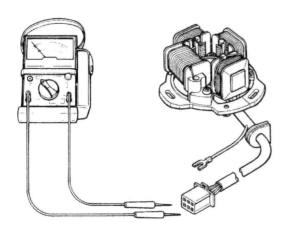

Fig. 7.11 Flywheel generator windings continuity test

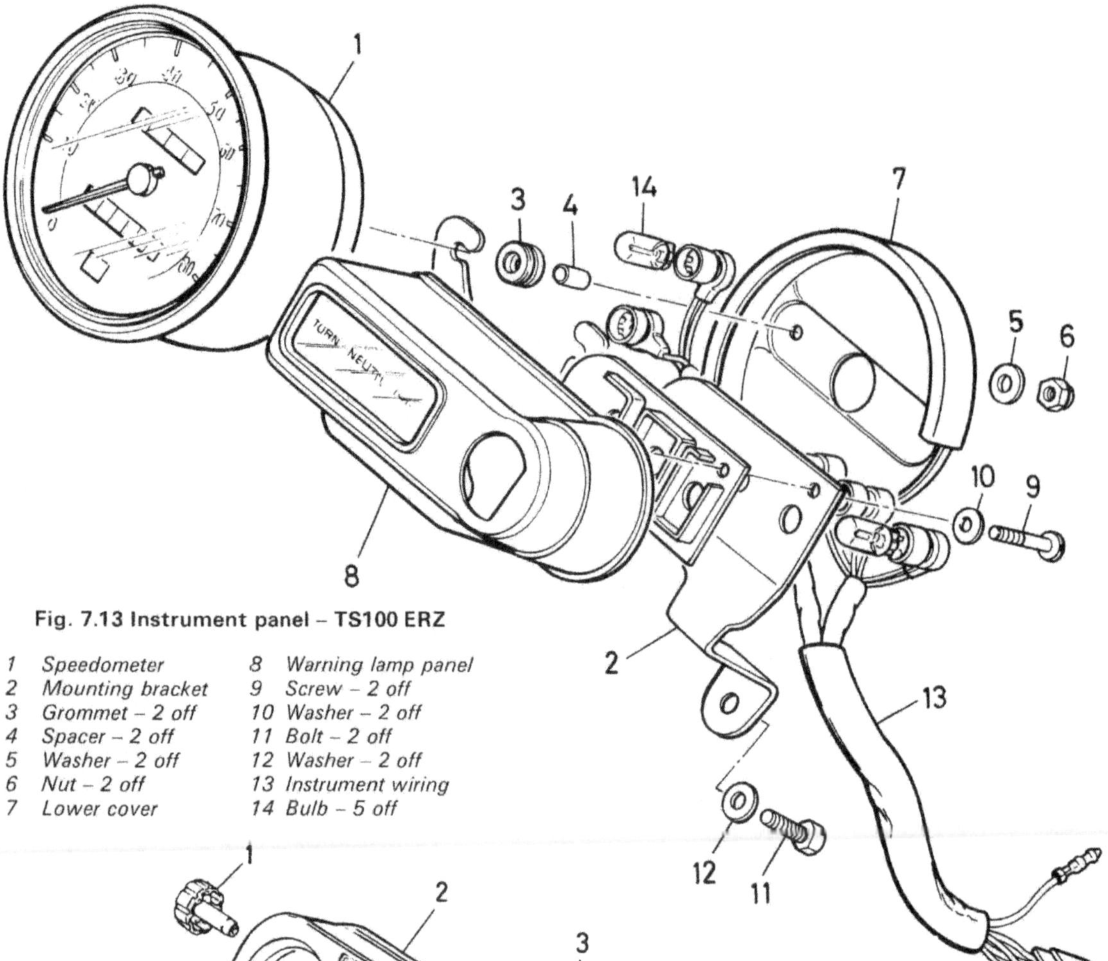

Fig. 7.13 Instrument panel – TS100 ERZ

1 Speedometer	8 Warning lamp panel
2 Mounting bracket	9 Screw – 2 off
3 Grommet – 2 off	10 Washer – 2 off
4 Spacer – 2 off	11 Bolt – 2 off
5 Washer – 2 off	12 Washer – 2 off
6 Nut – 2 off	13 Instrument wiring
7 Lower cover	14 Bulb – 5 off

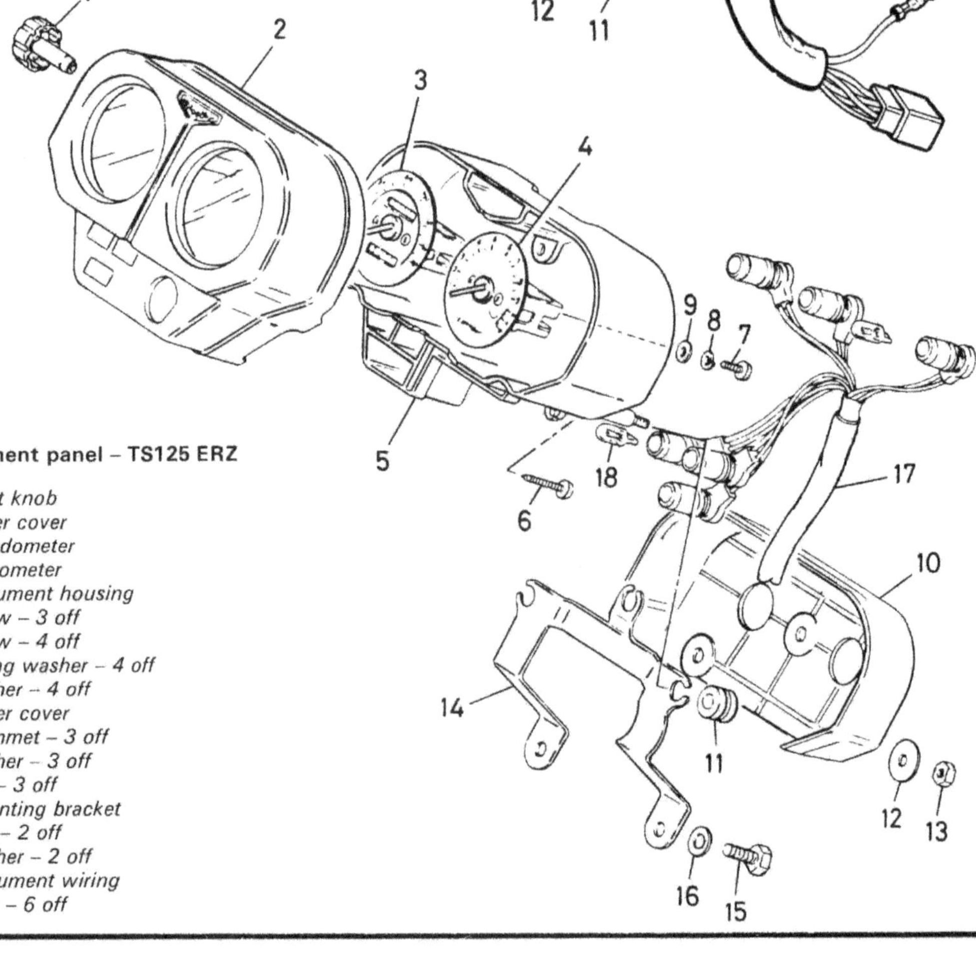

Fig. 7.14 Instrument panel – TS125 ERZ

1 Reset knob
2 Upper cover
3 Speedometer
4 Tachometer
5 Instrument housing
6 Screw – 3 off
7 Screw – 4 off
8 Spring washer – 4 off
9 Washer – 4 off
10 Lower cover
11 Grommet – 3 off
12 Washer – 3 off
13 Nut – 3 off
14 Mounting bracket
15 Bolt – 2 off
16 Washer – 2 off
17 Instrument wiring
18 Bulb – 6 off

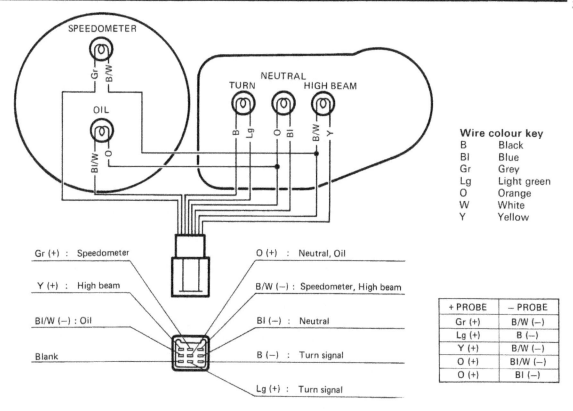

Fig. 7.15 Instrument panel test connections – TS100

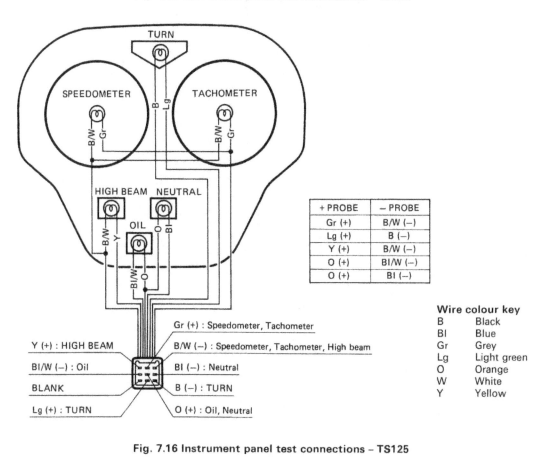

Fig. 7.16 Instrument panel test connections – TS125

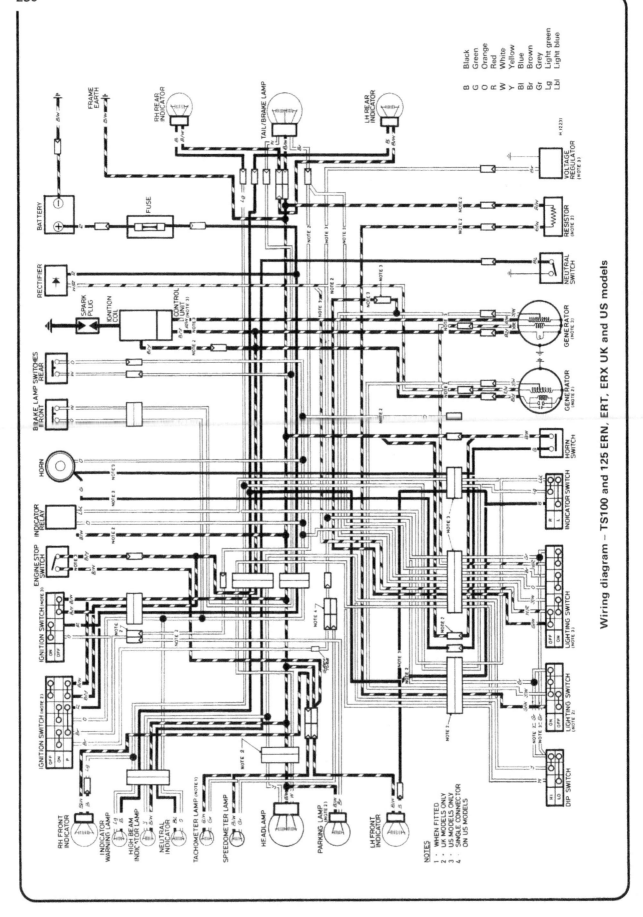

Wiring diagram – TS100 and 125 ERN, ERT, ERX UK and US models

B	Black
G	Green
O	Orange
R	Red
W	White
Y	Yellow
Bl	Blue
Br	Brown
Gr	Grey
Lg	Light green
Lbl	Light blue

NOTES
1 : WHEN FITTED
2 : UK MODELS ONLY
3 : US MODELS ONLY
4 : SINGLE CONNECTOR ON US MODELS

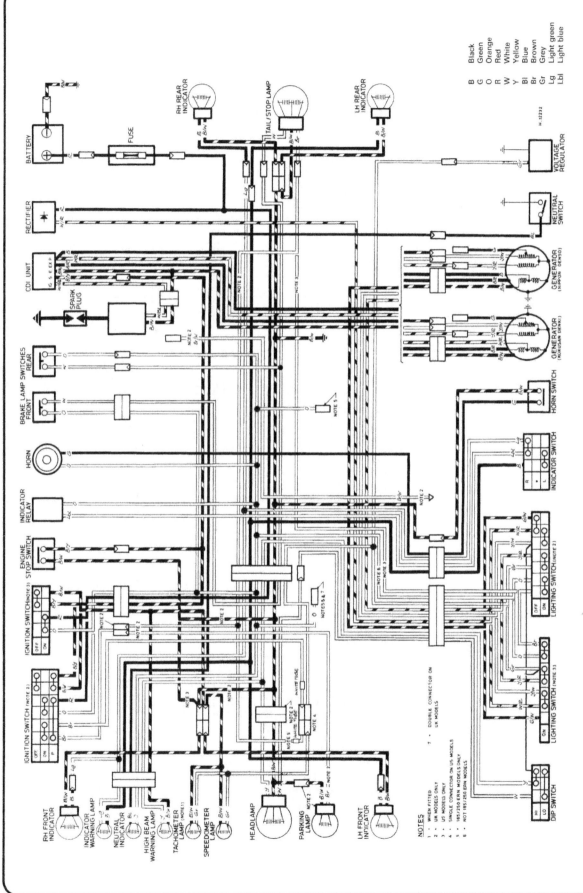

Wiring diagram – TS185 and 250 ERN, ERT, ERX UK and US models

231

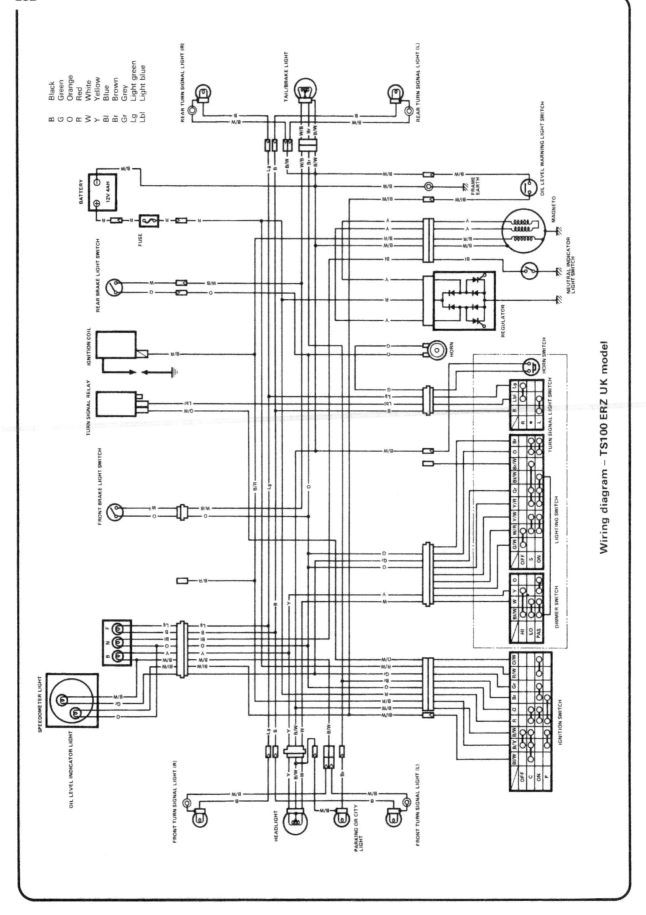

Wiring diagram – TS100 ERZ UK model

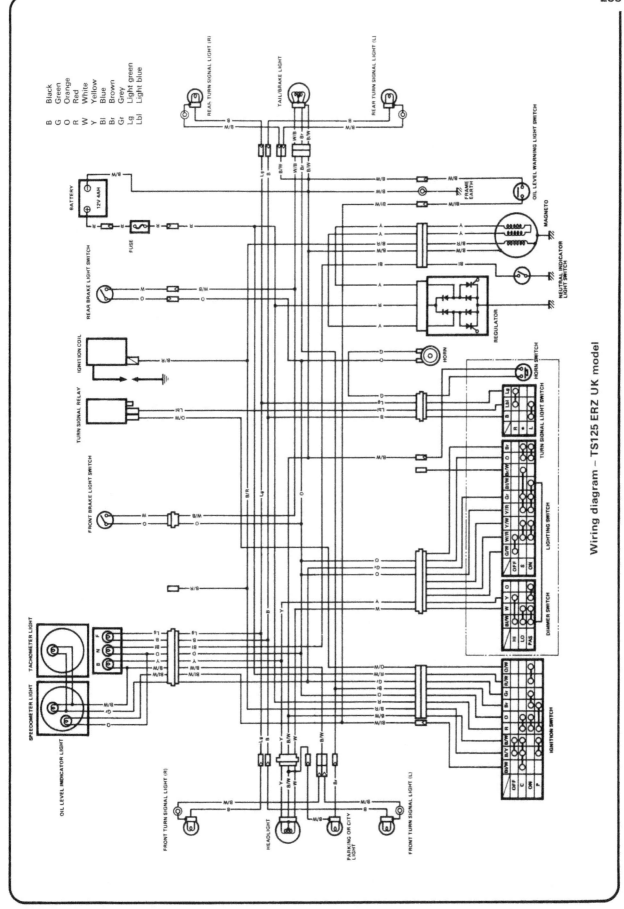

Wiring diagram – TS125 ERZ UK model

Conversion factors

Length (distance)
Inches (in)	X	25.4	= Millimetres (mm)	X	0.0394	= Inches (in)
Feet (ft)	X	0.305	= Metres (m)	X	3.281	= Feet (ft)
Miles	X	1.609	= Kilometres (km)	X	0.621	= Miles

Volume (capacity)
Cubic inches (cu in; in^3)	X	16.387	= Cubic centimetres (cc; cm^3)	X	0.061	= Cubic inches (cu in; in^3)
Imperial pints (Imp pt)	X	0.568	= Litres (l)	X	1.76	= Imperial pints (Imp pt)
Imperial quarts (Imp qt)	X	1.137	= Litres (l)	X	0.88	= Imperial quarts (Imp qt)
Imperial quarts (Imp qt)	X	1.201	= US quarts (US qt)	X	0.833	= Imperial quarts (Imp qt)
US quarts (US qt)	X	0.946	= Litres (l)	X	1.057	= US quarts (US qt)
Imperial gallons (Imp gal)	X	4.546	= Litres (l)	X	0.22	= Imperial gallons (Imp gal)
Imperial gallons (Imp gal)	X	1.201	= US gallons (US gal)	X	0.833	= Imperial gallons (Imp gal)
US gallons (US gal)	X	3.785	= Litres (l)	X	0.264	= US gallons (US gal)

Mass (weight)
Ounces (oz)	X	28.35	= Grams (g)	X	0.035	= Ounces (oz)
Pounds (lb)	X	0.454	= Kilograms (kg)	X	2.205	= Pounds (lb)

Force
Ounces-force (ozf; oz)	X	0.278	= Newtons (N)	X	3.6	= Ounces-force (ozf; oz)
Pounds-force (lbf; lb)	X	4.448	= Newtons (N)	X	0.225	= Pounds-force (lbf; lb)
Newtons (N)	X	0.1	= Kilograms-force (kgf; kg)	X	9.81	= Newtons (N)

Pressure
Pounds-force per square inch (psi; lbf/in^2; lb/in^2)	X	0.070	= Kilograms-force per square centimetre (kgf/cm^2; kg/cm^2)	X	14.223	= Pounds-force per square inch (psi; lbf/in^2; lb/in^2)
Pounds-force per square inch (psi; lbf/in^2; lb/in^2)	X	0.068	= Atmospheres (atm)	X	14.696	= Pounds-force per square inch (psi; lbf/in^2; lb/in^2)
Pounds-force per square inch (psi; lbf/in^2; lb/in^2)	X	0.069	= Bars	X	14.5	= Pounds-force per square inch (psi; lbf/in^2; lb/in^2)
Pounds-force per square inch (psi; lbf/in^2; lb/in^2)	X	6.895	= Kilopascals (kPa)	X	0.145	= Pounds-force per square inch (psi; lbf/in^2; lb/in^2)
Kilopascals (kPa)	X	0.01	= Kilograms-force per square centimetre (kgf/cm^2; kg/cm^2)	X	98.1	= Kilopascals (kPa)

Torque (moment of force)
Pounds-force inches (lbf in; lb in)	X	1.152	= Kilograms-force centimetre (kgf cm; kg cm)	X	0.868	= Pounds-force inches (lbf in; lb in)
Pounds-force inches (lbf in; lb in)	X	0.113	= Newton metres (Nm)	X	8.85	= Pounds-force inches (lbf in; lb in)
Pounds-force inches (lbf in; lb in)	X	0.083	= Pounds-force feet (lbf ft; lb ft)	X	12	= Pounds-force inches (lbf in; lb in)
Pounds-force feet (lbf ft; lb ft)	X	0.138	= Kilograms-force metres (kgf m; kg m)	X	7.233	= Pounds-force feet (lbf ft; lb ft)
Pounds-force feet (lbf ft; lb ft)	X	1.356	= Newton metres (Nm)	X	0.738	= Pounds-force feet (lbf ft; lb ft)
Newton metres (Nm)	X	0.102	= Kilograms-force metres (kgf m; kg m)	X	9.804	= Newton metres (Nm)

Power
Horsepower (hp)	X	745.7	= Watts (W)	X	0.0013	= Horsepower (hp)

Velocity (speed)
Miles per hour (miles/hr; mph)	X	1.609	= Kilometres per hour (km/hr; kph)	X	0.621	= Miles per hour (miles/hr; mph)

Fuel consumption*
Miles per gallon, Imperial (mpg)	X	0.354	= Kilometres per litre (km/l)	X	2.825	= Miles per gallon, Imperial (mpg)
Miles per gallon, US (mpg)	X	0.425	= Kilometres per litre (km/l)	X	2.352	= Miles per gallon, US (mpg)

Temperature
Degrees Fahrenheit = ($^\circ$C x 1.8) + 32

Degrees Celsius (Degrees Centigrade; $^\circ$C) = ($^\circ$F - 32) x 0.56

*It is common practice to convert from miles per gallon (mpg) to litres/100 kilometres (l/100km), where mpg (Imperial) x l/100 km = 282 and mpg (US) x l/100 km = 235

English/American terminology

Because this book has been written in England, British English component names, phrases and spellings have been used throughout. American English usage is quite often different and whereas normally no confusion should occur, a list of equivalent terminology is given below.

English	American	English	American
Air filter	Air cleaner	Number plate	License plate
Alignment (headlamp)	Aim	Output or layshaft	Countershaft
Allen screw/key	Socket screw/wrench	Panniers	Side cases
Anticlockwise	Counterclockwise	Paraffin	Kerosene
Bottom/top gear	Low/high gear	Petrol	Gasoline
Bottom/top yoke	Bottom/top triple clamp	Petrol/fuel tank	Gas tank
Bush	Bushing	Pinking	Pinging
Carburettor	Carburetor	Rear suspension unit	Rear shock absorber
Catch	Latch	Rocker cover	Valve cover
Circlip	Snap ring	Selector	Shifter
Clutch drum	Clutch housing	Self-locking pliers	Vise-grips
Dip switch	Dimmer switch	Side or parking lamp	Parking or auxiliary light
Disulphide	Disulfide	Side or prop stand	Kick stand
Dynamo	DC generator	Silencer	Muffler
Earth	Ground	Spanner	Wrench
End float	End play	Split pin	Cotter pin
Engineer's blue	Machinist's dye	Stanchion	Tube
Exhaust pipe	Header	Sulphuric	Sulfuric
Fault diagnosis	Trouble shooting	Sump	Oil pan
Float chamber	Float bowl	Swinging arm	Swingarm
Footrest	Footpeg	Tab washer	Lock washer
Fuel/petrol tap	Petcock	Top box	Trunk
Gaiter	Boot	Torch	Flashlight
Gearbox	Transmission	Two/four stroke	Two/four cycle
Gearchange	Shift	Tyre	Tire
Gudgeon pin	Wrist/piston pin	Valve collar	Valve retainer
Indicator	Turn signal	Valve collets	Valve cotters
Inlet	Intake	Vice	Vise
Input shaft or mainshaft	Mainshaft	Wheel spindle	Axle
Kickstart	Kickstarter	White spirit	Stoddard solvent
Lower leg	Slider	Windscreen	Windshield
Mudguard	Fender		

Index

A

About this manual 2
Accessories – fitting 12
Acknowledgements 2
Adjustments :-
    brake 36, 198
    carburettor 30, 132, 222
    clutch 35, 111, 220
    contact breaker points 33, 152
    final drive chain 26, 197, 226
    headlamp beam height 209
    horn 215
    ignition timing 34, 35, 155 – 157
    oil pump 28, 139, 222
    rear stop lamp switch 215
    rear suspension units 179
    spark plug 32, 162
    steering head bearings 37, 175
    throttle cable 31
Air filter element :-
    cleaning 31
    fault diagnosis 17
    removal, examination and refitting 134, 222

B

Balancing – wheel 202
Ballast resistor 207
Battery:-
    charging procedure 208
    examination and maintenance 27, 207
    fault diagnosis 23
    specifications 203
Bearings:-
    big-end 65
    gearbox 64
    main 64
    small-end 65, 104
    steering head 37, 40, 175
    swinging arm 40, 176
    wheel:
        front 190
        rear 196
Bleeding the oil pump 144
Brakes:-
    adjustment 36, 198
    dismantling 198
    fault diagnosis 23
    rear pedal 182
Bulbs:-
    fault diagnosis 24

flashing indicator lamps 212
headlamp 209, 210
specifications 204, 219
speedometer and tachometer lamps 213, 227
stop and tail lamp 211
warning light console 212, 227
Buying:-
    accessories 12
    spare parts 7
    tools 9

C

Cables:-
    clutch 35
    lubrication 26, 28
    speedometer and tachometer 38
    throttle 31
Carburettor:-
    adjustment 30, 132
    cleaning 39, 123
    dismantling, examination and reassembly 122
    emission control legislation 132
    fault diagnosis 17
    modifications 220
    removal and refitting 119
    settings 133
    specifications 114, 115, 218
Castrol grades 43
CDI unit 160, 222
Chain – final drive 26, 27, 197
Checks:-
    battery electrolyte level 27
    brake shoe wear 36, 200
    contact breaker points 33
    engine oil level 26
    float level 132
    flywheel generator 161, 205, 227
    ignition timing 34, 35, 155 – 157
    legal 25
    safety 25
    spark plug 32
    steering head bearings 37
    tyre pressures 25
Cleaning:-
    air filter element 31, 134
    carburettor 39
    fuel tap collector bowl 37
    metal and plastic components 41
Clutch:-
    adjustment 35, 111, 220
    cable 35, 111
    examination and renovation 74

fault diagnosis 19
modifications 220
refitting 98
removal 56
specifications 45, 47, 217
Coil – ignition 161, 222
Condenser 159
Connecting rod 65
Contact breaker assembly:-
adjustment 33, 152
gap 151
renewal 39
removal, renovation and refitting 158
Conversion factors 234
Crankcases:-
joining 89
modifications 219
separating 62
Crankshaft:-
examination and renovation 65
main bearings 64
refitting 89
removal 63
specifications 45, 47
Cush drive 196
Cylinder barrel:-
decarbonising 38, 66
examination and renovation 67
refitting 104
removal 52
specifications 45, 46, 217
Cylinder head:-
decarbonising 38, 66
examination and renovation 66
refitting 104
removal 52
specifications 45, 46

D

Decarbonising:-
cylinder head and barrel 38, 66
exhaust system 145
Descriptions – general:-
electrical system 204
engine, clutch and gearbox 48
frame and forks 165
fuel system and lubrication 116
ignition system 152
wheels, brakes and tyres 187
Dimensions and weights 6
Dual seat 185
Dust caps – tyre valves 201

E

Electrical system:-
ballast resistor 207
battery:
charging procedure 208
examination and maintenance 27, 207
fault diagnosis 23
flasher unit 215
flashing indicator lamps 212, 227
flywheel generator 205, 227
front brake stop lamp switch 215
fuse 208
general description 204, 226
handlebar switches 213
headlamp 209
horn 215
ignition switch 213

neutral indicator switch 215
rear brake stop lamp switch 215
rectifier 206, 227
stop and tail lamp 211
speedometer and tachometer heads 213, 227
testing 204
voltage regulator 207, 227
warning light console 212
wiring diagram 230 – 233
Engine:-
bearings:
big-end 65
main 64
small-end 65, 104
connecting rod 65
crankcases:
joining 89
separating 62
crankshaft 63, 65, 89
cylinder barrel 52, 66, 67, 104
cylinder head 52, 66, 104
decarbonising 38, 66
dismantling – general 51
examination and renovation – general 63
fault diagnosis 16
general description 48
gudgeon pin 52, 68, 104
lubrication 139
modifications 219, 220
oil pump 56, 79, 101, 139, 220
oil seals 63
piston and rings 52, 68, 104
reassembly – general 80
specifications 45, 46, 217
starting and running the rebuilt engine 112
tachometer drive 62, 80, 94
torque wrench settings 46, 47, 217
Emission control information 132
Exhaust system:-
decarbonising 145
fitting of aftermarket exhausts 14
removal and refitting 148

F

Fault diagnosis 15 – 24
Filters:-
air 31, 134, 222
fuel 37, 119
Final drive chain 26, 27, 197, 226
Flasher unit 215
Flashing indicator lamps 212, 227
Flywheel generator:-
output check 161, 205, 227
refitting 103
removal 55
Footrests 181, 226
Frame 180
Frame and forks:-
dual seat 185
fault diagnosis 22
footrests 181, 226
fork yokes 175
frame 180
front fork legs:
dismantling 166
modifications 223
removal 165
general description 165
kickstart lever 184
prop stand 180
rear brake pedal 182

rear suspension units 179
specifications 164, 218
speedometer and tachometer 175
steering head 172, 175
steering lock 175
swinging arm fork 176, 223
Front brake:-
adjustment 36, 198
dismantling 198
fault diagnosis 23
Front wheel:-
bearings 190
examination and renovation 187
fault diagnosis 21
removal and refitting 188
Fuel system:-
air filter 31, 134
carburettor 30, 39, 119 – 133, 220
fault diagnosis 17
fuel feed pipe 40, 119
fuel tank 116
fuel tap 37, 116, 222
general description 116
reed valve 133
specifications 114
Full floater rear suspension 223
Fuse 208

G

Gearbox:-
examination and renovation 69
fault diagnosis 20
lubrication 30, 116, 145
modifications 220
reassembly 81
removal 62
specifications 46, 47, 217
Gearchange mechanism:-
examination and renovation 73
fault diagnosis 19
reassembly 96
removal 60
specifications 46, 47
Generator – flywheel:-
output check 161, 205, 227
refitting 103
removal 55
Gudgeon pin 52, 68, 104

H

Handlebar:-
aftermarket handlebars 13
removal 173
switches 213
Headlamp 209, 227
High tension lead 162
Horn 215

I

Ignition switch 213
Ignition system:-
CDI unit 160, 222
condenser 159
contact breaker:
adjustment 33, 152
renewal 39
removal, renovation and refitting 158
fault diagnosis 18

flywheel generator:
output check 161, 205, 227
refitting 103
removal 55
general description 152, 222
high tension lead 162
ignition coil 161, 222
ignition timing 34, 35, 155 – 157
pulser coil 160, 222
spark plug 32, 162
specifications 151, 218

K

Kickstart:-
dismantling 62
examination and renovation 76
lever 184
modifications 220
refitting 92

L

Lamps:-
flashing indicator 212
fault diagnosis 24
headlamp 209
specifications 204, 219
stop and tail lamp 211
warning lamps 212
Legal obligations 25
Lubrication:-
brake cam shaft 38
control cables 26, 28
engine 26
final drive chain 26, 27
gearbox 30
general 28
speedometer and tachometer cables 38
Lubrication system:-
engine 139
gearbox 145
oil pump:
bleeding 144
removal and refitting 139, 220

M

Main bearings 64
Maintenance – routine 25 – 41

N

Neutral indicator switch:-
refitting 103
removal 56
testing 215

O

Oil pump:-
adjustment 28, 139, 222
bleeding 144
examination and renovation 79
refitting 101, 139, 220
removal 56, 139, 220
Oil seals 63
Ordering:-
spare parts 7
tools 9

P

Pedal – rear brake 182
Petrol feed pipe 40, 119
Petrol tank 116
Petrol tap 37, 116, 222
Piston and rings:-
    examination and renovation 68
    refitting 104
    removal 52
Prop stand 180
Pulser coil 160, 222

R

Rear brake:-
    adjustment 36, 198
    dismantling 198
    fault diagnosis 23
    pedal 182
Rear chain 26, 27, 197, 226
Rear suspension linkage 223 – 226
Rear suspension units 179
Rear wheel:-
    bearings 196
    cush drive 196
    examination and renovation 193
    fault diagnosis 21
    removal and refitting 193
    sprocket 196
Rectifier 206, 227
Reed valve 133
Refitting the engine/gearbox unit
in the frame 107
Regulator – voltage 207, 227
Removing the engine/gearbox unit
from the frame 48
Resistor – ballast 207
Rings and piston 52, 68, 104
Routine maintenance 25 – 41

S

Safety precautions 8, 25
Security bolts – wheel 202
Small-end bearing 65, 104
Spanner size comparison 11
Spark plug:-
    check 32, 162
    high tension lead 162
    operating conditions – colour chart 153
    renewal 37
    specifications 151, 218
Speedometer:-
    bulb renewal 213, 227
    drive 176
    drive cable 38, 176
    head 175
Specifications:-
    clutch 45, 47
    electrical system 203, 204
    engine 45, 46
    frame and forks 164, 165
    fuel 114, 115

    gearbox 46, 47
    ignition 151
    TS100 and 125 ERZ models 217
    lubrication 115, 116
    wheels, brakes and tyres 186
Sprocket:-
    front 56, 103
    rear 196
Standard torque wrench settings 11
Steering head:-
    bearings 37, 175
    lock 175
    removal and refitting 172
Stop lamp switches 215
Suspension units – rear 179, 223
Swinging arm fork 176, 223
Switches:-
    handlebar 213
    ignition 213
    stop lamp 215

T

Tachometer:-
    bulb renewal 213, 227
    drive 62, 80, 94
    drive cable 38, 176
    head 175
Throttle cable 31
Timing – ignition 34, 35, 155 – 157
Tools 9
Torque wrench settings 11, 46, 47, 165, 186, 217 – 219
Tyres:-
    colour instructions 189
    pressures 25, 186
    removal, repair and refitting 200
    security bolts 202
    valves 201

V

Valve – reed 133
Valve – tyre 201
Voltage regulator unit 207, 227

W

Warning light console 212, 227
Weights and dimensions 6
Wheels:-
    balancing 202
    fault diagnosis 21
    front:
        bearings 190
        examination and renovation 187
        removal and refitting 188
    rear:
        bearings 196
        cush drive 196
        examination and renovation 193
        removal and refitting 193
        sprocket 196
Wiring diagram 230 – 233
Working facilities 9